# Poverty Reduction in a Changing Climate

# Poverty Reduction in a Changing Climate

Hari Bansha Dulal

LEXINGTON BOOKS
*Lanham • Boulder • New York • Toronto • Plymouth, UK*

Published by Lexington Books
A wholly owned subsidiary of The Rowman & Littlefield Publishing Group, Inc.
4501 Forbes Boulevard, Suite 200, Lanham, Maryland 20706
www.rowman.com

10 Thornbury Road, Plymouth PL6 7PP, United Kingdom

British Library Cataloguing in Publication Information Available

**Library of Congress Cataloging-in-Publication Data**

Dulal, Hari Bansha, 1974–
    Poverty reduction in a changing climate / Hari Bansha Dulal.
        pages cm
    Includes index.
    ISBN 978-0-7391-6801-1 (hbk. : alk. paper) — ISBN 978-0-7391-7989-5 (electronic)
1. Poverty—Environmental aspects—Developing countries. 2. Climatic changes—
Economic aspects—Developing countries. 3. Sustainable development—Economic
aspects—Developing countries. I. Title.
    HC79.P6D85 2012
    339.4'6091724—dc23                                              2012044848

♾™ The paper used in this publication meets the minimum requirements of American
National Standard for Information Sciences—Permanence of Paper for Printed Library
Materials, ANSI/NISO Z39.48-1992.

Printed in the United States of America

Dedicated to the memory of my father, Shankar Prasad Dulal, and to my wonderful wife, Sanjana Dulal, who supported me through each step of the way.

# CONTENTS

Contents

# ABOUT THE CONTRIBUTORS

**Arup Mitra** is professor of economics at the Institute of Economic Growth, Delhi. His research interest includes urban development, labour and welfare, migration, industrial growth and productivity and gender inequality–areas in which he has several publications. He has also been a consultant to ILO, OECD, World Bank, UNDP and ADB. He held the Indian Economy Chair at Sciences Po. Paris in 2010. He did his PhD from Delhi School of Economics and post-doctorate from Northwestern University (USA).

**Bryan R. Bushley** is a doctoral candidate at the University of Hawai'i and a Graduate Degree Fellow at the East-West Center in Honolulu. He is currently conducting his dissertation research on the socioeconomic and governance implications of REDD+ for forest-dependent communities in Nepal, where he has been engaged in REDD+ readiness activities and participatory action research on forest certification and REDD+. Bushley also holds concurrent masters degrees in Forestry and Public Administration from the University of Washington. His academic and professional interests include the implications of market-based mechanisms for livelihoods, conservation and forest governance, as well as the local implications of, and responses to, climate change policies. He has worked on water resource management policy in Central Asia, ecotourism and payments for environmental services in Central America, and forest conservation in South Asia. Bushley has contributed to both academic and applied publications on a range of topics, including livelihoods and protected areas in Bangladesh, community conserved areas in India, and community forestry and climate change policy in Nepal.

**Dilys Roe** is a senior researcher in IIED's Natural Resources Group and leads their work on biodiversity. Since 2004, Dilys has coordinated the Poverty and Conservation Learning Group, a network of conservation, development and indigenous/local community organisations that is intended to improve dialogue on poverty-conservation linkages, promote better understanding of different perspectives and experiences of these links and help inform policy on these matters. While the majority of Dilys' work focusses on biodiversity-development/conservation-poverty issues, she also has a research interest in community-based natural resource management and community-based conservation; ecosystem-based adaptation and high biodiversity REDD+. In addition to her work at IIED, Dilys has worked periodically as a consultant biodiversity advisor to the UK Department for International Development and is also a doctoral candidate at the Durrell Institute for Conservation and Ecology (DICE)

ix

at the University of Kent.

**Enrique Hennings** is an applied economist with several years of international experience in economic analysis, agricultural finance and agribusiness. Currently, he is the manager of the Global Producer Finance Unit at Fairtrade International overseeing Latin America, Africa and Asia. In the past, he worked with ACDI/VOCA as Deputy Director for a Millennium Challenge Account Project in Honduras (Farmer Access to Credit). He worked at the Center for Farm and Rural Business Finance at the University of Illinois. He also worked for the World Bank, Inter-American Development Bank, and the Organization of American States. He holds a master degree and a PhD in Agricultural Economic from the University of Illinois.

**George Joseph** holds a PhD in Economics from Rutgers University in New Brunswick, NJ. Over the last few years, he has been working at the World Bank, including over the last two years in the Human Development Network. His areas of expertise include applied economics and experimental economics, and his recent work has focused on education analysis and policy in Africa, and vulnerability to climate change in both South Asia and the Middle East and North Africa regions.

**Jivanta Schöttli** is a lecturer in the department of political science, South Asia Institute, Heidelberg University, Germany where she completed her doctorate with a summa cum laude. Jivanta did her bachelor's and master's degrees in International Relations and Economic History respectively, from the London School of Economics and Political Science. Her research interests include Indian politics and foreign policy. Based on her doctoral thesis, a book titled, "Strategy and Vision in Politics. Jawaharlal Nehru's policy choices and the designing of political institutions," is due to be published shortly by Routledge, London. She has co-written a *Dictionary of Economics and Politics of South Asia* (Europa, London, 2006) and co-edited *State and Foreign Policy in South Asia* (Samskriti, New Delhi, 2010).

**John Weiss** is professor of Development Economics and associate dean, Research in the School of Social and International Studies, University of Bradford. He has over thirty-five years experience of work on economic development. He has written eight books and is the editor of a further seven books. Recent publications include the *Economics of Industrial Development*, Routledge, 2011; *Poverty Strategies in Asia: Growth Plus* (edited with H. A. Khan), Edward Elgar; 2006; and *Poverty Targeting in Asia* (edited) Edgar Elgar, 2005. He has extensive experience of practical development issues having worked as a consultant for, amongst others, the World Bank, the Asian Development Bank, the United Nations Industrial Development Organisation and the Department for International Development, UK. Whilst on long-term leave from the university he has been a staff member of the European Investment Bank and the Asian Development Bank, where for six years he was Director of Research at the Asian Development Bank Institute, Tokyo.

**Kate Bird** is a socio-economist with substantial experience generating policy-relevant research findings and providing policy advice to both southern governments and donor agencies. Her work has combined an interest in pro-poor growth and poverty analysis and recent work has included analysis of the factors contributing to economic progress in different country settings, a gendered analysis of the role of productive assets in wealth creation, the aid for trade agenda and chronic and intergenerationally transmitted poverty. Kate is interested in methodological innovation and her recent work on conflict and the intergenerational transmission of poverty builds on previous work to explore the dynamic inter-relationship between vulnerability contexts and household and intrahousehold factors. She is a research fellow with the Overseas Development Institute (ODI) and former head of ODI's Growth and Equity Programme and leads interdisciplinary and multi-country research within the Chronic Poverty Research Centre into the intergenerational transmission of poverty, and previously led its work on spatial poverty traps and remote rural areas.

**Kate Higgins** leads the Governance for Equitable Growth team at the North South Institute in Ottawa, Canada. Her research interests include growth, trade, equity, redistribution and poverty dynamics and she has a keen interest in global development frameworks such as the Millennium Development Goals and the post-2015 framework. Previously, she was a Research Fellow at the Overseas Development Institute and an officer at the Australian Agency for International Development (AusAID).

**Luis F. López-Calva** is lead economist for the Poverty and Gender Unit, Latin America and the Caribbean at the World Bank. During 2007–2010 he was chief economist at the Regional Bureau of Latin America and the Caribbean of the United Nations Development Programme (UNDP). In the 2006–2007 period he was a visiting scholar at the Stanford Center for International Development at Stanford University. Until December 2006 he was associate professor and chair of the Masters in Public Economics in the Graduate School of Public Affairs of the Tecnológico de Monterrey, Mexico City Campus. Between 2002 and 2006 he collaborated as Director of the Human Development Research Office in Mexico (UNDP-Mexico). López-Calva has also been professor of Economics at the Universidad de la Américas en Puebla and El Colegio de México. He has published in both national and international journals and edited volumes, on issues related to child labor, poverty, institutional economics and economic development. He has published four books, on *Economic Integration of the Mexico and California Economies, Child Labor in Latin America*, the *Measurement of Human Development in Mexico* (co-edited with Miguel Székely) and *Declining Latin American Inequality* (co-edited with Nora Lustig). López-Calva has a bachelor's degree in Economics from the Universidad de las Américas Puebla, has a master's in Economics from Boston University, and a master's and PhD in Economics from Cornell University.

**Markus Pauli** is a PhD candidate in the Cluster of Excellence "Asia and Europe in a Global Context" at Heidelberg University. Markus has studied Political Science with a focus on Political Economy and International Relations at the Free University of Berlin and the London School of Economics and Political Science. Thereafter he worked as a project coordinator for InWEnt - Capacity Building International, a non-profit organisation dedicated to human resource development and commissioned by the German Federal Government. His doctoral thesis is on individual freedoms as a concept and practice in development and the operationalization of the capability approach for assessing the impact of microfinance in India.

**Milo Vandemoortele** at the time of writing was a researcher at the Overseas Development Institute, she is currently a PhD candidate at the LSE, London. She is a development economist whose key research work lies at the interface between growth, equity and human development. She has extensive developing country experience. Milo has worked on projects examining the link between growth and equity; specifically she has examined polarisation indices' explanatory value in the link between growth and human development (in Pakistan), contributed to the design and roll out of training workshops on pro-poor growth policies (Tanzania); led a team of researchers to measure equitable progress in development (using DHS and MICS data); led three case studies on factors contributing to progress in economic conditions (Viet Nam, Mauritius, Malawi); and contributed to a health system strengthening projects (in South Africa, Ethiopia) and a drug access programmes (in South Africa, Morocco, China, Thailand).

**Nora Lustig** is Samuel Z. Stone Professor of Latin American Economics at Tulane University and a nonresident fellow at the Center for Global Development and the Inter-American Dialogue. Her research focuses on inequality, poverty, social policy, and development with an emphasis on Latin America. A sample of her publications includes: *Declining Inequality in Latin America. A Decade of Progress?; Thought for Food: the Challenges of Coping with Soaring Food Prices; The Microeconomics of Income Distribution Dynamics; and Shielding the Poor: Social Protection in the Developing World.* She was co-director of the World Bank's World Development Report 2000/1, *Attacking Poverty,* founding member and president of LACEA (Latin American and Caribbean Economic Association), and chair of the Mexican Commission on Macroeconomics and Health. She is currently the editor of the *Journal of Economic Inequality's* Forum and a member of the editorial board of *Feminist Economics* and *Latin American Research Review.* She is also a member of the board of directors of the Institute of Development Studies and the Global Development Network; and a member of the advisory board of the Center for Global Development and Columbia University's Earth Institute. She received her doctorate in Economics from the University of California, Berkeley. Her current research focuses on assessing the impact of fiscal policy on inequality and poverty in Latin America.

**Rishikesh R. Bhandary** is a graduate student at the Fletcher School of Law and Diplomacy at Tufts University. He is specializing in international environment and resource policy and negotiations and conflict resolution. He has consulted for the Government of Nepal's Ministry of Forests and Soil Conservation on different elements of the REDD readiness process and on assessing the feasibility of payments for environmental services in Western Nepal. Bhandary's latest relevant work is the co-edited *Ready for REDD? Taking Stock of Achievements* that brought together the latest scholarly work on REDD in Nepal. He has a particular interest in how climate finance mechanisms like REDD are negotiated.

**Roberto Foa** is a doctoral candidate in Government at Harvard University, and a consultant at the World Bank. His research focuses on the historical origins of contemporary social and political institutions, with a focus on legacies of colonialism and settlement. At the World Bank Roberto was the designer of the Indices of Social Development, and has published a range of articles and book chapters on topics such as social capital, democracy, and economic development. He has also served as a consultant to the Club of Madrid, the OECD, the World Values Survey, the Africa Progress Panel, the Higher School of Economics, and the International Institute of Social Studies. His teaching at Harvard includes the Politics of India and the Sophomore Tutorial in Government.

**Sergei Soares** is a researcher at the Institute for Applied Economic Research of the Brazilian Government. He has publications in the areas of poverty, inequality, income distribution, labor markets, education, and, most of all, social protection and conditional cash transfers. Ha has worked previously at the Ministry of Education and the World Bank.

**Subrata Mitra** is professor of Political Science at the South Asia Institute, Heidelberg University. Professor Mitra, PhD (Rochester) has taught at the universities of Hull, Nottingham in Great Britain and California, Berkeley. Governance, sub-nationalism, rational choice theory, quantitative applications in Political Science and South Asian politics and security, and citizenship are among his main areas of interest. Among his publications are *The Puzzle of India's Governance: Culture, Context and Comparative Theory* (Routledge, 2005), *Modern Politics of South Asia* (edited, 5 volumes, Routledge, 2008), *When Rebels become Stakeholders* (Sage, 2009) and *Politics in India: Structure, Process and Policy* (Routledge, 2010). His articles have appeared in the *Journal of Commonwealth and Comparative Politics*, the *British Journal of Political Science*, *Comparative Politics*, *World Politics*, *Comparative Studies in Society and History*, *The International Social Science Journal*, *West European Politics*, *Third World Quarterly*, and *Comparative Political Studies*.

**Quentin Wodon** is adviser in the Human Development Network at the World Bank. Upon completing business engineering studies, he worked first in Thailand as Laureate of the Prize of the Belgian Minister for Foreign Trade, and next for Procter and Gamble Benelux. In 1988, he decided to shift career to work on poverty and development policy issues. He joined the International Movement ATD Fourth World, a grassroots and advocacy NGO working with the extreme poor. He later completed his PhD in Economics at American University, taught at the University of Namur, and came to the World Bank in 1998, where he worked first on Latin America and next on sub-Saharan Africa before joining the Human Development Network in 2008. Dr. Wodon has published more than a dozen books and 200 papers.

**Terry Sunderland** is a Senior Scientist with CIFOR's Forests and Livelihoods programme, and leads the research domain "Managing trade-offs between conservation and development at the landscape scale". Prior to joining CIFOR, Terry was based in Central Africa for many years. Terry holds a PhD from University College London and has published extensively on conservation and livelihood issues.

# PREFACE

Even though there has been a consistent decline in the proportion of people living in poverty worldwide over the past decades, in many developing countries, poverty has become more widespread, making poverty reduction the core challenge for development in the twenty first century. Climate change adds another layer of complexity to poverty reduction challenge by compounding existing vulnerabilities and risks. Climate-induced food and fuel shortages, which have direct bearing on poverty, have become increasingly frequent over the years.

Why tackle climatic risks when there are other shocks and stresses that poor households and communities face on a regular basis? Given the impact climate change has on economic growth, endowments, entitlements, and health, etc., it could slow or possibly even reverse poverty reduction progress and development gains made so far. Many sectors providing basic livelihood services to the poor are struggling to cope even with today's climate variability and stresses. Poverty reduction will increasingly become difficult in future, as poverty-stricken areas are, often, also the ones that are most vulnerable to climate change and climate-induced extreme events. Given the fact that the welfare of communities is highly vulnerable to weather-related extremes, assessment and reduction of vulnerability to climate change should be an integral part of poverty reduction strategy. Failure to include climate change considerations in poverty reduction strategies and frameworks will make it impossible to achieve poverty reduction goals. Undermining of climate change risks could further enhance other shocks and stresses that governments in developing countries are struggling to avoid or minimize. Increase in both the frequency and duration of climate-induced food and fuel shortages could adversely impact poor's ability to save and invest for the future while keeping pace with day-to-day basic needs.

In this volume, a number of chapters look at thorny issues such as social inequality, weak governance, and fragile institutions that adversely affect poverty alleviation and climate change mitigation and adaptation outcomes. Impacts of food and fertilizer prices on poverty in climate-sensitive countries are analyzed. The role of official development assistance in reducing poverty is critically assessed. Along with challenges, climate change also provides an opportunity to reduce poverty and vulnerability to climate change. Through targeted social protection policies and payment for ecosystem services programs, developing countries can reduce both climate vulnerabilities and poverty. The authors examine the use of various instruments that have the potential to reduce both poverty and vulnerability to climate change. Prospects of climate change adaptation and the international aid regime policy convergence is analyzed. Country case studies are provided to illustrate the possibility of equitable growth, which is crucial for both climate vulnerability and poverty reduction. Through cases from various continents, the volume tries to tease out spatial similarities and differences in poverty reduction challenges and opportunities.

**1**

# MARKETS, THE STATE AND THE DYNAMICS OF INEQUALITY IN LATIN AMERICA: BRAZIL, MEXICO AND URUGUAY

*Nora Lustig and Luis F. Lopez-Calva*

Inequality in Latin America is very high—allegedly the highest in the world—and has been so for decades. In the last twenty-five years the distribution of income in Latin America has undergone two distinct trends. During the 1980s and early 1990s, income inequality increased in most countries for which comparable data is available.[1] Starting in the second half of the 1990s, inequality began to decline (figure 1.1). Between 2000 and 2009, while inequality was rising in other parts of the world, it declined in thirteen of the seventeen countries for which comparable data exist at an average pace of close to 1 percent per year (figure 1.2).

**Figure 1.1. Inequality in Latin America: 1990s–2000s (Gini coefficient)**

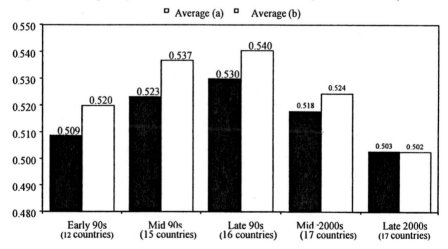

*Source*: Lustig et al. (2011), Figure 2.

*Note*: Unweighted averages. (a) for all countries; (b) for countries in which inequality declined in the 2000s

**Figure 1.2. Declining inequality in Latin America by country: 2000–2009 (annual % change in Gini)**

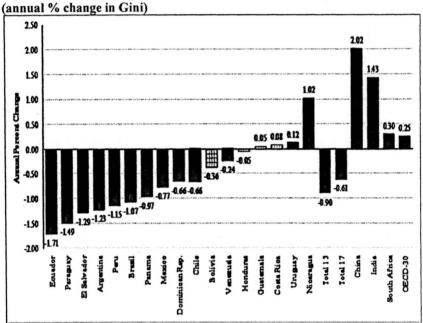

*Source*: Lustig et al. (2011), Figure 3

*Note*: Solid bars represent cases where changes are statistically significant based on SEDLAC's estimates. Data for Argentina and Uruguay are for urban areas only. In Uruguay, urban areas covered by the survey represent 80 percent of the total population; in Argentina, they represent 66 percent. The average change in the Gini for each country is calculated as the percentage change between the end year and the initial year divided by the number of years; the average for the total is the simple average of the changes by country (thirteen countries in which inequality fell). The years used to estimate the percentage change are as follows: Argentina (2009–00), Bolivia (2007–01), Brazil (2009–01), Chile (2009–00), Costa Rica (2009–01), Dominican Republic (2009–00), Ecuador (2009–03), El Salvador (2008–00), Guatemala (2006–00), Honduras (2009–01), Mexico (2008–00), Nicaragua (2005–01), Panama (2009–01), Paraguay (2009–02), Peru (2009–01), Uruguay (2009–00), and Venezuela (2006–00). Using the bootstrap method, with a 95 percent significance level, the changes were not found to be statistically significant for the following countries: Bolivia, Costa Rica, Guatemala, and Honduras (represented by grid bars in the figure). The years used in non-Latin American countries are as follows: China (1993–Mid'00s), India (1993–Mid'00s), South Africa (1993–08), and OECD-30 (Mid 80s–Mid'00s).

Widespread as the decline in inequality has been, in some countries the distribution of income became more unequal (figure 1. 2). Why did inequality decline in some countries while it increased in others? How important were market forces and state actions in explaining the disparate dynamics? Here we will address these questions by synthesizing the results of three case studies that were part of the UNDP-sponsored project "Markets, the State and the Dynamics of Inequality in Latin America," coordinated by the authors. We will focus on two countries in which inequality declined (Brazil and Mexico) and one in which it rose (Uruguay). Particular emphasis will be placed on the role played by returns to education and government transfers in accounting for the inequality dynamics.

The paper is organized as follows. The first three sections will present the results for Brazil, Mexico and Uruguay, respectively. The fourth section will conclude.

# BRAZIL[2]

After a few years with very little change, the Gini coefficient has been falling steadily since 1998. The steepest decline occurred after 2000 when Brazil's Gini coefficient declined to the tune of 1.3 percent per year[3] (figure 1.3). Extreme and moderate poverty declined as well[4] (table 1.1).

**Figure 1.3.–Evolution of the Gini coefficient for the Brazilian distribution of persons according to their family per capita income**

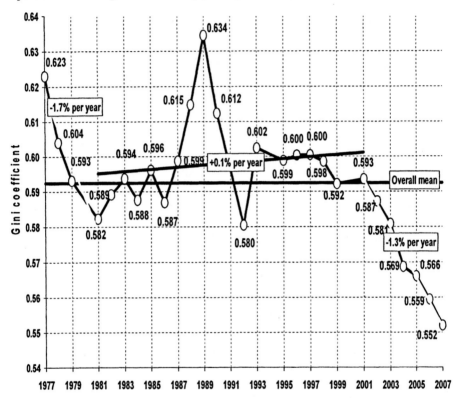

*Source*: Based on Pesquisa Nacional de Domicilios (PNAD), 1997 to 2007, Barros et al. (2009)

**Table 1.1. Poverty and Extreme Poverty in Brazil, 2001–2007**

| Indicators | 2001 | 2007 | Variation 2001-2007 (%) |
|---|---|---|---|
| **Poverty**[2] | | | |
| Headcount ratio | 39 | 28 | -28 |
| Poverty gap[1] | 18 | 12 | -34 |
| Poverty severity[1] | 11 | 7 | -37 |
| **Extreme poverty**[2] | | | |
| Headcount ratio | 17 | 10 | -42 |
| Poverty gap[1] | 7 | 4 | -40 |
| Poverty severity[1] | 5 | 3 | -37 |

*Source*: Estimates based on Pesquisa Nacional por Amostra de Domicílios (PNAD), 2001–2007.

*Notes*: 1. The poverty gap and severity are expressed in multiples of the poverty line.

2. Estimates made using regional poverty lines. The national average poverty line is equal to R$175 per month, and national average extreme poverty line is equal to R$88 per month.

*Source*: Barros et al. (2009).

Thus, based on the observed trends in poverty and inequality, Brazil's growth pattern could be defined as "pro-poor:" i.e., the growth of the income of the poor has been higher than the growth of the income of the rich. From 2001 to 2007, the per capita income of the poorest 10 percent grew 7 percent per year, a rate of growth nearly three times the national average (2.5 percent) while that of the richest 10 percent grew only 1.1 percent (figure 1.4). Two–thirds of the decline in extreme poverty can be attributed to the reduction in inequality.

**Figure 1.4. Annual growth rate for per capita income in Brazil, by percentile, 2001–2007**

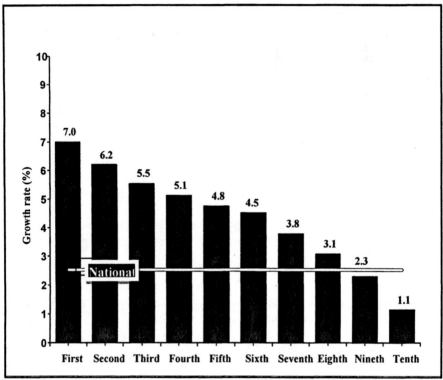

Based on Pesquisa Nacional de Domicilios (PNAD), 1977 to 2007
*Source*: Barros et al. (2009).

    The three main factors behind Brazil's decline in inequality: i) a reduction in education inequality and decreasing wage differentials by educational level; ii) increasing spatial and sectoral integration of labor markets; and, iii) higher government transfers for the poor.[5,6] Public transfers account for over 80 percent of families' non-labor income[7] and the percentage of the population in families with at least one beneficiary increased by 10 percentage points since 2001. In contrast to the episode of falling inequality in the late 1970s, demographic factors and the role of employment were not significant in either direction. For example, while not the case in the late 1970s, changes in the dependency ratio among the poor were of relatively little importance. In the past, a rise in unemployment among the poor prevented the fall of inequality from being more pronounced. In the current episode, however, unemployment has not played a role. All in all, the decline in labor income inequality has been primarily due to the reduction in wage inequality, and, because of the large expansion of access to education for the lower end of the distribution, the reduction in wage inequality is associated with the reduction in education inequality. Since 1995, labor earnings differentials by education level have declined at all levels. This reduction is much clearer after 2002, particularly for secondary and higher education.

As mentioned above, the decline in non-labor income inequality is also very important in explaining the reduction in overall household income inequality. While the size of the contribution varies depending on the methodology, for consistency purposes here we show the results that use the same method as the one used for estimating the contribution of the change in labor income inequality.[8] The decomposition exercises attempt to isolate the contribution of each source to the overall change in inequality: assets (rents, interest and dividends), private transfers, and public transfers.

In sum, in the case of Brazil the rapid decline in income inequality observed since 2001 may be attributed to the reaping of the benefits of the expansion of education, the changes in spatial patterns of labor demand and supply, and the larger size and increased progressivity in some public transfers, from both social security and social assistance, but more importantly from the former. However, the wage gap between formal and informal workers continued to increase and some government policies tempered the progress achieved in inequality reduction. In particular, it seems that raising the minimum wage—which in effect raises the social security benefits that are tied to the minimum wage—is less effective in reducing inequality and extreme poverty than targeted programs such as *Bolsa Família*.

# MEXICO[9]

After a period of rising household income inequality from 1984 until the mid-1990s, Mexico's inequality has been falling. In particular, between 2000 and 2006, the Gini coefficient dropped from close to .53 to close to .49 or by percentage points. This means a fall of 1.3 percent per year which is equal to the one observed in Brazil for the same time period.

Extreme poverty[10] has also been falling consistently since the mid-1990s following the spike in poverty caused by the 1994–95 peso crisis[11] (figure 1.5). In particular, extreme poverty fell by 43 percent between 2000 and 2006. This is especially remarkable given that during this period per capita GDP grew at a modest 2.5 percent or less per year. The latter emphasizes the role played by the reduction in inequality in explaining the reduction in poverty.

**Figure 1.5. Mexico: Incidence of extreme poverty (left axis): 1968–2006**

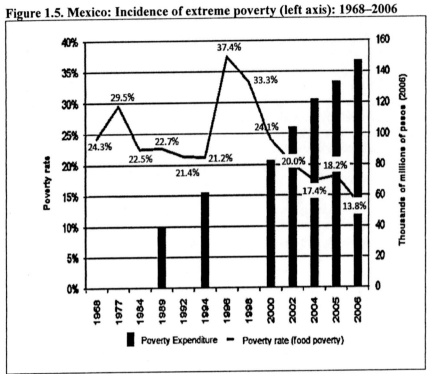

*Source*: Scott (2009).

The growth incidence curve[12] for 2006/2000 plotted in figure 1. 6 shows that the incomes of the poorest 40 percent grew faster than the mean of the growth rates for the entire distribution—the higher of the two horizontal lines in the figure.[13] Thus, during this period Mexico experienced "pro-poor" growth. The next question is which factors explain this growth pattern: changes in demographics, changes in employment patterns, changes in wage inequality, or changes in government transfers?

**Figure 1.6. Mexico: Growth Incidence Curves: 2006/2000**

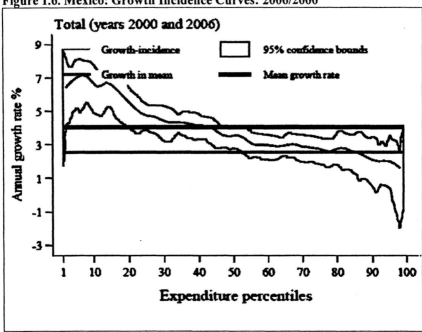

*Source*: Esquivel (2009).                    Expenditure Quantiles

Recent decomposition exercises of the change in inequality between the period 2000 and 2006 found the following results. Demographic changes as measured by the proportion of adults were equalizing and so were the changes in the proportion of employed adults. This means that the dependency ratio and the number of working adults per household "improved" relatively more for the poorer households than for the richer households. The inequality in the distribution of labor income and non-labor income fell, with both contributing to the reduction in overall household income inequality. Labor income includes wages and remunerations of the self-employed. Non-labor income includes incomes from property, own businesses and transfers. Transfers, in turn, can be private (remittances and gifts, for example) or public (pensions and conditional cash transfers, for example).

The fall in inequality of labor income is by far the most important factor in explaining the decline in overall household income inequality. Between 2000 and 2006, the Gini coefficient fell by 3.07 percentage points or by 5.8 percent. If the only factor that changed between 2000 and 2006 had been the distribution of labor income, the Gini would have fallen by 3.19 percentage points—that is, by even more than the overall decline in inequality.[14]

The decline in labor income inequality reflects the fall in the skilled/unskilled workers wage gap. This was one of the major components explaining the increase in overall income inequality between the mid–1980s and the mid-1990s. Since the mid-1990s, however, the upward trend of labor income inequality was reversed. Because it coincided with the implementation of the North American Free Trade Agreement (NAFTA) in 1994, there has been a lot of interest in determining to what extent this equalizing trend in relative wages was a product of NAFTA. So far, this question remains unanswered. With NAFTA there was an increase in demand for low-skilled workers for the "maquiladora" sector. However, during the same period there was also an increase in the share of workers with post-secondary education relative to those without. The share of less-skilled workers (those with less than secondary education) went from 55 percent in 1989 to 32 percent in 2006.

So, it seems that both demand (for example, increased employment in maquiladoras) and supply (changes in the relative abundance of low-skilled workers) factors may have played a role in reducing the wage gap between the skilled and the low-skilled. Figure 1.7 shows how, between 1996 and 2006, the wages of the less educated and less experienced—that is, the low-skilled—workers increased while the wages of the high-skilled workers fell slightly. This is consistent with the shape of the growth-incidence curve, and the large contribution, stemming from the fall in labor income inequality, to the decline in household income inequality.

**Figure 1.7. Mexico: Mean log wage of male workers by education and experience**

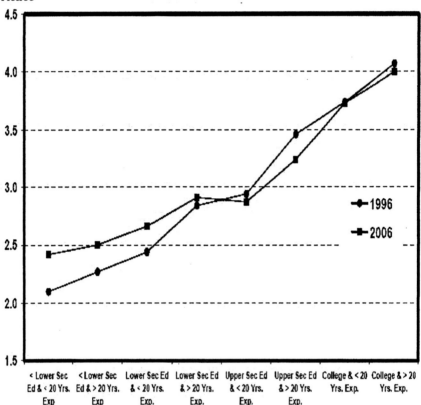

*Source*: Esquivel (2009).

In sum, the decline in household income inequality in Mexico appears to be determined by a relative (relative to workers with more skills, that is) increase in the demand for low-skilled workers and a relative fall in their supply. The latter must be the product of the progress made in education as more and more cohorts stay in school for more years (something that is confirmed by the steady increase in years of schooling).

Other factors that may have contributed to a rise in demand for workers at the bottom of the distribution could have been the increase in remittances and cash transfers from *Progresa/Oportunidades*, the Mexican government's signature anti-poverty program. However, since they all move more or less in tandem, the direct effect of remittances and cash transfers seems to affect the level but not necessarily the trend in income inequality. What might be more important is the indirect effect: that is, the spillover effect that remittances and cash transfers have on employment in poor local economies. Households which receive remittances tend to use them to build, expand or refurbish their dwellings. This generates demand for construction workers in the local economy, who in turn generate demand for other goods and services, and so on. One can think of remittances and cash transfers as myriad "stimulus packages" benefiting poor communities.

Mexico, thus, seems to be a case of lackluster overall growth in GDP and total factor productivity because a large portion of the employment generation occurs at the low-productivity/low-wage end instead of in the high-wage/high-productivity sectors. However, even if the new employment opportunities are low-wage, the wages (or remunerations) they pay are higher than what this group of low-skilled workers used to receive before 2000. In this sense, Mexico's growth pattern is "pro-poor." Although the launching of the anti–poverty conditional cash transfer program *Progresa/Oportunidades* made public spending more progressive, the bulk of transfers (pensions, in particular) is not.[15] By some estimates, without *Oportunidades* the Gini coefficient would be around one percentage point higher,[16] which is not insignificant. Nonetheless, public spending remains largely not pro-poor and in a number of cases it is plainly regressive. Thus Mexico's recent reductions in inequality, while important, remain limited because social policy still has serious shortcomings and inconsistencies. The good news is that this means that there are plenty of opportunities to further reduce poverty and inequality.

# URUGUAY[17]

In the early 1970s, Uruguay began a liberalization process that resulted in high growth rates and an increase of the manufacturing share in total exports (from 40 percent to 70 percent). However, in the early 1980s, the country faced a strong recession with GDP decreasing by 17 percent from 1981 to 1984. During the second half of this decade, as democracy was reestablished, export-oriented policies were applied, free trade agreements were encouraged, and the *Consejo de Salarios* was restarted, all of which led to a significant increase in real wages. In the 1990s, Uruguay joined MERCOSUR and a new stabilization policy based on the exchange rate was implemented. The joint result was a strong appreciation of currency, as well as the growth of investment and GDP. The *Consejo de Salarios* was abolished again in 1990–91 and labor contracts were liberalized instead. At the end of the 1990s, the unstable regional situation led to an economic recession that turned into a severe crisis in 2002: per capita GDP decreased 11 percent and unemployment rates were higher than they were in the crisis of 1982. Since the last quarter of 2003, motivated by the demand for primary goods, the economy has started to grow at an extraordinary rate. The centre-left party came to power in 2005 and launched several reforms such as the restoration of the *Consejo de Salarios* (Wage Council), the introduction of an income tax, the implementation of cash transfers schemes, and a health reform. Increasing inequality is the main feature of the period analyzed in the study (figure 1.8).

**Figure 1.8. Gini index and household per capita income, 1986–2009**

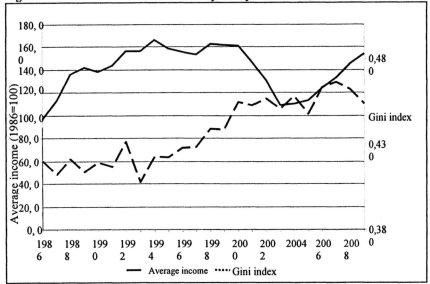

An analysis thorough Growth Incidence Curves (GIC) by sub-periods shows that from 1986 to 1994, per capita income expanded by 7 percent annually, with a higher increase for the poor than for the upper deciles.[18] However, per capita income for the upper deciles also increased at a higher rate than the mean and, as a result, overall inequality did not change substantially. Over 1994–2007, the Gini coefficient rose from 42.3 to 46.6 and average per capita income decreased at 2.7 percent annually, falling at a faster rate among the poor. Finally, over 2007–2009, inequality decreased as the growth of income for the poor was higher than the mean growth.[19] Several policy changes may explain the latest result: the minimum wage increased in 2005 and could have contributed to lower earnings inequality; ex ante analysis of the tax reform in 2008 also suggests equalizing effects (a decline between 1 and 2 percentage points in the Gini index); and the effect of cash transfer programs on inequality, although modest, was equalizing (with a decrease of 0.5 points in the Gini index).

To analyze the contribution of each income source to the overall change in inequality, the authors decomposed the Gini coefficient for the first year of each of the sub-periods analyzed: 1986, 1994, 2007 and 2009.[20] Overall, the results suggest that labor income has had an important role in explaining the inequality increase between 1994 and 2007. During 2007–09, the decline inequality indices experienced is mainly explained by the equalizing effect of both labor income and social benefits, specifically social transfers, which have been expanded in recent years.

To disentangle the driving forces explaining the changes in inequality, the authors estimated a parametric microsimulation model, focusing on the effects of labor market participation and education.[21] The microsimulation was carried out for the following periods: 1986–2009; 1986–94; 1994–2007; and, 2007–09.[22] For the period as a whole, 1986–2009, during which inequality rose, the leading factors were the unequalizing contribution of changes in returns to education and the earnings gender gap (around 70 percent of the increase is explained by the former and 20 percent by the latter). Between 1986 and 1994, inequality did not experience any change. During this period, changes in returns to education were unequalizing; however, this was more than compensated for by equalizing changes in the distribution of hours worked and a reduction in the regional gap between Montevideo and the rest of the urban areas in earnings.

The leading force explaining the increase in both labor income inequality and per capita household income inequality in 1994–2007 is the returns to education (approximately 80 percent). Labor market participation, the evolution of the gender gap, and the regional gap reinforced this trend (and changes in the distribution of the stock of education were equalizing but the effect was too small to countervail the unequalizing forces). During the small decline in inequality between 2007 and 2009, returns to education were the main driving force behind the decrease in inequality, both at the individual and the household level (78.5 percent). The increase in employment and the reduction in the regional gap also had an equalizing effect. The changes in the distribution of education was equalizing but with a negligible order of magnitude.

The results show that income inequality appeared to have peaked around 2007. Since then, inequality of both total per capita household income and hourly labor earnings has declined (figure 1.9). When analyzing the evolution of inequality by educational group, inequality was always higher among the higher skilled workers. Until 2006, inequality increased mainly among high-skilled workers, but after 2007 it increased among workers with six years of education or less and decreased for the rest. In 2009, the decrease of inequality occurred for all of the educational groups and the inequality gaps between groups were reduced. Participation rates increased for almost all educational groups during the post-crisis recovery, and unemployment reached its historical minimum, with a greater decrease for lower and medium skilled workers. In addition, results suggest changes in the structure of the labor force by education level, with the low (high) education workers becoming relatively more scarce (abundant), although this evolution has been milder than it has been in most Latin American countries. The main explanation for the mild evolution relies in the endemic drop-out rates affecting the secondary school system. These rates may explain why inequality, in contrast with other countries, continued to rise during most of the decade.

**Figure 1.9. Gini coefficient for household per capita income and labor hourly earnings, 1986–2009**

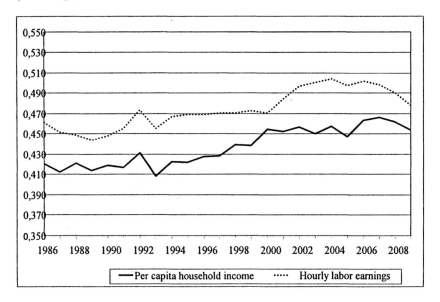

From the microsimulation model, the results suggest that the leading forces in the evolution of both household per capita income and labor income inequality were the returns to education, labor market participation and the residual factors. The latter was evident over 1994–2007 when inequality increased at its greatest rate. Meanwhile, the regional gap decreased, possibly a result of the increase in minimum wage (which were probably binding in the more deprived areas outside Montevideo), and the presence of the *Consejo de Salarios*.[23]

Summing up, the whole period 1986–2009 is marked by an increasing inequality trend. The 2002 crisis and the recovery were accompanied by increasing inequality. However, this trend was reversed in 2007. Overall, labor market forces and returns to education were the main causes for both the rise and the fall of inequality. Why has the skill premium declined since 2008? One plausible explanation may be the active labor market policies, in the form of higher minimum wages and the restitution of industry-wide collective bargaining processes, implemented by a left-leaning government. In addition, the government introduced a tax reform which increased the effective tax rate for top income earners (including high-wage earners) and, thus, reduced the skill premium. However, if it generates a higher supply of informal workers, the active labor market policies, however, may eventually result in unequalizing forces thereby increasing the gap between formal and informal labor's incomes. It is important to note that the upgrading of education does not appear to have played an important role, since Uruguay is one of the sad cases in which the pace at which the increase in average years of schooling has been particularly slow. The reduction in inequality since 2007, therefore, appears to depend more on pro-labor government actions and pro-poor transfers than on market forces.

## CONCLUSION

Since around 2000, inequality has been declining in the majority of countries in Latin America. It has also risen in some. In this chapter we analyzed the causes underlying the dynamics of inequality by focusing on two cases where inequality declined (Brazil and Mexico) and one in which it rose (Uruguay). The analyses presented above seem to indicate that both market forces and state action played important roles in explaining inequality dynamics. In particular, the decline in inequality in Brazil and Mexico seems to be strongly associated with a decline in the skill premium or returns to education.[24] In contrast, the leading force explaining the increase in income inequality in Uruguay during 1994–2007 is the increase in returns to education (skill premium rose). Due to particularly elevated high-school drop-out rates, the composition of its population by education level hardly changed in Uruguay.

State action contributed to the decline in inequality in Brazil, Mexico and Uruguay (2007–2009) in three main ways. First, greater access to basic education is the result of explicit efforts made by the governments of Brazil and Mexico. Second, government (net) transfers became more generous and progressive. The large-scale signature conditional cash transfer programs *Bolsa Familia* (Brazil) and *Oportunidades* (Mexico) reduced per capita household income inequality by between 10 and 20 percent. Third, in Brazil and in the recent period in Uruguay (2007–2009), state action let itself be felt through active labor market policies. In both countries, higher minimum wages compressed the wage distribution. In Uruguay, this process was reinforced by the reinstatement of industry-wide collective bargaining.

## NOTES

1. See, for example, Altimir (2008) and Londoño and Szekely (2000). As we shall see in the next section, in the 1980s only six of the countries had national data while the remaining seven covered urban areas (and in some countries only the large metropolis).

2. This section is based on Barros, de Carvalho, Franco and Mendonça (2009b and 2010).

3. The decline in income inequality in Brazil fulfills the "Lorenz dominance" test and it is statistically significant. Barros et al. (2009).

4. Barros et al. (2009).

5. Barros et al. (2009).

6. If one decomposes the change in household income inequality, it is found that roughly half of the decline can be attributed to a reduction in labor income inequality and half to a reduction in non-labor income inequality. Labor income includes wages and remunerations of the self-employed. Non-labor income includes incomes from property, own businesses and transfers. Transfers, in turn, can be private (remittances and gifts, for example) or public (pensions and conditional cash transfers, for example).

7. Household surveys usually do not accurately capture households at the very top of the distribution whose income is likely to come from assets. This is true for all countries.

8. See Barros et al. (2009).

9. This section is based on Esquivel, Lustig and Scott (2010).

10. The incidence of poverty is measured using the headcount ratio. In Mexico, extreme poverty is measured using the official "food poverty" line.

11. The 2005 figures should be taken with caution because surveys may not be comparable to surveys in the rest of the years.

12. A "growth incidence curve" plots the changes in household per capita income (or expenditure) for each quantile of income (or expenditure), from poorest to richest households. Quantiles are usually percentiles or smaller.

13. The mean of the growth rates of the entire distribution was slightly above 2.5 percent.

14. Alejo et al (2009).

15. Scott (2009).

16. Scott (2009).

17. This section is based on Alves et al. (2011).

18. ECH gathers after-tax current income. The household income aggregate included in this study adds up labor earnings for salaried workers, self-employed

and entrepreneurs (both in kind and in cash), capital income (coming from interests, rents, royalties, financial assets), public and private transfers (including remittances) and imputed owner's occupied housing. This income definition is different from the one used by INE as ours excludes the imputation for health insurance coverage. INE adds to household income the market value of health insurance for formal workers. Since 2007, when a significant reform in the health system took place and coverage was expanded to the offspring aged 18 years or younger, our average income series does not show the same trend as the INE ones. The Gini index also varies when these imputations are included. In spite of this, the inequality trends are the same when removing these imputations.

19. Overall, if the whole time span is considered, the GIC curve is U shaped, reflecting higher rates of income growth for the lower and upper tiers of the income distribution.

20. The decomposition of the Gini index by income source follows the methodology developed by Lerman and Yitzhaki (1985).

21. See Bourguignon and Ferreira (2005).

22. In the model, labor income is designed following two equations: i) an hourly earnings equation using Heckman, explained by sex, age, square of age, region of residence and a set of dummy variables for education levels (the selection equation also includes a variable reflecting whether the person was cohabiting with a partner and a variable reflecting the presence of children aged 5 years or less); and ii) an equation for worked hours using a Tobit model, estimated separately for heads, spouses and the rest of the members, and explained by sex, schooling, age, region and non-labor income for the household head. To estimate counterfactuals for education and unobservable factors, the authors estimated an equation using the years of schooling as the dependent variable. From this model, they distinguished the effect on total inequality of the returns to education, returns to experience, changes in the gender gap, returns to residing in Montevideo, hours worked, educational structure and unobservable factors.

23. From 1985 to 1991, private, public and rural wages were set in collective agreements among the government, entrepreneurs and unions. This wage setting mechanism was suppressed during 1990–91, when wage councils were no longer convened. From then on, wage bargaining took place mainly at firm level, either by means of a negotiation union-firm or, most of the times, through a direct agreement between the worker and the employer. In 2005 collective wage bargaining was reinstalled.

24. Similar dynamics in the labor market were found in Argentina and Chile; see Gasparini and Cruces (2010) and Eberhardt and Engel (2008).

## REFERENCES

Alejo, Javier, Marcelo Bergolo, Fedora Carbajal, and Guillermo Cruces. "Cambios en la desigualdad del ingreso en América Latina. Contribución de sus principales determinantes (1995–2006). Informe Final." *Markets, the State and the Dynamics of Inequality in Latin America* co-ordinated by Luis Felipe Lopez-Calva and Nora Lustig. Background paper prepared for the UNDP project, 2009. http://undp.economiccluster-lac.org/.

Altimir, Oscar. "Distribución del ingreso e incidencia de la pobreza a lo largo del ajuste." *Revista de la CEPAL* no. 96, December 2008.

Alves, Guillermo, Verónica Amarante, Gonzalo Salas, and Andrea Vigorito. (2011). "The Evolution of Inequality in (Urban) Uruguay in the Last Three Decades (1986–2009)." Discussion paper prepared for the UNDP Project *Markets, the State and the Dynamics of Inequality: How to Advance Inclusive Growth,* co-ordinated by Luis Felipe Lopez-Calva and Nora Lustig.

Barros, Ricardo, Francisco H. G. Ferreira, José R. Molinas Vega, and Jaime Saavedra Chanduvi. *Measuring Inequality of Opportunities in Latin America and the Caribbean.* Washington, DC: World Bank, 2009.

Barros, Ricardo, Mirela de Carvalho, Samuel Franco, and Rosane Mendonça. "Markets, the State and the Dynamics of Inequality: Brazil's Case Study," prepared for the project *Markets, the State and the Dynamics of Inequality* UNDP, 2009b.

Barros, Ricardo, Mirela de Carvalho, Samuel Franco, and Rosane Mendonça. "Markets, the State and the Dynamics of Inequality in Brazil." In *Declining Inequality in Latin America: A Decade of Progress?* Ed. Luis F. López Calva and Nora Lustig. Washington DC: Brookings Institution and UNDP, 2010.

Bourguignon, Francois and Francisco G. H. Ferreira. "Decomposing Changes in the Distribution of Household Incomes: Methodological Aspects." In *The Microeconomics of Income Distribution Dynamics in East Asia and Latin America,* ed. Bourguignon, Francois, Francisco G. H. Ferreira and Nora Lustig. New York: Oxford University Press, 2005.

Esquivel, Gerardo. "The Dynamics of Income Inequality in Mexico since NAFTA." Background paper prepared for the UNDP project *Markets, the State and the Dynamics of Inequality: How to Advance Inclusive Growth,* co-ordinated by Luis Felipe Lopez- Calva and Nora Lustig. UNDP, 2009. http://undp.economiccluster-lac.org/.

Eberhard, Juan and Eduardo Engel. "Decreasing Wage Inequality in Chile." Discussion paper prepared for the UNDP Project *Markets, the State and the Dynamics of Inequality: How to Advance Inclusive Growth*, co-ordinated by Luis Felipe Lopez-Calva and Nora Lustig. UNDP, 2008. http://undp.economiccluster-lac.org/.

Esquivel, Gerardo, Nora Lustig, and John Scott. "A Decade of Falling Inequality in Mexico: Market Forces or State Action?" In *Declining Inequality in Latin America: A Decade of Progress?* Chapter 6, ed. Luis F. López Calva and Nora Lustig. Washington DC: Brookings Institution and UNDP, 2010.

Gasparini, Leonardo and Guillermo Cruces. "A Distribution in Motion: The Case of Argentina." In *Declining Inequality in Latin America: A Decade of Progress?* Chapter 5, ed. Luis F. López Calva and Nora Lustig. Washington DC: Brookings Institution and UNDP, 2010.

Lerman, R. and S. Yitzhaki. "Income Inequality Effects by Income," *The Review of Economics and Statistics*, MIT Press 67, no.1 (1985): 151-56.

Londoño and Szekely. "Persistent Poverty and Excess Inequality: Latin America, 1970-1995." *Journal of Applied Economics* no. 1 (May 2000): 93-134.

Lustig, Nora. "Is Latin America Becoming Less Unequal?" in *Vision for Latin America 2040. Achieving a More Inclusive and Prosperous Society*, prepared for CAF (Andean Development Corporation) by Centennial Group, Washington DC, 2010.

Lustig, Nora, Luis F. López-Calva and Eduardo Ortiz-Juarez. "The Decline in Inequality in Latin America: How Much, Since When and Why." Working Paper no. 1118, Tulane University, 2011.

López-Calva, Luis F. and Nora Lustig. "Declining Inequality in Latin America: A Decade of Progress?" Washington, DC: Brookings Institution, 2010.

Scott, John. "Gasto Público y Desarrollo Humano en México: Análisis de Incidencia y Equidad." Working Paper for *Informe de Desarrollo Humano de México 2008/2009*. México: PNUD, 2009.

————. SEDLAC (Socio-Economic Database for Latin America and the Caribbean), CEDLAS y Banco Mundial. La Plata, Argentina y Washington DC, 2009. http://www.depeco.econo.unlp.edu.ar/cedlas/sedlac/.

## 2

## POLITICS AND THE ROOM TO MANEUVER: DEMOCRACY, SOCIAL OPPORTUNITY AND POVERTY IN INDIA

*Subrata K. Mitra, Jivanta Schöttli, Markus Pauli*

Ownership is the essence of economic citizenship. Beyond actual possession, a sense of personal welfare and proprietorship, or at least the hope of achieving them, constitute a necessary and important complement to being stakeholders in a society. Together with efficacy and legitimacy, these are necessary attributes of political agency. In this chapter we examine the interplay among democracy, social opportunity and economic security, drawing upon survey data to explore popular perceptions of India's new economic policy. The chapter first situates itself within the context of new theoretical literature on the multi-dimensional nature of poverty and how to measure it, then identifies aggregate indicators of the performance of Indian states and the overall achievements and failings of India in terms of poverty alleviation. In the second half, the chapter identifies what seems to be the lack of a "politics of poverty" in India and the various cultural, historical, and political explanations that have been proffered for this apparent anomaly. Finally, the impact of democracy on poverty reduction is examined through the programmes that have been launched and which aim at creating a level playing field but which nevertheless have the potential to degenerate into highly populist measures.

Students of poverty measurement and alleviation find contemporary India puzzling. The economy thrives yet poverty persists. This sends mixed signals. As for the former, enthusiastic supporters of neo-liberal economic policies find in India a suitable context for the enthusiastic endorsement of liberalisation. On the other hand, persistent poverty—the dark under-belly of India's thriving corporate sector—is exactly the cue that India's radical Left needs to unleash a barrage of anti-colonial, anti-liberalisation and anti-capitalist policies. In this chapter, we take a middle position between the radical rejection of India's past, and its mechanical acceptance. Instead we argue that it is politics that provides the linkage among poverty, democracy and social policies.

In terms of absolute numbers, India, it is often pointed out has more poor people than any other country. This is true on every count, no matter the method of measurement applied. Furthermore, India's high levels of poverty have

21

persisted despite more than a half-century of policies and rhetoric promising to eradicate mass poverty. In his inaugural speech at Independence in 1947, Jawaharlal Nehru endorsed Mahatma Gandhi's goal of removing "every tear from every eye." For much of the period since Independence, government policy sought to reduce poverty by constraining the freedom of the private sector. Poor farmers were strongly encouraged to form cooperatives. The state intervened to dictate which sections of the industrial sector were to be reserved for small and micro or "cottage" industries. While policies and debates since the 1950s have seen reversals and re-evaluations, committee reports[2] and institutional innovations, the question of how best to tackle and alleviate poverty remains to a large extent, elusive.

Furthermore, the issue of mass poverty brings to the fore the core challenge for India's political economy: how to balance economic growth with redistribution. Scholarly opinion remains divided. Many critics of the Indian model of development consider the continued existence of mass poverty as evidence of the shortcomings of Indian democracy and its political economy of development. Others point in the direction of relative improvements in India's infrastructure, GDP, rate of growth, and improved trade figures as signs of progress. In theoretical and methodological terms, mass poverty raises issues of incredible complexity, pitting quantitative methods against the qualitative and problems of politics and public policy versus moral issues of poverty amidst increasing displays of prosperity and plenty. Some core questions that are going to be raised in this chapter include, how can poverty be measured? Is a cross-cultural, objective measure of poverty possible? Is poverty merely the lack of money or is it a "state of mind"? Is poverty necessarily culture-and context-specific? Why does mass poverty co-exist with spectacular growth in India? Is liberalization "good" for the poor? What has the government done, or is doing for poverty alleviation?

## Conceptualizing and Measuring Poverty

What is poverty? A renowned glossary on poverty identifies twelve distinct definitions.[3] Poverty as a material concept: (1) *Need* constituted by lack of material goods and services—such as basic human needs like safe drinking water, food, shelter, sanitation, health, education or information.[4] (2) *Patterns of deprivation* taking into account the combinations, seriousness and duration of deprivations. (3) *Limited command over resources* as every need is a need for something. Poverty as economic circumstances: (4) *Standard of living* measured as income or consumption. (5) *Inequality* as an intrinsic conceptual part of poverty when poverty is linked to a minimum standard of living tolerable in a society. (6) *Economic position* in society, abstracted as "class" and reflecting the inequality of a social structure. Poverty as social circumstances: (7) *Social class* attaching a socio-economic status to an economic position in society—e.g. "underclass" or in the Indian context "low castes" or "untouchables." (8) *Dependency* mainly conceptualised in the relationship towards the state as following from being dependent on social benefits. (9) *Lack of basic security* leading to vulnerability to social risks. (10) *Lack of entitlement* as the underlying reason for lack of resources. (11) *Exclusion* from the "minimum acceptable way of life" due to limitations in material, social and cultural resources.[5] (12) Poverty as moral judgement: pointing to the element of *unacceptable hardship* of poverty, implying that something ought to be done about it.

Poverty is however, multidimensional. Decisive in the acknowledgement of the multiple dimensions of poverty has been the human development approach spearheaded by the UNDP. Ever since the launching of the human development index in 1990 by Mahbub ul Haq and Amartya Sen, which aimed at putting health and education in addition to income at the centre of poverty conceptualisation, more dimensions of poverty have been acknowledged as essential. A milestone in the multidimensional conceptualisation of poverty has been Jean Drèze's and Amartya Sen's (1989) notion of poverty as a lack of entitlement, pointing out that a lack of essential resources echoes a lack of entitlements.[6] Or in other words it is not the lack of food that produces famines but the inability of people to buy existing food, it is not the lack of housing that constitutes homelessness but the lack of access to housing. Developing this idea further, Sen's capability approach defines development as the expansion of peoples' individual freedoms and their overall capability to enjoy a life they have reason to value. One of the central contributions of the capability approach has been that it draws attention to the importance of agency and freedom of choice. Sen argues: "With adequate social opportunities, individuals can effectively shape their own destiny and help each other. They need not to be seen primarily as passive recipients of the benefits of cunning development programmes. There is indeed a strong rationale for recognizing the positive role of free and sustainable agency—and even of constructive impatience."[7]

This leads to the question of how poverty can be measured. The most pervasive international measures, the poverty lines set by the World Bank, are $1.25 and $2 using 2005 purchasing-power parity (PPP). These poverty lines refer to an underlying definition of poverty as "the inability to attain a minimal standard of living."[8] Such an absolute measure of living standard, usually measured as income or consumption, defines a specific and, to a certain extent arbitrary, threshold under which one is regarded as being poor. It has to be distinguished from a relative standard of living in a society, which conveys inequality and which is the common way to conceptualise and measure poverty in advanced economies (e.g. poverty line at 60 percent of average household income). The beauty of attaining a single, allegedly precise number of poor people brings with it a high price. First, it defines the standard of living narrowly in terms of income, not taking into account other central aspects like life expectancy or access and quality of healthcare and education. Second, non-monetary income and income transfers like free or subsidized public services do not feature due to data limitations. Third, the way income is adjusted for household size and type is rather arbitrary. Fourth, the differences in cost-of-living between different regions and social groups are not captured appropriately.[9] Fifth, inequality within the household is usually not captured. To gain a more comprehensive and instructive understanding of poverty, it seems inevitable that there is a need to look into further dimensions beyond income or consumption.

An innovative measurement, the multidimensional poverty index (MPI) was recently launched in the 20th anniversary Human Development Report.[10] The MPI assesses the "nature and intensity of poverty at the individual level."[11] The MPI goes beyond a mere headcount of poor people and identifies the depth of poverty, meaning the intensity of deprivation across different dimensions of poverty. So far the MPI takes into account three dimensions: (1) health, measured as nutrition and child mortality; (2) education with the indicators being years of schooling and children enrolled; and (3) standard of living, comprising electricity, drinking water, sanitation, type of floor, cooking fuel and assets. However, Alkire and Santos suggest going further. "Missing dimensions," such as quality of work, physical safety, empowerment, psychological well-being as well as the ability "to go about without shame" could be integrated into the MPI.[12] One core feature of the MPI is that it enables the black box of "poverty" to be opened as findings can be analysed according to indicators (e.g. how much the indicators of sanitation or cooking fuel contribute to overall poverty), by groups of the population (e.g. in what dimensions groups like scheduled castes experience more or less deprivation, and to what intensity) and by regions (e.g. which states or even regions within states are exposed to what kind and what intensity of poverty).

The following two tables serve to compare the status of poverty in India in relation to its regional neighbours (Bangladesh and Pakistan) as well as its international peers classified through the BRICS nomenclature. Table 2.1 illustrates the broad range of people classified as poor depending on which method of measurement is applied. In the case of India, Bangladesh and

Pakistan, the percentage of people classified as poor under the MPI is situated in between the two poverty lines proposed by the World Bank (US$1.25 and US$2 a day), ranging from 51 percent to 58 percent. For China and South Africa however, the MPI identifies less people as poor compared to both World Bank poverty lines, at the level of 13 percent for China and 3 percent for South Africa.

According to World Bank´s US$1.25 and US$2 poverty lines as well as to the MPI proportion of poor (in other words the headcount H) the ranking regarding the percentage of people living in poverty is as follows: Bangladesh having the highest percentage of poor, followed by India and then Pakistan. This is not so when using the national poverty lines, as defined by the states themselves, according to which there are some percentage less poor people in India than in Pakistan. Looking at the MPI (last column in table 2.1) one finds that India has the worst performance among the three South Asian states. This is due to the way the MPI is computed: it takes into account the average intensity of deprivation (A) and multiplies it with the proportion of poor (H).[13] The average intensity of deprivation "reflects the proportion of dimensions in which households are, on average, deprived."[14] The dimensions being health, education and standard of living with the (so far) ten indicators, introduced above. On average, poor households in India are deprived across more dimensions than in Bangladesh, whereas Pakistan has an even slightly higher average intensity of deprivation than India (see second last column in table 2.1). Due to this higher average intensity of deprivation in India (despite having some percentage less poor people than Bangladesh), India performs slightly less well than Bangladesh concerning the measurement by the MPI.

In table 2.2 the three dimensions of education, health and living standards are showcased, comparing their indicators among the selected countries. This decomposition allows for a far more comprehensive policy advice since it highlights shortcomings in specific areas. For example in the regional context of South Asia, Bangladesh outperforms India and Pakistan concerning the enrolment of children whereas Pakistan is far better in providing electricity to its citizens. Given the weights of the different indicators the biggest contributor to multidimensional poverty in India is the health sector, with inadequate nutrition contributing disproportionately. Within the living standard dimension cooking fuel and sanitation can be pinpointed as major contributors to multidimensional poverty in India.

Applying the MPI raises several interesting insights. First, the huge variation in multidimensional poverty between the Indian states ranges from 16 percent being poor in Kerala, 32 percent in Tamil Nadu, 58 percent in West Bengal to 81 percent in Bihar (see table 2.3). Second, decomposition along social classes confirms the common impression that marginalised social groups are disproportionately affected by poverty. For example, 81.4 percent of Scheduled Tribes, 65.8 percent of Scheduled Castes and 58.3 percent of Other Backward Classes in India are classified as poor by the MPI.[15] A special feature of the MPI is that it highlights the fact that poverty needs to be examined both in terms of its spread as well as its depth. As seen in table 2.3 the average intensity of poverty, in other words the seriousness of deprivation along the different dimensions, tends to rise in those states with a greater proportion of people living in poverty.

**Table 2.1. Comparison of Poverty Lines & MPI – BRICS plus Pakistan and Bangladesh**

| | **Comparative Poverty Measures** | | | | **Poverty Depth** | **Index** |
|---|---|---|---|---|---|---|
| | US$ 1.25 a day % | US$2 a day % | National Poverty Line % | MPI – Proportion of Poor (H)[a] % | MPI – Average Intensity (A) % | MPI (= H x A) |
| India [b] | 42 | 76 | 29 | 55 | 53.5 | 0.296 |
| Bangladesh[c] | 50 | 81 | 40 | 58 | 50.4 | 0.291 |
| Pakistan | 23 | 60 | 33 | 51 | 54.0 | 0.275 |
| China [d] | 16 | 36 | 3 | 13 | 44.9 | 0.056 |
| Brazil | 5 | 13 | 22 | 9 | 46.0 | 0.039 |
| South Africa | 26 | 43 | 22 | 3 | 46.7 | 0.014 |
| Russia | 2 | 2 | 20 | 1 | 38.9 | 0.005 |

Values have been rounded.
[b] Survey Data for India: DHS 2005.
[c] Survey Data for Bangladesh and Pakistan: DHS 2007.
[d] Survey Data for China, Brazil, South Africa and Russia: WHS 2003.
*Source*: Oxford Poverty and Human Development Initiative (OPHI), Country[a] Briefing – India, Bangladesh, Pakistan, China, Brazil South Africa, Russia, 2010 & Human Development Report, 2010.

| States | Education[a] | | Health[a] | | Living Standard[a] | | | | | |
|---|---|---|---|---|---|---|---|---|---|---|
| | Schooling % | Enrolment (Children) % | Mortality (Children) % | Nutrition % | Electricity % | Sanitation % | Drinking Water % | Floor % | Cooking Fuel % | Assets % |
| India | 18 | 25 | 23 | 39 | 29 | 49 | 12 | 40 | 52 | 38 |
| Bangladesh | 24 | 9 | 24 | 37 | 39 | 48 | 2 | 54 | 57 | 45 |
| Pakistan | 19 | 34 | 30 | N/A | 9 | 33 | 8 | 36 | 42 | 26 |
| China | 11 | 0 | 0 | 3 | 0 | 8 | 3 | 3 | 9 | 2 |
| Brazil | 8 | 0 | 0 | 1 | 0 | 4 | 2 | 2 | 7 | 0 |
| South Africa | 3 | 0 | 0 | 1 | 0 | 2 | 1 | 1 | 2 | 1 |
| Russia | 1 | 0 | 0 | 0 | 0 | 0 | 0 | 0 | 0 | 1 |

Table 2.2. MPI Deprivation for Each Sector – BRICS plus Pakistan and Bangladesh
[a]Values have been rounded.

Source: Oxford Poverty and Human Development Initiative (OPHI), Country Briefing – India, Bangladesh, Pakistan, China, Brazil South Africa, Russia, 2010.

Table 2.3. Decomposition of Multidimensional Poverty across Indian States

| MPI Rank | States | Population (millions) 2007 | Number of MPI Poor (millions) | Proportion of Poor (H) (%) | Average Intensity (A) (%) | MPI (H x A) |
|---|---|---|---|---|---|---|
| 1 | Kerala | 35.0 | 5.6 | 15.9 | 40.9 | 0.065 |
| 2 | Goa | 1.6 | 0.4 | 21.7 | 43.4 | 0.094 |
| 3 | Punjab | 27.1 | 7.1 | 26.2 | 46.0 | 0.120 |
| 4 | Himachal Pradesh | 6.7 | 2.1 | 31.0 | 42.3 | 0.131 |
| 5 | Tamil Nadu | 68.0 | 22.0 | 32.4 | 43.6 | 0.141 |
| 6 | Uttaranchal | 9.6 | 3.9 | 40.3 | 46.9 | 0.189 |
| 7 | Maharashtra | 108.7 | 43.6 | 40.1 | 48.1 | 0.193 |
| 8 | Haryana | 24.1 | 10.0 | 41.6 | 47.9 | 0.199 |
| 9 | Gujarat | 57.3 | 23.8 | 41.5 | 49.2 | 0.205 |
| 10 | Jammu & Kashmir | 12.2 | 5.4 | 43.8 | 47.7 | 0.209 |
| 11 | Andhra Pradesh | 83.9 | 37.5 | 44.7 | 47.1 | 0.211 |
| 12 | Karnataka | 58.6 | 27.0 | 46.1 | 48.3 | 0.223 |
| 13 | Eastern Ind. States | 44.2 | 25.5 | 57.6 | 52.5 | 0.303 |
| 14 | West Bengal | 89.5 | 52.2 | 58.3 | 54.3 | 0.317 |
| 15 | Orissa | 40.7 | 26.0 | 64.0 | 54.0 | 0.345 |
| 16 | Rajasthan | 65.4 | 41.9 | 64.2 | 54.7 | 0.351 |
| 17 | Uttar Pradesh | 192.6 | 134.7 | 69.9 | 55.2 | 0.386 |
| 18 | Chhattisgarh | 23.9 | 17.2 | 71.9 | 53.9 | 0.387 |
| 19 | Madhya Pradesh | 70.0 | 48.6 | 69.5 | 56.0 | 0.389 |
| 20 | Jharkhand | 30.5 | 23.5 | 77.0 | 60.2 | 0.463 |
| 21 | Bihar | 95.0 | 77.3 | 81.4 | 61.3 | 0.499 |
| | **India** | **1,164.7** | **645.0** | **55.4** | **53.5** | **0.296** |

*Source*: Based on Oxford Poverty and Human Development Initiative (OPHI), Country Briefing India, 2010.

# Mass Poverty, Growth versus Redistribution: The Dilemma of Development in Post-Colonial States Revisited[16]

A number of early books on the subject of India's potential for political and economic development were pessimistic about the ability to maintain democratic institutions while enabling mass mobilization and delivering governance and development. Barrington Moore for instance was highly negative about the Indian model, attributing poor performance to a dysfunctional "trickle-down, felt-needs" model. Democracy in Moore's eyes simply complicated the matter. In his words, "Only one line of policy that seems to offer real hope, which, to repeat, implies no prediction that it will be the one adopted. In any case, a strong element of coercion remains necessary if a change is to be made. Barring some technical miracle that will enable every Indian peasant to grow abundant food in a glass of water or a bowl of sand, labor will have to be applied much more effectively, technical advances introduced, and means found to get food to the dwellers in the cities. Either masked coercion on a massive scale, as in the capitalist model including even Japan, or more direct coercion approaching the socialist model will remain necessary. The tragic fact of the matter is that the poor bear the heaviest costs of modernization under both socialist and capitalist auspices" (emphasis added).[17]

Similar sentiments have been expressed by Dandekar and Rath in Poverty in India (1971) who, at the peak of the period of the populist counterattack, suggested that poverty alleviation needed higher taxation and employment generation through public works. The same moral imperative seems to have been at work in the concepts of entitlement made famous by Amartya Sen. Going against widespread scholarship on the link between economic growth and poverty, Sen and Drèze pinpointed the low levels of social indicators in India as the pre-eminent determinant of the country's poor performance in the past, advocating that the state needs to contribute to a people-centred economic development aimed at the expansion of human capabilities.

Examining and contrasting development across Indian states, Atul Kohli has pointed out that while rates of economic growth might predict where poverty is likely to come down, the same unit of economic growth has drastically different effects from state to state. Hence, one unit of growth in Kerala or West Bengal has been four times more "efficient" in reducing levels of poverty than for examples in Bihar or Madhya Pradesh. Rather than examine aspects like irrigation infrastructure, enhanced farm yields, and access to credit which Kohli regards as proximate determinants, he points to the form of social and political power as an answer. As a hypothesis therefore, Kohli has proposed that poverty comes down the fastest in those states where effective governmental power has been founded upon a broad political base, where the hold of upper classes have been minimised, and middle and lower classes have successfully organised and used these resources to reach the power. Democracy may indeed have enabled the Indian state to be a more responsive state but even coupled with higher growth rates, during the 1990s and 2000s, it has not led to dramatic reductions in poverty. This leads Kohli to propose that the fault lies in "the state's limited institutional capacities and faulty policies at the central, state and local levels."[19]

Kaushik Basu has countered this position by arguing that bureaucracy and governmental overactivity have in fact fettered the Indian economy.[20] Basu goes as far as to say that bureaucratic control emanates from the particular nature of India's democracy which is "a system of overlapping rights" where every one has the right to decide on every matter as opposed to a system of "partitioned rights" where everyone has a domain over which she or he has the full right to decide. As a result Indian-type democracy allows for conditions where many can exercise veto power, impairing policy design, flexibility and, critically, implementation.

As a country where most of the population lives in villages and depends on agricultural for their livelihood, one of the greatest challenges to the Indian state has been agrarian modernization. In fact it is estimated that more than three-fourths of India's poor live in its villages. Here again, extensive critiques have been mounted against Indian policy makers and their failure to allocate enough attention, funds and resources to this sector, mesmerized as they were by projects of steel plant and dam—construction in the 1950s, establishing the country's "temples of modernity" as Jawaharlal Nehru once famously put it. Agrarian modernization it can be argued requires a deep paradigm shift involving both a structural and cultural change.

## The Perception of the "Poor" in Indian Culture and History: Poverty as Karma?

Poverty in India is a socially meaningful concept although it lacks powerful political articulation. One explanation for this puzzle draws upon the power of religion and the construction of the self in India. May Weber for instance shows how the overall hegemony of the upper castes and their values provides a hegemonic structure where the caste system functions as a transmission belt, siphoning off economic creativity and political resentment into investments in spirituality, intended only to reinforce the power and dominance of the Hindu social structure.[21] Furthermore, in his 1904 *Protestant Ethic and the Spirit of Capitalism*, Weber was certain that Asia was doomed to centuries of economic stagnation because of the entrenched structural and ideational incompatibilities between all the great philosophies of the East (Hinduism, Buddhism and Confucianism) with the requirements of modern economic rationality. This therefore could be used as an explanation for why the poor have never been a national political force in India. Both the Naxalites in 1960s Bengal and the *Garibi Hatao* (Remove Poverty) rhetoric of Indira Gandhi failed to takeoff as a nation-wide mass movement.

However, Weber's privileging of culture as the determinant of human action, and his privileging of the Calvinist ethos as giving rise to a distinctive behaviour that was both rationalized and puritan, has met with a wide range of criticism. Not least is the charge that Weber's model of causality is unable to explain why and how cultural attitudes change or persist over time as underlying conditions alter. Furthermore it has been proposed that such an interpretation is misleading as it assumes a highly compartmentalized approach to behaviour, assuming that the actor chooses his or her actions one at a time and is able to assess the rationality of each move. Instead it is argued that action is embedded within wider strategies or "assemblages" of action and that "culture" is to be seen not as an endogenous but rather an independent variable. Hence, the focus and emphasis shifts away from the end goals that a culture may or may not prioritize to the context that has produced a certain "cultural tool-kit."

## Contextualising the Debate: How Did India's Poor Get into the Political Radar Screen?

A historical explanation for what can be described as the "absence of the politics of poverty" is provided by Barrington Moore. He formulates an answer with reference to the historic roles of British rule, Gandhi and the Congress party. The departure of the British deprived the Congress of their legitimate basis for a "unite to oppose" strategy and the democratic empowerment of the socially marginal groups made the demand for both growth and redistribution all the more articulate as regular, competitive and largely fair elections became a normal part of politics. Interestingly, even the Karachi Resolution of 1931, considered to be the first major statement of the Indian National Congress on social and economic policy, and the pet project of Jawaharlal Nehru, contains no explicit reference to the problem of endemic poverty. Furthermore, though the more radical presidents of the INC prior to independence, such as Nehru in Faizpur and Subhas Chandra Bose in Haripura made references to poverty in their presidential addresses, little emerged in the form of resolutions or policy agendas. As a result, although by the mid-1950s there was a seeming consensus on the direction towards a "socialistic pattern of society,"[22] little agreement had been reached on the methods and mechanisms of targeting and alleviating poverty, leaving the field open to wild swings between the rhetoric of the left and right as represented within the catch-all umbrella of the Congress. The following brief sketch of development discussions and policies within India captures the oscillations (see table 2.4).

By the mid-1950s, Jawaharlal Nehru's model, the "socialistic pattern of society" had gained precedence. The Socialist goal was to be attained through measures such as land reform. However, this early momentum soon met its roadblock in the form of rural landlords who were important king-makers in local party politics.[23] As a result, it has been documented by various scholars that land reform remained mostly as rhetoric, making little headway in terms of actual implementation. It has also been pointed out that there was simply a dearth of land available for distribution to the landless.[24] India's Five Year Plans directed public funds towards private enterprise, infrastructure building, not employment generation. Egalitarian measures such as land reform eventually gave way to more populist and direct measures of poverty alleviation that at the same time did not involve confronting the landed elite. This was done through government subsidies, preferential credit in the form of programmes such as the Small Farmers' Development Agency (SFDA) programme (1971–79), the Integrated Rural Development Programme (IRDP) (1979–99), and the Swarnjayanti Gram Swarozgar Yojana (SGSY) (1999).

**Table 2.4. Development Discussions and Policies in India**

| Period | Environment surrounding policy discussion. | Policy decisions / Directions |
|---|---|---|
| 1947–51 | Policy debate within Congress party and in the Constituent Assembly → "Socialistic Pattern of Society" | Nehru's mixed economy emerges victorious over Gandhian ideas about Community Development. |
| 1952–63 | Planning emerges as the primary tool of government policy formulation and implementation. Political control over resources, import-substitution, public sector, industry as leading sectors: "the commanding heights of the economy." | Public Distribution System as a mechanism for providing price support to producers and providing food subsidy for consumers. |
| 1963–69 | The policy debate is revived and institutional reforms re-emerge. | Green Revolution and indications of a shift towards the right. |
| 1969–73 | The populist comeback and counterattack. | Land Reforms, Price Policy Reforms, Nationalization of Commercial Banks and Administrative reforms. |
| 1974–84 | Surreptitious and incremental liberalization. | Direct action privileged: launching of *Integrated Rural Development Program.* |
| 1985–91 | Half-hearted liberalization or "liberalisation by stealth." | Innovations such as induction of elected village *panchayats* in financing, planning & implementation functions & the creation of Self-Help Groups as recipients of micro-loans. |
| 1991–2004 | Towards "non-reversible" liberalization. | New paradigm of "growth taking care of distribution." |
| 2005– | The "India Shining" reality check. | National Rural Employment Guarantee Act. |

Nehru's model of import substitution, industrialization, modernization of agriculture, planning, was a model based on the "felt needs, trickle-down theory." During the first three five year plans over the years, 1951–1966, the prime emphasis was placed on the need to achieve higher growth rates in the belief that capital accumulation and enhanced savings/ investment would create a "trickle-down" effect of growth. However, the plans were over-ambitious, misguided and quickly ran into bottlenecks, particularly during the third plan when inflation, war with Pakistan, and drought created massive dissatisfaction. By the late 1960s the land situation had become polarized. Bullock capitalists on the one side and radicalized peasantry on the other were both contributing to an environment of hostility and resentment that many thought would be ripe for a Maoist revolution. The split in the Communist Party of India, giving birth to the Communist Party of India (Marxist), rise of Naxalite violence and political instability in many Indian States, indicated deep, inherent problems within the Indian model of development although the much-heralded revolution did not materialise.

What followed the radical '60s was a spate of reformist legislation, nationalization and some conspicuous programmes for instance, the Twenty Point Programme, land to the landless, homestead land, target group programmes—measures that were introduced by Prime Minister Indira Gandhi during the eighteenth month of the Emergency. Many of these social-democratic policies were put on hold when the Janata party came to power after the end of the Emergency and the fall of Indira Gandhi. However, the general tendency towards direct action programmes continues, for instance, through the Integrated Rural Development Programme (IRDP) which aimed at providing assets to the asset-less (small and marginal farmers, agricultural labourers, rural artisans) through income-generating activities. During the 1980s this scheme was extended to cover schedule castes and tribes, women and rural artisan. Various structural problems plagued the IRDP. For instance unskilled landless labourers were offered credit to develop entrepreneurship without being provided the experience to manage and enterprise, as a result of which banks were disinterested in providing credits to the poor. Recognising the failings of the IRDP, the government launched Swarnjayanti Gram Swarozgar Yojana (SGSY) in 1999 that aimed at creating "self-help groups" rather than focusing on individuals in the bid to develop micro-enterprises. The strengths of such an approach included the linking with existing banking institutions, providing banks also with the opportunity to penetrate into rural areas.

Wage employment programmes have aimed at providing rural poor with a livelihood during a lean agricultural season as well during drought and floods. Continuing into the post-1991 reform era, these programmes have been revised and re-launched. For instance a new emphasis has been placed on the need to create economic assets and infrastructure for villages with the idea that the creation of employment will follow as a by-product. The Public Distribution System (PDS) has been modified as a result of which it adopts a much more targeted approach, identifying households below the poverty line and providing them with subsidised food grains. A number of problems have dogged the PDS including costing challenges, wastage, pilferage and diversion to the open market that occurs at different stages from procurement to distribution.

Purchasing power of rural people was a major hurdle to poverty alleviation programmes such as the PDS, where people were simply unable to purchase the grains even at subsidised prices. In response to this, the government's National Rural Employment Guarantee Act, enacted by legislation on August 25, 2005, sought to ameliorate the problem of purchasing power by providing a legal guarantee for one hundred days of employment in every financial year to adult members of any rural household. As the various programmes briefly outlined above have demonstrated, the government has faced a whole host of structural and technical problems. What has, however, emerged centre stage also in terms of government-formulated programmes is a focus on the political and social dimensions of poverty. Hence the attention given to Panchayati Raj Institutions (PRI) and the function they can play in financing, planning and implementation of poverty-alleviation schemes. Moving away from the emphasis on income or entrepreneurship is the understanding that economic, social and political aspects of poverty alleviation are interlinked to one another. Hence economic upliftment alone cannot alleviate poverty but must also be connected to social and political empowerment.

## Democracy, Social Opportunity and Economic Security in India

Does democracy give rise to both social opportunity and economic security? How successful has the Indian state been in reconciling the goals of removing mass poverty, extending economic democracy and, at the same time, maintaining democratic and political competition?

The record remains a highly disputed area of discussion. Even writers like Rajni Kothari and Morris–Jones, originally convinced that the flexibility of India's political institutions would deliver, seem less sure in later works. Scholars like Atul Kohli have used regime as a variable to identify minor gains in egalitarianism.[25] Hence, an interventionist state, backed up by a leftist coalition has been seen as undertaking effective anti-poverty mechanisms. Alternative interpretations include the Rudolphs' who highlight the role of demand groups and state-dominated pluralism. Mitra, on the other hand focuses on two-track strategies where popular protest is as much part of "normal" politics in India as is participation.[26] The state, in fact, manages to draw upon protests as a means through which it can legitimise its authority by creating new institutions, responding to local demands and recruiting protestors into the political arena. Mitra has depicted this dynamic in the form of a new institutional model that focuses on elite strategies in the form of institutional reforms, law and order management or constitutional adjustments (see figure 2.7).[27]

## CONCLUSION: The Neo-Institutional Model

The poor have neither disappeared nor formed themselves into a political party or movement, but they continue to exist as a demand group whose presence is a brake on rapid and radical liberalization. However, as the flow diagram below demonstrates, India has found a mechanism of coping. Critical room to manoeuvre is provided through the constitution, and the self-reinforcing dynamic of a system based on political accountability makes it necessary for elites to engage in purposeful social intervention. As a result institutional reforms and policy change can in fact add to stability rather than undermine it. Hence, the response of decision-making elites to challenges takes the form of law and order management, strategic reform and redistributive policies and constitutional change which are likely to affect the perception of local actors.

**Figure 2.1. A dynamic neo-institutional model of governance based on elite strategies**

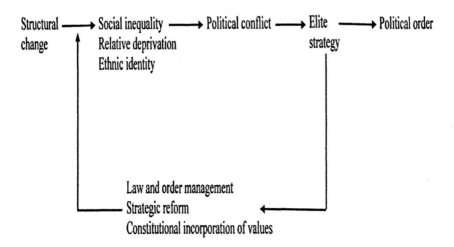

*Source*: Mitra, Subrata K., *The Puzzle of India's Governance: Culture, Context and Comparative Theory*. London: Routledge, 2006.

Nevertheless, Ashutosh Varshney laments the fact that poor democracies do not seem to have an inherent advantage in their being democracies to tackle poverty[28] and offers an important and useful explanation for why this is so. Distinguishing between class and ethnicity, where class is essentially an economic category, but ethnicity is defined in terms of ascriptive identities, Varshney argues that the ethnic politics of subaltern groups are not normally framed in terms of poverty. In a multi-ethnic democracy, Varshney posits that political mobilization tends to happen along ethnic fault lines rather than economic or class issues. Hence, if the poor belong to very different groups, defined by caste, language, race, religion, then the likelihood of collective action converging solely around an agenda of poverty removal is unlikely. This, combined with the fact that democracies tend to favour direct and more popular methods of poverty removal, has led to the country's sustained disappointing results. Furthermore, while agreeing with Sen that democracy cuts down the worst excesses of poverty, such as famines, Varshney has pointed out that it creates clienteles around interventionist-direct action policies.[29]

As seen from the above theoretical, empirical and methodological overview, every measurement of poverty in India has its specific politics behind it. The intent of the chapter has been to argue that poverty alleviation is a policy issue that inevitably involves a politics–policy nexus. Extending the statistical data from the World Bank and other international agencies for cross-cultural comparison needs to take into account the value premises (liberal, institutionalism, democratic legitimacy as based on entitlement) on which the data are based. Sen, for example, while emphasizing quantitative measures, leaves open the scope for subjective attributes such as legitimacy. In fact from survey research[30] one comes up with inexplicable evidence of a sense of qualified optimism, even among the poor. This would be explained by the neo-institutional model in figure 2.1 that refers to the fact that perception is a key variable in the dynamic model. Such a finding, it is argued, is not mere evidence of the efficacy of democratic rhetoric and confirmation of the low expectations, karma-bound, fatalistic Indians. Rather, it is posited that there is something seriously wrong with the left-leaning, structural-reformist approaches to poverty research that pervades the field, especially when it comes to the case of India. These perspectives tend to overlook the flexibility and resilience of structures like the Indian constitution that are capable of bringing about changes in the rules of the game and create new policy arenas.

The macro co-existence of poverty and plenty in a country like India, which has had its share of mass movements, can only be explained by disaggregating India. Different regions have had hugely different success rates in targeting and tackling poverty. However, yet another avenue that needs further exploration is whether liberalization and globalization can lead to poverty reduction. The state of Bihar may in fact be showing the way, for recent election results revealed voting patterns based on the government's performance and output rather than caste acting as the prominent deciding factor. Furthermore, as a result of, or in conjunction with, this political revival, are impressive economic growth rates that promise to change the reputation of Bihar being one of the infamous BIMARU states. The case of Bihar therefore may indeed herald a new dynamic, a "race to the top" perhaps, as states within the federal union compete for foreign and domestic investment and the hope of attracting both wealth and skilled labour.

## NOTES

1. Subrata K. Mitra is Professor of Political Science, South Asia Institute, Heidelberg University. Markus Pauli is Doctoral Fellow, Graduate Program in Trans-cultural Studies, Cluster of Excellence, Heidelberg University and Jivanta Schöttli is Lecturer in Comparative and International Politics, South Asia Institute, Heidelberg University.

2. The latest being the Tendulkar Committee which was set up in 2009 to look into the methodology of measuring poverty.

3. Spicker Paul, Sonia Alvarez Leguizamón, and David Gordon, eds. *Poverty: An International Glossary* (2nd ed.: Zed Books Ltd, 2007), 230-240.

4. United Nations Department of Economic and Social Affairs. *Copenhagen Declaration on Social Development 1995* (URL: http://www.un.org/esa/socdev/wssd/ copenhagen_declaration.html, last visited February 28th 2011), 57.

5. European Economic Community. *On Specific Community Action to Combat Poverty* (85/8/EEC, Official Journal of the EEC, 2/24, 1985).

6. Drèze, Jean and Amartya Sen. *Hunger and Public Action* (Oxford: Clarendon Press; New York: Oxford University Press, 1989)

7. Sen, Amartya. *Development as Freedom* (Oxford: Oxford University Press, 1999), 11.

8. World Bank. *World Development Report 1990* (Oxford, New York [etc.]: Oxford University Press, 1990), 26.

9. Spicker (et al.). *Poverty: An International Glossary*, 61.

10. UNDP. *Human Development Report 2010* (New York: UNDP 2010).

11. Alkire, Sabina and Maria Emma Santos. *Multidimensional Poverty Index* (Oxford Poverty and Human Development Initiative, July 2010).

12. See also the homepage of the Oxford Poverty and Human Development Initiative (OPHI), URL: http://www.ophi.org.uk/research/missing-dimensions, last visited February 28th 2011.

13. Concerning the methodology see: Alkire, Sabina and Maria Emma Santos, *Acute Multidimensional Poverty: A New Index for Developing Countries* (OPHI Working Paper 38) as well as UNDP, Human Development Report 2010.

14. Alkire and Santos. *Multidimensional Poverty Index*, July 2010.

15. Oxford Poverty and Human Development Initiative (OPHI). Country Briefing: 2010 for India.

16. The social origins of India's post-independence state and the redistributive efforts of this state have been well studied. See for example: Frankel (2005), Chibber (2003), Nayar (1989), Kohli (1987). Bardhan (1984), and Herring (1983).

17. Moore, Barrington. *Social Origins of Dictatorship and Democracy: Lord and Peasant in the Making of the Modern World* (Boston: Beacon Pr., 1966), 410.

18. Jean Drèze and Amartya Sen. *India: Economic Development and Social Opportunity* (Oxford University Press, 1995).

19. Kohli, A. Chung-in Moon, Georg Sørensen. *States, Markets, and Just Growth: Development in the Twenty-first Century* (United Nations University Press, 2003).

20. Kaushik Basu, "The Impact of Structural Adjustment on Social Sector Expenditure: Evidence from Indian States." In: Rao, C. H. Hanumantha and Linnemann, Hans (eds.) *Economic Reforms and Poverty Alleviation in India* (New Delhi: Sage Publications, 1996).

21. See Max Weber. *The Religion of India: the Sociology of Hinduism and Buddhism* (New York, Free Press, 1958).

22. Passed within the Congress party on August 14/15, 1957 as part of the Avadi Resolution.

23. For a detailed discussion of Jawaharlal Nehru's early policy choices and the long-run impact on India's political development, please see Schöttli, Jivanta, *Vision and Strategy in Indian Politics: Jawaharlal Nehru's Policy Choices and the Designing of Political Institutions* (London: Routledge, 2011 forthcoming).

24. V. M. Dandekar and Nilakantha Rath. *Poverty in India* (Poona: Indian School of Political Economy, 1971).

25. Kohli, Atul. *The State and Poverty in India: The Politics of Reform* (Cambridge: Cambridge University Press, 1987).

26. Mitra, Subrata K. *Power, Protest, and Participation: Local Elites and the Politics of Development in India* (New York: Routledge, 1992).

27. Mitra, Subrata K. *The Puzzle of India's Governance: Culture, Context and Comparative Theory* (London: Routledge, 2005).

28. Varshney, Ashutosh. *Democracy, Development, and the Countryside: Urban-Rural Struggles in India* (Cambridge: Cambridge University Press, 1995).

29. Varshney, Ashutosh. *Why Haven't Poor Democracies Eliminated Poverty?* In: Varshney, A. (ed.) *India and the Politics of Developing Countries: Essays in Memory of Myron Weiner* (New Delhi: Sage, 2004).

30. Mitra, Subrata K. and Singh, V.B. *When Rebels Become Stakeholders. Democracy, Agency and Social Change in India* (New Delhi: Sage, 2009), 155-156.


## BIBLIOGRAPHY

Alkire, Sabina and Maria Emma Santos. "Acute Multidimensional Poverty: A New Index for Developing Countries." Oxford Poverty and Human Development Initiative, Working Paper 38, 2010.

———. "Multidimensional Poverty Index." Oxford Poverty and Human Development Initiative, July 2010.

Bardhan, Pranab Kumar. *The Political Economy of Development in India.* Oxford: Blackwell, 1984.

Basu, Kaushik. "The Impact of Structural Adjustment on Social Sector Expenditure: Evidence from Indian States." In *Economic Reforms and Poverty Alleviation in India,* ed. Rao, C. H. Hanumantha and Linnemann, Hans. New Delhi: Sage Publications, 1996.

Chibber, Vivek. *Locked in Place: State-building and Late Industrialization in India.* Princeton: Princeton University Press, 2003.

Dandekar, V.M. and Nilakantha Rath. *Poverty in India.* Poona: Indian School of Political Economy, 1971.

Drèze, Jean and Amartya Sen. *India: Economic Development and Social Opportunity.* Oxford: Oxford University Press, 1995.

———. *Hunger and Public Action.* Oxford: Clarendon Press; New York: Oxford University Press, 1989.

European Economic Community. "On Specific Community Action to Combat Poverty," In: 85/8/EEC, *Official Journal of the EEC,* 2/24, 1985.

Frankel, Francine R. *India's Political Economy, 1947–2004: The Gradual Revolution.* 2nd ed., New Delhi: Oxford University Press, 2005.

Kohli, Atul, Chung-in Moon and Georg Sørensen. *States, Markets, and Just Growth: Development in the Twenty-first Century.* New York: United Nations University Press, 2003.

Kohli, Atul. *The State and Poverty in India: The Politics of Reform.* Cambridge: Cambridge University Press, 1987.

Herring, Ronald J. *Land to the Tiller: The Political Economy of Agrarian Reform in South Asia.* New Haven: Yale University Press, 1983.

Mitra, Subrata K. and Singh, V.B. *When Rebels Become Stakeholders. Democracy, Agency and Social Change in India,"* New Delhi: Sage, 2009.

Mitra, Subrata K. *The Puzzle of India's Governance: Culture, Context and Comparative Theory.* London: Routledge, 2005.

Mitra, Subrata K. *Power, Protest, and Participation: Local Elites and the Politics of Development in India.* New York: Routledge, 1992.

Moore, Barrington. *Social Origins of Dictatorship and Democracy: Lord and Peasant in the Making of the Modern World.* Boston: Beacon Pr., 1966.

Nayar, Baldev Raj. *India's Mixed Economy: The Role of Ideology and Interest in Its Development.* Bombay: Popular Prakashan, 1989.

Oxford Poverty and Human Development Initiative (OPHI). "The Missing Dimensions of Poverty Data," accessed February 28, 2011. http://www.ophi.org.uk/research/missing-dimensions.

Schöttli, Jivanta. *Vision and Strategy in Indian Politics: Jawaharlal Nehru's Policy Choices and the Designing of Political Institutions.* London: Routledge, 2011 (forthcoming).

Sen, Amartya. *Development as Freedom..* Oxford: Oxford University Press, 1999.

Spicker, Paul, Sonia Alvarez Leguizamón, and David Gordon (eds.) *Poverty: An International Glossary.* 2nd ed., New York: Zed Books, 2007.

United Nations Department of Economic and Social Affairs. "Copenhagen Declaration on Social Development," 1995, accessed February 28, 2011. http://www.un.org/esa/socdev/wssd/copenhagen_ declaration.html

United Nations Development Programme. *Human Development Report 2010. The Real Wealth of Nations.* New York: UNDP, 2010.

Varshney, Ashutosh." Why Haven't Poor Democracies Eliminated Poverty? "In: Varshney, A. (ed.) *India and the Politics of Developing Countries: Essays in Memory of Myron Weiner.* New Delhi: Sage, 2004.

Varshney, Ashutosh. *Democracy, Development, and the Countryside: Urban-Rural Struggles in India.* Cambridge: Cambridge University Press, 1995.

Weber, Max. *The Religion of India: the Sociology of Hinduism and Buddhism.* New York: Free Press, 1958.

World Bank. *World Development Report 1990.* Oxford, New York: Oxford
   University Press, 1990.

## 3

# THE IMPORTANCE OF PLACE AND SPACE: THE SPATIAL DIMENSIONS OF POVERTY AND DEVELOPMENT POLICY

*Kate Bird and Kate Higgins*

The relationship between geography and development has sparked significant interest in recent decades. Some have focused on interactions among geography, the economy and development and contributed to a field known as economic geography. For example, Paul Krugman's Nobel Prize winning work explored the spatial dimensions of economic activity.[1]

Others have focused more specifically on the relationship between geography and poverty. For example, early work by Jyotsna Jalan and Martin Ravallion investigated the relationship between geographic endowments and well-being.[2] More recently, the World Bank published *Reshaping Economic Geography—the 2009 World Development Report*, putting the spatial dimensions of development in the development policy spotlight.[3]

This chapter explores the role of "place" and "space" in development, with a particular focus on poverty. Specifically, what are the spatial dimensions of poverty? Why do they matter? What policies and strategies have attempted to overcome them? And what lessons can we glean for policy and practice?

## THE SPATIAL DIMENSIONS OF POVERTY

Even when controlling for individual and household characteristics, such as age, household composition or ethno-linguistic group, the endowments of an area explain a substantial proportion of the poverty that people living in it experience.[4] Essentially, when it comes to explaining why people are poor, location matters.

The endowments of an area can be defined as an area's "geographic capital"—the physical, natural, social, political and human capital of a location. The endowments may be "first nature" geographic characteristics, such as topography or proximity to the coast. But they may also be "second nature" geographic characteristics, such as the geographical distribution of infrastructure and public services.[5] According to the Chronic Poverty Research Centre (CPRC), locations that will experience particular spatial disadvantage will be:

- *geographically remote areas*, which are far from the centre of political and economic activity;
- *low potential or marginal areas*, which are ecologically disadvantaged and therefore have poor agricultural potential and low levels of natural resource endowments;
- *less favored areas,* which are politically or socially excluded or disadvantaged; and
- *weakly integrated areas*, which are poorly linked in terms of markets and communication.[6]

Theoretical and conceptual literature establishing the spatial dimensions of poverty has been accompanied by numerous empirical investigations. Following Jalan and Ravallion's original work in China,[7] several studies empirically demonstrate the relationship between geography and poverty. Bird and Shepherd, in their study of semi-arid zones in Zimbabwe, identify a clear link between high levels of remoteness, low levels of public and private investment and high incidence of chronic poverty.[8] Bird, McKay and Shinyekwa find a strong correlation between isolation and poverty in Uganda, and find that households in remote areas have lower levels of market participation, make less use of public services (which are often further from users) and spend more time fetching wood and water.[9] Escobal and Torero find similar results in Peru: they identify a strong association between spatial inequality and variation in private and public assets.[10] Christiaensen et al. in their cross-Africa study, find that the impact of economic growth on poverty reduction depends on how remote households are from economic centers, and how well they are served by public infrastructure.[11] Burke and Jayne find that in rural Kenya, spatial factors are a substantial determinant of wealth, explaining a relatively similar share of the total variation in wealth as household-specific factors.[12] Fafchamps and Moser find, based on their research from Madagascar, that crime increases with distance from urban centers and conclude that crime and insecurity are associated with isolation. They argue that isolated regions tend to have more banditry and are more likely to harbor armed terrorist or insurgent groups than better connected areas, and crime and insecurity may contribute to the well-being, or poverty, of an area.[13] While much of the evidence focuses on the link between remoteness and poverty, as Grant reminds us, the spatial dimensions of poverty are not only found in remote rural areas. Given rapid urbanization in developing countries, and rising levels of urban poverty, the spatial dimensions of poverty are becoming more apparent in urban contexts. Urban areas, despite their proximity to economic and political centers, can indeed be politically and social excluded, and weakly integrated into markets and communication.[14]

Some argue that geographical constraints can "trap" people in poverty and make the case for the existence of "spatial poverty traps." For example, Minot et al. argue explicitly that spatial poverty traps exist in Vietnam, because interventions have been unable to address the small number of agro-climatic and market access variables that explain roughly three-quarters of rural poverty.[15] Daimon describes the presence of spatial poverty traps in Indonesia, where spatial factors, including quality of public goods in a district, remoteness and rural residence, are statistically significant in determining levels of per capita expenditure and poverty rates.[16] Burke and Jayne have challenged the conception of spatial poverty "traps," however. This conclusion is based on their research in rural Kenya, where they find that that while there is a strong correlation between spatial factors and welfare, not all households in areas characterized as spatial poverty traps are chronically poor. In fact, they find that poor households are often surrounded by others who manage to remain above the bottom tercile, or even rise out of poverty, indicating that spatial factors are not the only determinants of poverty. In addition, the proportion of households that have climbed out of poverty is not greatly different between areas of low and high mean wealth, indicating that location does not necessarily function as a poverty "trap." They also emphasize that household-specific factors are of considerable importance in explaining variation in household wealth across the nationwide sample and that the relative explanatory power of spatial factors, though substantial, is slightly less than that of household-specific factors. This leads them to suggest that "spatial disadvantage" is a more accurate way of describing the spatial dimensions of poverty in Kenya, and highlight the need for policies that address household, as well as spatial, drivers of poverty.[17]

# WHY DO THE SPATIAL DIMENSIONS OF POVERTY MATTER FOR DEVELOPMENT POLICY?

## *Why Do the Spatial Dimensions of Poverty Deserve Attention?*

First, regardless of whether one speaks of the spatial dimensions of poverty, spatial poverty traps, or spatial disadvantage, there is a clear and compelling body of evidence demonstrating that low levels of geographic endowments can partially explain why households are poor. If we are seeking entry points for tackling poverty, spatial disadvantage must be one of them.

Second, many people live in spatially disadvantaged, or geographically un-endowed, areas. *Reshaping Economic Geography* states that there are 1 billion people living in slums in the cities of the developing world, 1 billion people living in fragile regions and 1 billion people living in regions that are divided, distant from markets and lack a large local economy.[18] On top of this, intra-country inequality is on the rise in many countries. In India, China, Russia, Mexico and South Africa, as well as most other developing and transitional economies, spatial and regional inequalities—of incomes, economic activity and social indicators—are increasing.[19] Most national household survey data continues to show a significant regional dimension to the incidence of poverty[20] and spatial disadvantage will be found even where a country has experienced economic growth and aggregate reductions in headcount poverty.[21] So, if reducing poverty and facilitating the more equitable distribution of opportunities and benefits relating to growth are desired outcomes, understanding and addressing spatial disadvantage is crucial.

Third, it is likely that the poverty in spatially disadvantaged areas is characterized by multiple forms of disadvantage: low returns on all forms of investment, partial integration into fragmented markets, social and political exclusion and inadequate access to public services. This means that people in these areas are not only more likely to be income poor (which is typically measured by poverty headcount), but also chronically poor (which is typically measured by poverty depth and poverty duration). Chronic poverty is the most difficult form of poverty to escape. For this reason, it is more likely that chronic poverty will be passed on to the next generation.[22]

Fourth, the "bad neighborhood effect" constrains the opportunities of people living in spatially disadvantaged areas, limiting opportunities for poverty exit. Even if an individual has the entrepreneurial skills, the investment capital and the will to invest in a business, the returns on their investment will be lower than in better-connected areas with higher levels of geographic capital. The "bad neighborhood effect" also extends its blighting effect to investments in human capital. In spatially disadvantaged areas, it is likely that there will be a poorer return on investment in education, for example. Even if a child attends a good school, and receives a good education, the absence of local and accessible successful role models and good entry-level employment opportunities will make success harder to come by. Tackling elements of spatial disadvantage may enable some households to relatively quickly escape poverty. We know that individual and household characteristics have significant explanatory power when it comes to poverty, but in cases where these are not prominent, and spatial disadvantages are the key factors holding a household back, tackling these constraints may be an efficient way to help households move out of poverty.

## SPATIAL DISADVANTAGE AND THE DEVELOPMENT POLICY LEXICON

While spatial inequality within countries appears to be on the rise, it is certainly not new, and there is no doubt that governments, and donors, have sought to tackle issues relating to geographic exclusion and spatial disadvantage. But in the current policy lexicon, what attention is paid to spatial disadvantage?

### *Government Policy*

A review of the poverty reduction strategy papers (PRSPs) and national development plans of developing countries is one way to discern how the spatial dimensions of development, and poverty, are integrated into national planning policies. We conducted a review of PRSPs from fourteen purposively selected countries.[23] We did not attempt a comprehensive discourse analysis of these policies, but rather focused very specifically on identifying whether the language of spatial poverty—such as "spatial disadvantage," "regional inequality," and "lagging regions"—had penetrated the PRSPs. We found that this language featured prominently in the selected PRSPs.

We also examined the recent national development plans of China and India: China's 11th Five–Year Plan 2006–2010 and Plan for National Economic and Social Development and India's 11th Five–Year Plan 2007–2012. These plans sit outside the conventional PRSP framework and—in contrast to many early PRSPs—have been country-led initiatives.[24]

The goal of promoting balanced development among regions is a central theme in China's 11th Five–Year Plan, with specific attention given to "improving the mechanism of regional balance and interaction."[25] A recent report on progress on China's Plan for National Economic and Social Development clearly states the goal of "mak[ing] development among different regions more balanced."[26] While the plan does not discuss the implementation of specific policies for each region, it gives special attention to policies and programs to support the development of the western region. It envisions a strong role for the state in ensuring economic and social development in old revolutionary areas, ethnic minority areas, border areas and economically depressed areas, through increased transfer payments from the state and other unspecified forms of "strong support."[27]

India's 11th Five–Year Plan devotes an entire chapter to spatial disparity between regions. A chapter entitled "Spatial Development and Regional

Imbalances" acknowledges the challenges presented by a pattern of growth and poverty reduction where:

> ... high growth rates have led to a spiral of commercial and service sector activity in the already developed regions, [while] the backward areas continue to lack even basic amenities such as education, health, housing, rural roads, drinking water and electricity.[28]

At the inter-state level, the importance of India's national project, and national unity, provides the impetus for central government action:

> Redressing regional disparities is not only a goal in itself but is essential for maintaining the integrated social and economic fabric of the country, without which the country may be faced with a situation of discontent, anarchy and breakdown of law and order.[29]

Further, the plan suggests that the development efforts of States need to be supplemented by national efforts in order to "minimize certain distinct geophysical and historical constraints." These supplementary policies and initiatives include General Purpose Resource Transfers, the Backward Regions Grant Fund, the Hill Area Development Programme, the Western Ghats Development Programme and the Border Area Development Programme.[30]

The review of the PRSPs and the national plans of China and India reveal that these national-level plans recognize the need to address spatial disparities and spatial disadvantage. Of course, language use in national policy documents does not tell us anything about the effectiveness of policy implementation. There could well be a divergence between policy and practice. Programs seeking to deliver on policy pronouncements may be hampered be a range of budgetary, political and institutional obstacles. Conversely, practice may be ahead of policy directives. What our analysis does reveal, however, is that spatial disadvantage does feature in the policy lexicon of a number of developing country governments.

## Donor Policy

How does spatial disadvantage feature in the policy lexicon of donors? We · reviewed the policies of selected donors using the same approach we applied to the review of PRSPs above. We examined the DFID 2009 White Paper, AusAID's 2006 White Paper, the EU's 2005 Consensus on Development, USAID's 2004 White Paper, UNICEF's 2008 State of the World's Children,

Irish Aid's 2006 White Paper and CIDA Sustainable Development Strategy: 2007–2009.

Among the organizations included in the survey, the Australian Agency for International Development (AusAID) is the only organization that specifically and directly referred to spatial or geographic disadvantage. Using the language of "lagging and vulnerable regions," AusAID's 2006 White Paper links spatially disadvantaged areas in the southern Philippines and eastern Indonesia to regional growth and infrastructure. Language relating to spatial disadvantage did not feature in any of the other donor policy statements, illustrating that, from a policy perspective, the spatial dimensions of poverty are not high on the agenda of donor agencies. These results offer a contrast to the results of the PRSP review. While this analysis is relatively superficial, and we must take care with the type of conclusions we draw from it, it does appear to indicate that national governments give greater weight to spatial concerns than their international development partners.

Why may this be the case? One explanation could be that sectoral (e.g. health) or thematic (e.g. gender) programs subsumes donor efforts to address spatial disadvantage. Alternatively, differing sources of accountability may influence policy priorities. The domestic state–citizen contract means that, ideally, developing country governments are accountable to their citizens. In the interest of re-election, but also social cohesion and stability, governments may want to appease disadvantaged areas through policy promises and investments. In contrast, donors are ultimately accountable to their own parliaments or to their own resource base, which is typically in richer countries. As a result, they may be more inclined to frame issues in ways that gain political, or fundraising, traction at home. As well, their obligation to deliver measurable results and evidence of efficiency may influence programmatic design and spending behavior. For example, investments may be made in areas closest to the poverty line or in places where dramatic results can be expected. This may be at the expense of investing in more difficult to reach areas, or more entrenched and structural inequalities.

# A GLOBAL NARRATIVE: THE 2009 WORLD DEVELOPMENT REPORT

The World Bank's 2009 World Development Report *Reshaping Economic Geography* brought the spatial dimensions of development into the development policy spotlight. As the World Bank's flagship report, the World Development

Report offers a global narrative on an issue worthy of exploration, attention and action.

*Reshaping Economic Geography* highlights uneven economic development at local, national and international levels, and argues this poses one of today's biggest development challenges. It may seem that the obvious solution is to spread economic development as far as possible, to reach those who are currently excluded. But the report challenges this approach, and drawing on the historical experience of "developers," such as Europe, East Asia and the United States, the report argues that successful development requires spatial *concentration,* coupled with economic integration.

According to the report, development requires three transformations: a greater density of population to facilitate economic integration, as seen in the growth of cities; shorter distances (through transport infrastructure, for example), to encourage businesses and workers to migrate towards dense areas; and fewer divisions, through thinner economic borders and greater access to world markets. To achieve these transformations, the report argues, requires three strategies: urbanization, territorial development and regional integration. Urbanization is needed to facilitate higher population density, territorial development can integrate nations (and shorten distances), and regional integration can lead to fewer inter-country divisions, through greater access to global markets. The report argues there are three policy instruments available for supporting these transformations and strategies: institutions, infrastructure and interventions. "Institutions" is shorthand for policies that are spatially blind in their design and universal in their coverage. Examples include policies to deliver social services such as health, education and water and sanitation, and regulations affecting land, international trade and labor. "Infrastructure" is shorthand for investments and policies that connect geographic spaces, such as roads, airports, railways and communication systems that move goods, people, services and ideas. "Interventions" is shorthand for programs that are targeted at particular spaces, such as slum clearance, fiscal incentives for manufacturing investment and preferential trade access for developing countries. Importantly, it is argued that "interventions" too often dominate policy discussions on spatial disparities. The report therefore calls for a "rebalancing" of debates to include all three groups of instruments to achieve successful economic integration and genuine development results.

# INITIATIVES TO TACKLE SPATIAL DIMENSIONS OF POVERTY

How do the ideas and policy pronouncements play out in practice? Case studies enable us to explore different approaches in a concrete way. Eight initiatives, each which relate to spatial disadvantage, are explored below.

## Case Study 1: Focal Area Development in the Lao People's Democratic Republic

Poverty has a regional dimension in Lao People's Democratic Republic (PDR), with poverty concentrated heavily in remote upland areas inhabited largely by ethnic minority communities. In these areas, living conditions are poor and human capabilities are low—particularly in the more isolated locations. These areas correspond with the seventy-two districts that the government has identified as poor and the forty-seven it has identified as being the poorest.[31] .

The Government of Lao PDR (GoL) recognizes that inter-region inequalities exist across human development dimensions. Focal Area Development is one of the policies in the GoL's National Growth and Poverty Eradication Strategy (NGPES).[32] The policy involves "consolidating" and/or moving dispersed villages to areas identified for development. The aim is to improve service delivery to the rural population (for example, education, health, agricultural and electricity services). The policy also seeks, by relocating upland villages to lowland areas, to help "stabilize" shifting cultivation, eradicate opium production, extend administrative control (by consolidating villages into larger units), foster the cultural integration of ethnic minorities into Lao society and resettle villagers with a history of armed rebellion.[33]

All resettlement in Lao PDR is nominally voluntary rather than forced. But banning shifting cultivation and opium cultivation, and destroying forests, reduce the livelihood options of upland communities. They are left with little choice but to move. To put pressure on communities to move, membership of "mass organizations" is withheld, the village head is no longer recognized and public infrastructure (e.g. schools and health centers) is dismantled. Communities are not involved in identifying Focal Area sites—district and provisional authorities choose these. Upland communities are moved to lowland areas, alongside roads, and many have better access to schools, clinics, electricity and marketing opportunities. However, below 80 percent of the "receiving" sites have sufficient land for production and housing, road access to markets and/or an area for school construction. Some sites are in poor, politically sensitive areas, with low development potential.[34]

Communities rarely receive compensation, advice or support to help them adjust to their new environment. Relocation without compensation or support, along with the paucity of the new locations, has had adverse consequences. The first participatory poverty assessment (PPA) in Lao PDR found that, in many cases, rural people described themselves as newly poor and explained that their worsening situation was a result of village relocation and land and forest allocation.[35] The adverse consequences and impacts of relocation include:

- *Loss of assets:* This included loss of land, forest resources and livestock (through disease).
- *Increased poverty, food insecurity and vulnerability:* Resettled households found that the abrupt change in livelihood strategy, from semi-subsistence to being forced into cash crop production (maize, soybean),[36] increased their vulnerability. People had a poor understanding of the new agro-ecological zone,[37] their new context and production systems, and this led to low levels of productivity and food security. The change from upland to lowland rice production reduced total rice production and extended the "hungry season" poor producers experienced. A drop in farm-gate prices for maize and soybean and seasonal price variability exacerbated this. The second round of the Lao PDR PPA found that women who had experienced impoverishment from relocation hired out their labor to domestic employers, migrated to Thailand in search of work and were drawn into prostitution.
- *Higher mortality rates:* Changed livelihoods and lack of access to non-timber forest products led to a worsening diet. Despite access to roads and markets, resettled villagers were found to have a poorer diet than those who remained in their remote upland homes.
- *Reduced social capital and community cohesion:* In some cases, resettled villagers formed the lowest social strata when they joined established villages with other ethnic groups (e.g. Khmu, Xiengmoon and Hmong joining Lao and Phou Thai villagers). Without adequate land, the resettled groups became casual laborers for Lao and Phou Thai farmers and were not represented in village committees and other public bodies. Adverse incorporation, marginalization and community-level conflict increased as a consequence of resettlement and intensified conflict between new settlers and their "host" communities.[38]

## Case Study 2: Model Secondary Schools in Ghana

The regional dimension of poverty is significant in Ghana. Poverty has declined in Accra and around the Rural Forest area, but is still very widespread in the northern regions (Northern, Upper East and Upper West).[39] The developmental divide between north and south in Ghana arises from a combination of circumstances and policies. These include the geographical concentration of agricultural resources and activities (such as cocoa, minerals and forest resources) in the southern regions; the British colonial legacy of investing more heavily in regions where exploitable resources, such as diamonds, gold, timber and cocoa, were available and cheap to produce and export; and the transfer of out of, and lack of educational investment into, the north.[40] Northern regions of Ghana have essentially lagged behind the rest of the country and, as a result, Ghana has missed out on significant opportunities to increase the country's average income and foreign exchange earnings, as well as to further bed down national stability.[41]

Reducing inter-area differences is recognized as central to improving Ghana's economic performance and human development indicators.[42] Reducing inter-area differences in education is one dimension of this. Persistent geographic disparities in education are recognized in the second Ghana Poverty Reduction Strategy (GPRS II). To address this, the Government of Ghana (GoG) has instituted its model secondary school policy. This involves upgrading one secondary school in each district to "model" secondary school status, "to address the issues of geographic disparities in access to quality education."[43] The objective is to provide each district with a school that could be compared favorably with the leading schools in the country. The program was a feature in the first GPRS (2003–2005), in which enhancing the delivery of social services to ensure locational equity and quality was a priority. The initial phase identified thirty-one schools for upgrading after assessment of physical infrastructure and academic requirements. In 2004, 66 percent of work on the first batch of thirty-one schools was complete, and the second phase—which would upgrade twenty-five additional schools—was in planning.[44]

As yet, no studies have been conducted to assess the impact of the model secondary school policy on reducing inter-district disparities in education outcomes.[45] The preliminary observations of Ghana education experts vary. On the one hand, there has been criticism of the over-emphasis on equality in distribution under this policy. The approach of establishing one model school in each district reflects a tendency in Ghana to distribute public goods equally between districts (irrespective of needs), rather than specifically targeting disadvantaged areas and marginalized populations to achieve pro-poor and geographically equalizing outcomes.[46] In addition, there is concern that this policy further entrenches vertical, or class-based, divisions. To gain acceptance to a model secondary school, students must meet the senior secondary school admission criteria, which is more challenging for those from poor households and poor districts. Therefore, the policy does not address fundamental, underlying inequalities, and those with more economic and social capital can manage and benefit from the system.[47]

On the other hand, the potential for this policy to reduce regional inequalities in senior secondary education is recognized. While the policy may not address underlying structural inequalities in the short term, it may be able to address these in the long term. Properly resourced schools may attract teachers to remote and rural areas, motivate teaching staff and therefore improve education outcomes. The policy may have a circular effect: students who reap the benefits of model secondary schools may become teachers and be motivated to return to their districts and contribute through education. In addition, the facilities and resources model secondary schools offer give students something to aspire to. Essentially, the policy will enable students to have access to good quality senior secondary education across the country regardless of district, reducing the current bias towards a small number of prestigious government-funded and urban-based senior secondary schools. It may also improve the chances of students from remote and rural districts gaining admission to competitive universities.[48]

## Case Study 3: Universal Primary Education in Northern Uganda

Uganda has made significant development progress over the past two decades. According to a recent Uganda National Household Survey, the national poverty rate has declined to 31 percent for 2005/06, from 56 percent in 1992/93 and 38 percent in 2002/03.[49] Prudent macroeconomic policies have generated robust growth, adult HIV/AIDS prevalence has declined significantly from around 18 percent in the early 1990s to 6.4 percent in 2005 and primary level net enrolment rates have increased from 62.3 percent in 1992 to 92 percent for girls and 94 percent for boys in 2006.[50] The poverty reduction has not been experienced uniformly across the country, however. In the northern region of Uganda, poverty decline has been modest at only approximately 17 percent since 1992/93. This is compared to substantial progress made in the West and Central regions of Uganda, where poverty has declined by around 60 percent since 1992/3.[51] Census and household survey data indicate that the Northern region has the highest poverty rate, the highest annual average population growth, the highest fertility levels, the highest proportion of people living in a hut (as opposed to a permanent house) and the lowest proportion of people owning a mobile phone. These data indicate that the North is lagging behind the rest of Uganda across a range of development indicators.[52]

Lackluster progress in the North can be attributed to a vicious cycle of conflict, massive displacement, thwarted economic activity, complications in service delivery and the immense poverty that has gripped the northern region of the country.[53] The conflict between the Lord's Resistance Army (LRA) and Ugandan Government's Uganda People's Defence Forces (UPDF)—ongoing for more than two decades—is the most prominent driver of poverty in the North, and the a key reason for continued disparity between northern Uganda and the rest of the country.

Education is no exception to this trend. While the Northern region fares relatively well in terms of education access, when it comes to education quality, competition and performance, it is well below national average. Table 3.1 outlines these trends.

**Table 3.1. Education Quality and Performance Indicators in Uganda**

| District/ region | Pupil teacher ratio | Pupil classroom ratio | National Examination Division 1 Scores (%) | National Examination Division U Scores (%) | Schooling status – no formal schooling (15+) 2005/06 (%) | Literacy (18+ years) 2005/06 (%) |
|---|---|---|---|---|---|---|
| Districts in the Northern region (2005) | | | | | | |
| Gulu | 67 | 94 | 1.1 | 28.4 | - | - |
| Kitgum | 85 | 98 | 2.1 | 15.7 | - | - |
| Pader | 91 | 137 | 1.2 | 22.7 | - | - |
| Regions | | | | | | |
| North | 58 (2004) | 92 (2004) | - | - | 27 | 59 |
| Central | 43 | 63 | - | - | 23 | 80 |
| East | 55 | 90 | - | - | 20 | 61 |
| West | 73 | 73 | - | - | 26 | 66 |
| National | 50 (2004 & 2005) | 74 (2005 | 5 | 15.2 | 20 | 69 |

*Sources*: Ministry of Education and Sports (2005); National Household Survey 2005/06 and National Service Delivery Survey 2004 (in Regional Forecasts, 2007); Uganda National Examinations Board Primary Leaving Examinations (PLE) 2005; Uganda National Examinations Board PLE 2005.

A number of factors drive this regional disparity in education. Conflict has had a pronounced negative impact on education infrastructure, staff, resources and systems. Many schools, and their teaching materials and resources, have been destroyed, and a number of schools had to relocate during the LRA war. Teachers are not attracted to teaching in the north. There are high rates of teacher absenteeism, because of inadequate teacher housing, long commutes to schools and low salaries without hardship supplementation.[54] The poverty experienced in the north also affects the capacity of children to learn. The conflict displaced many people and their access to productive livelihoods was extremely limited. While Uganda's policy of Universal Primary Education (UPE) means that families do not have to pay school fees, they still must purchase uniforms and school materials, make Parent and Teacher Association (PTA) contributions and forgo children's labor when they are in school. Children are also hungry and some are still traumatized by war. Combined, these factors limit school enrolment, attendance and children's ability to learn and complete their primary education.

The Government of Uganda's (GoU's) central primary education policy is UPE, which it introduced in 1996. UPE involved the abolition of tuition and other costs, such as contributions to parent teacher associations and to school building funds. President Museveni's decision to implement UPE made Uganda the first African country in the post-structural adjustment era to institute a policy that made primary education free at the point of delivery. This removed a significant obstacle to education for poor families and sent a signal to the nation as to the importance of education. Since the early 1990s, the education budget has increased from 1.6 percent to 3.8 percent of gross domestic product (GDP), boosted by funds made available by donors and resources from the Poverty Reduction Fund and the Heavily Indebted Poor Countries (HIPC) initiative.[55] Education is a priority of the human development pillar of Uganda's Poverty Eradication Action Plan (PEAP), reflecting the centrality of education in Uganda's development agenda.

The impact of UPE on primary school enrolment across Uganda has been significant. In the Northern region, between 1992/93 and 2002/03 girls' and boys' enrolment rose from 39.7 percent to 72.5 percent and from 54.4 percent to 73.5 percent, respectively. However, while UPE has had an equalizing effect in terms of education access, it has not had the same effect when it comes to education quality and performance. Essentially, the universal approach that underpins UPE means that the policy does not address the particular challenges facing the north in terms of education delivery, and therefore has not proved effective in reducing disparities in education quality and performance.

Analysis of the flow of public resources to the north of Uganda sheds some light on the constraints to reducing regional educational disparity, showing that these resources are limited and potentially inadequate. Public resource flows include central government transfers to districts through unconditional and conditional transfers, GoU development expenditures (such as the Northern Uganda Peace and Recovery Plan), donor on- and off-budget projects (such as the Northern Uganda Social Action Fund [NUSAF]), UN agencies' resources and NGO resources. An analysis of government spending shows that per capita central government transfers to the wider north are approximately equal to the Ugandan national average. However, the region receives little government funding to compensate for the effects of a long-duration conflict. While an equalization grant is in place, and 85 percent of this flows to the Northern region, the size of this is small and therefore ineffective in achieving its objective of equalizing disparities.[56]

In the education sector specifically, the main outputs from the GoU's primary education budget are the Primary Teachers' Wage Bill, the UPE Capitation Bill and the School Facilities Grant.[57] Teachers' wages are standard, and no supplementation exists for teaching in difficult contexts. As such, this budgetary mechanism does not take regional educational disparities into account. UPE capitation grants are paid for on the basis of the number of students enrolled and their level of education.[58] Again, regional education disparities are not taken into account. The School Facilities Grant is based on need, however, with allocation based on pupil-to-permanent classroom ratios. As a result, allocations to the north are well above average, potentially with some regional inequality-reducing impacts.[59]

Districts can supplement primary education through the district budgeting process. However, conflict-affected districts of Northern Uganda have an eroded revenue base, limiting their ability to increase expenditure through local taxation. Preliminary analysis of the Northern Uganda Public Expenditure Review (NUPER) shows that local revenue mobilized in the north is roughly 30 percent below the national average (excluding Kampala). This is unsurprising given the long-term impact of conflict, insecurity and displacement on the economy. Additionally, prior to the abolition of the graduated tax, internally displaced persons (IDPs) were exempt. Thus, graduated tax compensation transfers to Northern Uganda are lower than the national average. It might be assumed that humanitarian agencies and NGOs fill this gap, but this does not appear to be the case. In fact, the NUPER did not find a significant difference between per capita donor and NGO support to the districts of the wider north and districts elsewhere in Uganda.[60]

This analysis of public resource flows shows that, on the whole, central government unconditional and conditional grants do not do enough to acknowledge and/or redress regional inequalities. With a low local revenue base, this leaves local government with limited room to invest in measures to "level the playing field." Donor and NGO funding does not appear to sufficiently fill the gap left by inadequate central government transfers and the eroded revenue base, making post-conflict recovery and the removal of inter-area inequality in primary education difficult.

## Case Study 4: Targeting Lagging Regions through the 8-7 National Poverty Reduction Program in China

In the three decades since the reform and opening-up period began, China has made widely recognized progress in absolute poverty reduction, measured according to both the $1 a day (purchasing power parity [PPP] and national poverty lines.[61] But this poverty reduction has been accompanied by increases in inter-area and within-area inequality across several economic and social indicators.[62] For example, there has been a divergence in infant mortality rates and illiteracy across the rural–urban and inland–coastal divides. The type of development strategy pursued in the post-1979 reform era—one focused on expanding and facilitating foreign direct investment (FDI) and export-led growth—has contributed to this trend.[63] Prioritization of the eastern coastal region, along with fiscal decentralization and the removal of redistributive mechanisms of fiscal transfers from rich provinces to poor ones, has allowed the concentration of wealth and capital in particular geographic regions.[64]

The poor development performance of lagging regions has not gone unnoticed by the Chinese government, as their 11th Five-Year Plan discussed above demonstrates. The 8–7 National Poverty Reduction Program, introduced in 1994, was an initiative specifically designed to target areas where poverty was persisting.[65] The program targeted 592 "poor" counties and was designed around three components: subsidized loans for households and enterprises, food for work programs drawing in surplus farm labor for infrastructure projects, and government budgetary grants for investment. Spending on the 8–7 Program accounted for 5–7 percent of total government expenditure from 1994–2000, with RMB 124 billion allocated to the 592 counties.[66]

It is difficult to isolate the effects of the 8–7 Program from other important changes in the institutions and policies guiding social and economic development. While the rate of poverty reduction accelerated during the period of implementation, with the number of poor rural people declining from 80 million to 32 million between 1993 and 2000, this could be attributed to a combination of different policies and interventions.[67] For instance, important reforms in procurement increased farm-gate prices and represented a shift away from previously urban-biased pricing policies. This may have been more important in driving down rural poverty than programmes such as the 8–7 Program.[68]

But the 8–7 Program did make a noticeable contribution to growth in poor areas, and evidence suggests that resources from the program were well-targeted at the county level. Also, during the life of the 8–7 Program (1994–2000), designated "poor" counties experienced higher than average growth in agricultural production and household net income. The growth rate of agricultural GDP in the poor counties was 7.5 percent, compared with a national average of 7 percent. Household net income per capita increased from RMB 648 to RMB 1,337, growing at an annual rate of 12.8 percent, two percentage points higher than the national average. This echoes findings of higher rates of growth in geographically targeted poor counties in earlier Chinese poverty reduction programs.[69]

While these results are encouraging, closer examination reveals some key structural deficiencies, which may have reduced the program's effectiveness in alleviating spatial poverty and inequalities. While the 8–7 Program's investments led to higher rates of growth in poor counties, we must consider how this growth was distributed within counties. In particular, implementation of the 8–7 Program's subsidized loans was complicated by the leakage of benefits to non-poor villages, township and village enterprises and non-poor households. Weiss notes that because of the real financial constraints these counties faced, local officials had incentives to divert funds to projects capable of generating revenue, rather than funding projects that would achieve the greatest poverty impact. This may also have resulted in the diversion of funds away from longer-term poverty reduction goals.[70]

## Case Study 5: Presidential Instruction Programme for Less–Developed Villages in Indonesia

Even though Indonesia has made significant progress in growth and poverty reduction in recent decades, significant regional and sub-regional disparities remain. Intra-country welfare disparities are in part the result of spatial concentrations of both resource endowments and major industrial activities. For example, Jakarta and West Java—the nation's major industrial centers—account for roughly 30 percent of national GDP. Investments in infrastructure and human capital are located primarily on Java, with additional significant investments on North Sumatra. In contrast, the eastern provinces—where people rely primarily on agriculture—produce less than 1 percent of Indonesia's GDP. They also have far higher rates of poverty.[71] So, despite Indonesia's growth and reductions in poverty, high levels of poverty persist in remote areas. Indeed, evidence shows that spatial factors, including quality of public goods in a district, remoteness and rural residence, are statistically significant in determining levels of per capita expenditure and poverty rates in Indonesia.[72]

In 1994, the Presidential Instruction Programme for Less-Developed Villages (known in Indonesian as Instruksi Presiden Desa Tertinggal or IDT) came into effect. The program was intended to target poor villages and reduce the poverty headcount from 25 million (the official headcount in 1993) to 12 million (by the end of Replita VI, or the sixth five–year plan). It was expanded in 1996, but then abandoned in 1997 in the wake of the Asian financial crisis.[73] The program sought to reduce poverty through redistributive centre-to-region fiscal transfers. In response to criticisms of the overly centralized administration of a previous program (called Inpres), the IDT was designed to allow local needs and expertise to take precedence over the broad application of national goals and strategies. The implementation of the IDT involved two distributional stages. The first involved the selection of poor villages, by classifying each village according to a range of socioeconomic indicators. A village was classified as "poor" based on its position relative to the provincial average, as well as on a subjective assessment from a field inspection by local officials. Based on this approach, 31 percent of villages in Indonesia were classified as poor or neglected.[74] Each targeted village received identically sized transfers from the central government, regardless of population or severity of poverty. The second stage involved the disbursement of funds to *pokmas* (community groups), which made proposals for the use of funds to the sub-district office. This was in contrast with previous Inpres programs, which had increased the financial capacity of regional governments to deliver infrastructure projects.[75]

Conclusions about the effectiveness of the IDT are mixed. It has been found that the IDT program has contributed to reducing regional inequality.[76] It also helped develop self-employment activities among young men, which has reduced unemployment and increased household expenditure on non-food items of housing and clothing. In doing so, it has supported two of the program goals: increasing employment opportunities and improving welfare.[77] These findings must be interpreted with caution, however. All poor villages received grants, so there was no control group. The program may also have supported social cohesion in Indonesia. Research shows that people in poorer areas had a positive perception of the program and think of the funds as coming straight from "Pak Harto" (President Suharto).[78] There are critiques of the program, however. For example, it has been argued that IDT stifled entrepreneurship and innovation in receiving villages, created a system of "dependency" in poorer regions. It has also been suggested that IDT fuelled feelings of exploitation in resource-rich regions, with implications for national cohesion.[79]

## *Case Study 6: Vietnam's Programme for Socioeconomic Development in Communes Faced with Extreme Difficulties*

Vietnam is widely described as a development success story. In recent decades, it has experienced rapid economic growth and significant reductions in poverty. Social and economic development indicators show large and in some cases increasing spatial disparities, however. Headcount poverty rates are highest in the mountainous border regions in the northeast and northwest of the country, as well as in the interior areas of the central coast and the northern Central Highlands. Depth and severity of poverty display similar spatial concentrations.[80] While the areas around the major cities of Hanoi, Ho Chi Minh City and Haiphong have captured a greater portion of industrial production, the less-favored areas remain largely agricultural and relatively excluded from the nation's increasingly rapid economic growth. During the period since the inception of Doi Moi economic reforms, this regional divergence has only increased, with foreign investment concentrated in the industrial centre.[81] Work by Minot et al. suggests the presence of spatial poverty traps:

> The analysis of the geographic determinants of poverty reveals that three-quarters of the variation in rural poverty at the district level can be explained by a small number of agro-climatic and market access variables. This finding is somewhat troubling because it is not possible to design policy interventions that directly influence the agro-climatic variables. So those living in districts with steep slopes and poor soils may be caught in spatial poverty traps from which it is difficult to escape.[82]

As part of a larger effort to address issues of persistent poverty, the Government of Vietnam (GoV) initiated the Programme for Socioeconomic

Development in Communes Faced with Extreme Difficulties (also known as Programme 135 or P135) to address some of the challenges facing disadvantaged areas. The first phase of the program (1999–2004) reached 2,374 communes.[83] Communes targeted were in ethnic minority and mountainous areas. These areas were geographically remote, experienced harsh agro-ecological and climatic conditions and had few opportunities to participate in and contribute to Vietnam's economic growth.[84] Program management was decentralized to the commune level. Funding came largely from the central government, although some provinces provided funding too. Funds were allocated equally to each "poor" district and disbursed to each commune. Investment focused on five areas:

1.  Infrastructure at the village and commune level (e.g. roads, health centers, schools, irrigation systems, water supply systems and markets);
2.  Infrastructure at commune–cluster level (e.g. inter-commune roads, clinics and markets);
3.  Settlement of ethnic minorities;
4.  Agricultural and forestry extension programs; and
5.  Training of commune-level cadres.

The program allocated roughly 96 percent of funds to infrastructure at either the village and commune level (76 percent) or the commune–cluster level (19 percent). As a result, most local authorities treated the P135 as a channel for receiving central government investment in commune infrastructure.[85]

Between 1999 and 2004, P135 supported more than 20,000 small infrastructure projects and training for over 155,000 community staff. The program is credited with making important contributions to poverty reduction in targeted regions. Indeed, between 2002 and 2006, the Central Highlands experienced an annual 5.8 percent reduction in poverty and the North West region experienced an annual 4.8 percent reduction in poverty.[86] Qualitative analysis of the effectiveness of P135 infrastructure programs confirms this positive assessment. In a study conducted by the Ministry of Labor, Invalids and Social Affairs (MOLISA) and the UN Development Programme (UNDP), 85 percent of respondents said road construction under P135 had a "significantly positive impact on their lives," whereas roughly 75 percent noted that irrigation projects had a "significantly positive impact on their agricultural production." In both cases, the main reason given for a lack of satisfaction was poor quality. The real value in P135 seems to have come from the sheer quantity of its investments and the focus on better connecting remote communities to services and national growth processes.

The MOLISA/UNDP study concluded that improvements could be made to the second phase of P135 by increasing the focus on capacity building to enable further decentralization, which was thought to be crucial to meeting local needs and ensuring that P135 targeted poor people, rather than wealthier people living in "poor" communes.[87]

## Case Study 7: Mexico's PROGRESA

In Mexico, a north–south divide pervades a range of economic and social indicators, with the more industrialized north performing consistently better than the more agricultural south. Mexico's economic growth, liberalization and integration into world markets have further entrenched this divergence. For example, Rodríguez-Pose and Sánchez-Reaza find that while liberalization has led to increased benefits for states bordering on the US, as well as Mexico City, the agricultural and natural resource-dependent states in the south (including Chiapas, Oaxaca, Guerrero and Tlaxcala) have suffered relative declines and experienced poor rates of growth. Economic liberalization and integration have been connected to economic divergence and to a widening of the gap between a relatively rich north of the country and an increasingly poor south. The gap is along economic indicators, such as income, as well as social indicators, such as education. Poor social and geographic capital may contribute to keeping these southern regions trapped in poverty, especially given the likely future need for skilled labor in the Mexican economy.[88]

The National Program for Education, Health and Nutrition (Programa Nacional de Educación, Salud y Alimentación, or PROGRESA) began in 1997, with the objective of reducing long-term poverty by providing incentives for households to invest in human capital.[89] From August 1997 to early 2000, the program expanded from roughly 140,000 households in 3,369 localities to nearly 2.6 million rural households in 72,345 localities. The annual budget was around US$1 billion. The program was highly centralized in nature, representing a significant shift from its predecessor, the National Solidarity Programme (Programa Nacional de Solidaridad, or PRONASOL), which ran between 1988 and 1994, and the general shift towards decentralization in developing country poverty reduction. It was hoped that recentralization would make PROGRESA less vulnerable to local political influences. This appeared to be effective: research shows that PROGRESA's targeting mechanisms were extraordinarily robust in resisting elite capture and reaching the poorest households.[90]

PROGRESA provided conditional cash transfers of, on average, 20 percent of total household expenditures. There were two types of conditional transfers— one for education and one for health and nutrition—indexed for inflation and adjusted twice each year. Educational transfers sought to help offset the direct and opportunity costs of schooling that contribute to low levels of educational attainment in poor areas. They were provided for each child aged seven years or above with 85 percent school attendance. Households with older students or female students who met this criterion received a transfer of greater value. Health and nutrition transfers sought to address food poverty. They were linked to attendance at health clinics and nutrition and hygiene information sessions, and resulted in an additional nutritional supplement for households with children under three years old. Targeting occurred in two stages. The first stage used a "marginality index" to identify the most marginal rural localities. The selected localities were then visited to ensure they had access to the necessary infrastructure (e.g. schools and health clinics) to make sure the conditions attached to the transfers were viable. The second stage targeted households within eligible localities, using census data to identify "poor" households.[91]

Fuentes and Montes note that several studies of PROGRESA show that the program has been successful in reducing poverty in the short term, as well as building human capital to contribute to poverty reduction over the longer term. For example, PROGRESA localities were found in 1997 to have lower levels of poverty increase than "control localities" (localities not receiving benefits). This suggests the PROGRESA cash transfers protected poor communities from increases in poverty, even over the short term. School enrolment rates in targeted areas increased, suggesting children in households receiving PROGRESA transfers would obtain average earnings around 8 percent higher when they reached adulthood. Both children and adults in PROGRESA communities demonstrated improved health outcomes, including a 19 percent reduction in days of absence from work in the 18–50 age bracket, lower incidence of illness for under fives, greater participation in prenatal care and a 16 percent per annum improvement in mean growth between twelve and thirty-six months. Positive spill-overs to non-beneficiary households were also found to exist, raising the level of human capital throughout PROGRESA communities.[92] PROGRESA became *Oportunidades* (Oportunidades Human Development Program) in 2002 and has continued to expand and generate positive results.

## *Case Study 8: Circular Migration in India*

Circular migration is emerging as a dominant form of migration among poorer groups in India. Circular migration is when individuals, families or groups migrate temporarily or seasonally, for work or to access services, before returning to their "home" area. The number of short-term out-migrants in India has been estimated at 12.6 million, but recent micro studies documenting large and increasing numbers of internal migrants suggest that the true figure is 30 million and rising.[93] Circular migration rates are high in remote rural areas, especially among chronically poor people. Rates are particularly high in drought-prone areas with low agro-ecological potential, poor access to credit and high population densities. For example, an estimated 300,000 laborers, driven by a range and combination of push and pull factors, migrate from drought-prone Bolangir district in Western Orissa every year.[94] And research from Andhra Pradesh and Madhya Pradesh finds that circular migration is highest among chronically poor households in remote villages.[95] While acknowledging the presence of some negative impacts (e.g. acute shortage of labor and high dependency ratios in sending areas), there is overwhelming evidence that internal migration can lead to positive change in both sending and receiving areas, helping reduce, or halt the slide into, poverty.[96]

Despite the large numbers of people migrating, migration-related issues rarely get onto the local, state or national policy agenda in India. The economic benefits of migration are not recognized and migration tends to be viewed as an economically, socially and politically destabilizing process, because it overburdens urban areas, deprives rural areas of productive members, destabilizes family life, leads to labor exploitation by the informal sector and causes administrative and legislative headaches.[97]

There have essentially been two forms of policy response to migration in India, neither of which have enabled, or controlled, migration. The first has been to increase rural employment, in an attempt to stem the flow of migrants out of rural areas. This reflects an assumption that deteriorating agriculture leads to out-migration, and that improved natural resource bases and employment opportunities in rural areas can reduce or reverse migration. The suite of policies in place that aim to increase the availability of rural employment include the National Rural Employment Guarantee Scheme, which promises 100 days of wage labor to one adult member in every rural household that volunteers for unskilled work, numerous development programs to improve agricultural productivity and programs to develop small and medium towns, to help arrest migration to urban areas. The other policy response is essentially a "non-response." India does not recognize the centrality of migrant labor to economic growth, and perceptions of negative economic, political and social effects mean that local, state and national governments remain hostile towards migrants, while employers routinely disregard laws designed to protect their rights and needs. In many cases, policies remain predicated on a supposedly "sedentary" population.[98]

These policy responses—encouraging rural employment and ignoring and excluding migrants—do not stop migration. On the whole, opportunities are still better in urban areas and in high productivity rural areas. Meanwhile, these responses do mean that migrants and their children are poorer and more marginalized at their destination than they might otherwise be. This is largely because migrants are not recognized politically—they are not entitled to vote at their destination and they have limited access to public services and entitlements. This means that migrants:

- Have limited political agency, and the political process rarely recognizes or addresses their concerns.
- Often live in illegal settlements, where they face constant threats of eviction, disease, sexual abuse, underpayment and police harassment.
- Have inadequate access to essential services and amenities such as water and sanitation, electricity, health care and education.
- Experience harassment and exclusion, as local police and bureaucrats regulate the informal economy.
- Cannot access subsidized food through the Public Distribution System, because they cannot use their cards outside their home local authority. This means they spend a considerable proportion of their wages on basic food supplies and rents and, when whole families migrate, children often do domestic chores while their parents work, missing out on an education.
- Cannot easily access state schools, cheap housing or government health care.[99]
- Experience discrimination and sometimes exploitation, and are generally paid less than non-migrant workers.

Labor unions, donors and non-governmental organizations (NGOs) are attempting to tackle these problems. For example, mobile ration cards for 5,000 migrants are being piloted in small and major towns in Rajasthan, India. In Madhya Pradesh, the UK Department for International Development (DFID) funded a comprehensive migrant support program in eight tribal districts, which aims to provide information on opportunities and improve bargaining power by enhancing skills. Several NGOs, such as the Gramin Vikas Trust in Madhya

Pradesh and Adhikar in Orissa, have migrant support programmes to improve the efficiency, safety and cost of remittance mechanisms. Identity cards have been used with very positive results in Madhya Pradesh under a migrant support program that the Gramin Vikas Trust has implemented.[100] As Deshingkar argues, circular migration should be supported - it is not going away and on balance, benefits poor people. Policy makers need to recognize migration as an important poverty interrupter and develop policies and programs that maximize benefits for migrants, sending areas and receiving areas.[101]

# LESSONS FOR POLICY AND PRACTICE

What can we glean from this and other analyses? What are the lessons for development policy and practice about addressing the spatial dimensions of poverty?

## *Spatial Disadvantage as a Driver of Poverty*

There is a compelling body of evidence showing that spatial disadvantage is a driver of poverty. This means that if analysis and policy is to understand and address poverty, it must be spatially informed, and address a location's spatially specific characteristics as well as its wider integration into national, regional and global economies. Policies that support economic integration and attempt to reduce transaction costs will benefit aggregate growth. However, it is likely that place and space will retain their relevance in determining economic growth and poverty outcomes, even in this relatively connected world.

## *The Form and Processes of Integration into the Economy Matter*

Spatial integration is good. But spatially informed policymaking should not only facilitate integration, but also help manage the *nature* of the integration. Excluded areas (and households in those areas) may not always benefit from economic integration. For example, labor migration is a critical part of the economic integration process, but some people may not be equipped to migrate. This suggests that it is not enough simply to enable integration and agglomeration, with the assumption that labor markets will function in such a way as to move labor efficiently to where it is needed. Government policies must also deliver services to hard-to-reach and excluded populations in order to build their capabilities and agency so they can not only migrate, but also benefit from labor migration. As well, locations (and households in those locations) that are connected, but incorporated in an *adverse* way, may find it difficult to benefit from economic integration.[102] Policies need to address adverse impacts resulting from processes of integration, so that people can best make use of the opportunities that increased integration brings. Deshingkar's work on circular migration in India is particularly suggestive of the ways in which spatially integrative processes (i.e. the mobility of labor) can produce damaging outcomes if the policy environment either actively constructs barriers to productive inclusion or, through a non-response, simply fails to mitigate the negative consequences of adverse inclusion.[103] The case of resettled villagers in Lao PDR further emphasizes the need to account for the new challenges faced as a result of policies designed to facilitate population density and integration—even where there are acknowledged advantages. So while economic integration should be encouraged, the form and processes of integration need to be taken into account, so that poorer households, and poorer locations, can best benefit from the processes of change.[104]

## *Balance Universalism and Targeting*

Universal policies seeking to achieve universal outcomes need to be balanced with targeted initiatives that take into account the particular constraints faced by households in disadvantaged areas. Experiences of implementing universal education policies in Ghana and Uganda suggest that such policies may fail to achieve truly universal outcomes if they do not take into account the additional challenges faced in disadvantaged locations. Some of these challenges may be entrenched and structural, and the result of historical processes of investment and economic development, social exclusion and marginalization or conflict. In the case of Uganda, policies such as the Ugandan equalization grants may need to be scaled up in order to provide lagging regions with the additional resources they need to emerge from poverty. Programs such as the 8–7 Program in China, Vietnam's P135, the IDT in Indonesia and Mexico's PROGRESA show the value in specifically targeting spatially disadvantaged areas and households. In all cases, there is evidence of progress in terms of growth and/or poverty reduction in targeted areas. Targeting does not come without challenges, however. For example, in China, the leakage of benefits to non-poor locations and households complicated the program. In Mexico, PROGRESA dealt with this through second-round targeting (i.e. identifying households in poor locations), and this proved effective in stemming leakages and targeting intended beneficiaries. The challenge for policymakers in situations like these will be to balance the costs of additional targeting with potential benefits.

In sum, attempts to support integration and agglomeration should be balanced with policies to tackle the root causes of poverty and promote the accumulation of capabilities in spatially disadvantaged areas. Addressing spatial disadvantage often requires a multi-pronged approach: supporting economic integration between spatially disadvantaged and productive areas; mitigating the adverse impacts that stem from economic integration; and addressing the particular challenges faced by spatially disadvantaged areas. A balanced approach, which addresses specific locational needs, as well as supporting the positive integration of disadvantaged areas into broader development processes, is required. The relational dimension is key here, and an overly decentralized approach risks losing a wider perspective, and as a result, does not give concerns of equity, inequality or other relational concepts enough weight. Rodríguez-Pose and Gill have raised this concern, arguing a shift in global policy towards decentralization and devolution:

> reflects a subtle, but profound, renunciation of the traditional equalization role of national government, in favor of conditions fostering economic and public competition, and leading to greater development of initially rich and powerful regions to the detriment of poorer areas.[105]

## *Adopt Short-Term and Longer Term Policy Solutions*

In addressing spatial disadvantage, there appears a need to combine shorter term policies that address immediate deprivations with longer term policy solutions that address the structural issues that lead to disadvantage. For example, based on her research from Orissa, Shah argues:

> While resource transfers through wage employment or other subsidies are crucial to making a dent in chronic poverty in such regions, long-term solutions lie in addressing structural problems, such as the failure of entitlements and the integration of forest management into the larger framework of development in a number of forest-based economies.[106]

Ex–post evaluations of the 8–7 Program in China confirm this, and subsequent anti-poverty programs aimed at lagging regions in China, such as the Southwest Poverty Reduction Project and the Western Development Initiative, have taken a multi-sectoral approach that packages longer term investments in health and education with infrastructure expenditures, food for work programs and other more immediate return investments in economic performance. In the same sense, the achievements in health and education under PROGRESA in Mexico need to be matched with opportunities for productive employment that takes advantage of the additional human capital.

## *Respond to Different Scales and Settings of Spatial Disadvantage*

Policies must address the different scales on which spatial disadvantages are manifest, and the different settings in which spatial disadvantage might be found. Understanding context, and working at a sufficient level of disaggregation to identify areas or pockets of spatial disadvantage, is critical. It is not enough to provide aggregate growth or poverty reduction figures. We need to go behind these figures to understand how the benefits are distributed. For example, where are people living in income poverty, where do people not have access to markets and decent work, where are people unable to engage in political decision-making processes, and where do people not have access to good quality public services? Spatial disadvantage will be found in a range of contexts, at a range of scales, within countries but also within regions and cities. Policy responses need to be cognizant of this variation and respond to a specter of settings, ranging from remote rural areas through to overcrowded urban slums.

# CONCLUSION

We have argued that when it comes to poverty, the spatial dimensions matter. Despite progress on economic growth in many countries, increasing welfare disparities between regions and areas represent a prominent trend. This has led us to argue that if looking for entry points for tackling poverty, spatial disadvantage must be one of them. Our policy review leads us to suggest that, at least from a policy perspective, developing country governments have done a better job of acknowledging and grappling with this than donors.

We believe that inclusive growth should remain at the front of our minds. Attempts to kick-start and drive economic growth and economic integration need to be balanced with concerns as to how the benefits of this growth and integration are distributed. Medium-term economic change and progress need to be balanced with poverty reduction in the short-term. We cannot assume the presence of a developmental state and effective redistribution mechanisms that will take care of the current and very real needs of populations in spatially disadvantaged areas.

Traditional policy instruments, which seek to address pockets of poverty through universal approaches are not adequate. They need to be complemented by area-based interventions that consciously seek to address, in a context-specific way, low levels of geographic capital and its human development implications. Given the institutional weakness of many low-income developing countries, this is an ambitious agenda. Many governments struggle to deliver the most basic of public services to more privileged urban areas, let alone those in remote and isolated communities, or urban slums. But if governments are to move their populations out of poverty, and avoid social and political tensions, these spatial disadvantages need to be overcome. While universal and sectoral policies will be necessary, in many cases they may not be sufficient. Additional attention will need to be made to the spatial dimensions of poverty to ensure that households in spatially disadvantaged locations can move out of poverty.

## NOTES

1. Paul Krugman, "Increasing returns and economic geography," *Journal of Political Economy* (1991): 99, 3, 483-499.

2. Jyotsna Jalan and Martin Ravallion, "Spatial poverty traps?," Policy Research Paper, 1798 (Washington DC: World Bank, 1997); Jyotsna Jalan and Martin Ravallion, "Geographic poverty traps? A micro model of consumption growth in rural China," *Journal of Applied Econometrics* (2002): 17, 4, 329-346; and Ravi Kanbur and Tony Venables, "Spatial poverty and development," *Journal of Economic Geography* (2005): 5, 1, 1-2.

3. World Bank, *Reshaping Economic Geography – the 2009 World Development Report* (Washington DC: World Bank, 2008).

4. Jalan and Ravallion, "Spatial poverty traps?" and Martin Ravallion and Quentin Woden, "Poor Areas, or Only Poor People?," *Journal of Regional Science* (1997): 34, 4, 689-711.

5. Kanbur and Venables, "Spatial poverty and development."

6. Chronic Poverty Research Centre (CPRC) *Chronic Poverty Report 2004-05* (Manchester: Chronic Poverty Research Centre, 2004).

7. Jalan and Ravallion, "Spatial poverty traps?" and Jalan and Ravallion, "Geographic poverty traps? A micro model of consumption growth in rural China."

8. Kate Bird and Andrew Shepherd, "Livelihoods and Chronic Poverty in Semi-Arid Zimbabwe," *World Development* (2003): 31, 3, 591-610.

9. Kate Bird, Andy McKay Isaac Shinyekwa, "Isolation and poverty: the relationship between spatially differentiated access to goods and services and poverty," *ODI/CPRC Working Paper Series,* ODI WP322, CPRC WP162 (London: ODI and Manchester: CPRC, University of Manchester: 2010).

10. Javier Escobal and Maximo Torero, "Measuring the impact of asset complementarities: the case of rural Peru," *Cuandernos de Economia (Latin American Journal of Economics* (2005): 42, 125, 137-164.

11. Luc Christiaensen, Lionel Demery and Stefano Paternostro, "Macro and micro perspectives of growth and poverty in Africa," *World Bank Economic Review* (2003): 17, 3, 317-347.

12. Bill Burke and Thom Jayne, "Spatial disadvantage or spatial poverty trap. Household evidence from rural Kenya," *ODI/CPRC Working Paper Series,* ODI WP327, CPRC WP167 (London: ODI and Manchester: CPRC, University of Manchester, 2010).

13. Marcel Fafchamps and Christine Moser, "Crime, isolation and law enforcement," *Journal of African Economies* (2003): 12, 4, 615-671.

14. Ursula Grant, "Spatial inequality and urban poverty traps," *ODI/CPRC Working Paper Series,* ODI WP326, CPRC WP166, (London: ODI and Manchester: CPRC, University of Manchester, 2010).

15. Nicholas Minot, Bob Baulch and Michael Epprecht, "Poverty and inequality in Vietnam: spatial patterns and geographic determinants" (Washington DC and Brighton: IFPRI and IDS, 2003).

16. Takeshi Daimon, "The spatial dimension of welfare and poverty: lessons from a regional targeting programme in Indonesia," *Asia Economic Journal* (2001): 15, 4, 345-367.

17. Burke and Jayne, "Spatial disadvantage or spatial poverty trap. Household evidence from rural Kenya."

18. World Bank, *Reshaping Economic Geography–the 2009 World Development Report.*

19. Ravi Kanbur and Anthony Venables, *Spatial Inequality and Development* (Oxford: Oxford University Press, 2005).

20. Bird et al., "Isolation and poverty: the relationship between spatially differentiated access to goods and services and poverty."

21. Chronic Poverty Research Centre (CPRC), *Chronic Poverty Report 2004-05.*

22. Chronic Poverty Research Centre (CPRC), *Chronic Poverty Report 2004-05.*

23. Countries were purposively selected to represent a range of regions (sub-Saharan Africa, Latin America and the Caribbean, South East Asia and Central Asia) and income levels. The PRSPs reviewed were from Cote D'Ivoire; Djibouti; Tajikistan; Afghanistan; Albania; Armenia; Benin; Haiti; Lao PDR; Liberia; Maldives; Mali; Moldova; Niger; Rwanda; Uzbekistan.

24. The donor-led nature of PRSPs has been a prominent critique.

25. National Development and Reform Commission (NDRC), "Outline of the Eleventh Five-Year Plan for the Economic and Social Development of the People's Republic of China" (2007).

26. Government of China, "Report on the Implementation of the 2007 Plan for National Economic and Social Development and on the 2008 Draft Plan for National Economic and Social Development" (Beijing: Government of China, 2008).

27. Government of China "Report on the Implementation of the 2007 Plan for National Economic and Social Development and on the 2008 Draft Plan for National Economic and Social Development."

28. Government of India Planning Commission, "Eleventh Five Year Plan (2007-2012): Inclusive Growth, Volume 1" (New Delhi: Government of India, 2008).

29. Government of India Planning Commission, "Eleventh Five Year Plan (2007-2012): Inclusive Growth, Volume 1."

30. Government of India Planning Commission, "Eleventh Five Year Plan (2007-2012): Inclusive Growth, Volume 1."

31. Hans Dieter Bechstedt, "Impact of public expenditures on ethnic groups and women–Lao PDR," Poverty and Social Impact Assessment Final Report (2007).

32. Government of Lao PDR (GoL), "National Growth and Poverty Eradication Strategy (NGPES)" (Vientiane: GoL, 2003).

33. Bechstedt, "Impact of public expenditures on ethnic groups and women – Lao PDR."

34. Bechstedt, "Impact of public expenditures on ethnic groups and women – Lao PDR."

35. State Planning Committee, National Statistics Centre and Asian Development Bank, "Participatory Poverty Assessment, Lao PDR" (Vientiane: State Planning Committee, 2001).

36. State Planning Committee, National Statistics Centre and Asian Development Bank, "Participatory Poverty Assessment, Lao PDR."

37. State Planning Committee, National Statistics Centre and Asian Development Bank, "Participatory Poverty Assessment, Lao PDR."

38. Bechstedt, "Impact of public expenditures on ethnic groups and women – Lao PDR."

39. Harold Coulobme and Quentin Woden, "Poverty, livelihoods and access to basic services in Ghana," Ghana CEM: Meeting the challenge of accelerated and shared growth (Washington DC: World Bank, 2007).

40. Arnim Langer, Abdul Raufu Mustapha and Frances Stewart, "Horizontal inequalities in Nigeria, Ghana and Cote d'Ivoire: Issues and Policies," CRISE Working Paper, 45 (Oxford, CRISE, 2007).

41. Charles Jebuni, Andy McKay and Andrew Shepherd, "Economic growth in Northern Ghana," Paper presented at International Workshop on Spatial Poverty Traps, Stellenbosch, 29 March (2007).

42. International Development Association (IDA) and International Monetary Fund (IMF), "Growth and Poverty Reduction Strategy Joint Staff Advisory Note," (Washington DC: IDA and IMF, 2006) and World Bank, "Ghana Country Brief" (Washington DC: World Bank, 2007).

43. National Development Planning Commission, "Growth and Poverty Reduction Strategy (GPRS II) (2006-2009)," (Accra: National Development Planning Commission, 2005).

44. National Development Planning Commission, "Ghana Poverty Reduction Strategy 2004 Annual Progress Report."

45. Kwame Akyeampong, University of Sussex, pers. comm., 2007.

46. Don Taylor, DFID Ghana, pers comm., 2007.

47. Kwame Akyeampong, University of Sussex, pers. comm., 2007 and Don Taylor, DFID Ghana, pers.comm., 2007.

48. Leslie Casely-Hayford, Associates for Change, pers. comm., 2007 and Kwame Akyeampong, University of Sussex, pers comm., 2007.

49. Uganda Bureau of Statistics (UBOS), "Uganda National Household Survey, 2005/06" (Kampala: UBOS, 2006).

50. World Bank, "Uganda – Country Brief," (Kampala: World Bank, 2007).

51. World Bank, "Republic of Uganda Joint IDA-IMF Staff Advisory Note on the Poverty Reduction Strategy paper Annual Progress Report" (2007).

52. Regional Forecasts, "Uganda Multi-Donor Group Northern Uganda Public Expenditure Review (NUPER)," Final Report (2007). Data based on UBOS Population and Housing Censes 1991 and 2002, National Household Survey 2005/06 and Uganda Demographic Household Survey 2000.

53. World Bank, "Republic of Uganda Joint IDA-IMF Staff Advisory Note on the Poverty Reduction Strategy paper Annual Progress Report."

54. Women's Commission for Refugee Women and Children, "Learning in a war zone" (New York: Women's Commission for Refugee Women and Children, 2005).

55. World Bank, "Poverty and Vulnerability Assessment. Report No. 36996-UG," (Washington: World Bank, 2006).

56. Regional Forecasts, "Uganda Multi-Donor Group Northern Uganda Public Expenditure Review (NUPER)."

57. Ministry of Education and Sports, "The Education and Sports Sector Annual Performance Report," October (Kampala, 2007).

58. Lawrence Bategeka and Nathan Okurutm, "Universal Primary Education," *Policy Brief*, 10 (London: Inter-Regional Inequality Facility, 2006).

59. Regional Forecasts, "Uganda Multi-Donor Group Northern Uganda Public Expenditure Review (NUPER)."

60. Regional Forecasts, "Uganda Multi-Donor Group Northern Uganda Public Expenditure Review (NUPER)."

61. Wang Sangui, Li Zhou, Ren Yanshu, "The 8-7 National Poverty Reduction Program in China – The National Strategy and Its Impact," Conference on Scaling Up Poverty Reduction: A Global Learning Process, Shanghai, 25-27 May (2004).

62. Azizur Rahman Khan and Carl Riskin, *Inequality and Poverty in China in the Age of Globalization* (New York: Oxford University Press, 2001).

63. Xiaolan Fu, "Limited linkages from growth engines and regional disparities in China," *Journal of Comparative Economics* (2004): 32, 1, 148-164.

64. Xiaobo Zhang and Ravi Kanbur, "Spatial Inequality in Education and Health Care in China," *China Economic Review* (2005): 16, 2, 189-204.

65. Sangui Wang, "Poverty Targeting in the People's Republic of China," ABD Institute Discussion Paper, 4, (Tokyo: ADB Institute, 2004).

66. Wang Sangui, Li Zhou and Ren Yanshu, "The 8-7 National Poverty Reduction Program in China – The National Strategy and Its Impact," Conference on Scaling Up Poverty Reduction: A Global Learning Process, Shanghai, 25-27 May (2004).

67. Sangui et al., "The 8-7 National Poverty Reduction Program in China – The National Strategy and Its Impact."

68. John Weiss, "Reaching the Poor with Poverty Projects: What is the Evidence on Social Returns?," ADB Institute Discussion Paper, 9 (Tokyo: ADB Institute, 2004).

69. Sangui et al., "The 8-7 National Poverty Reduction Program in China – The National Strategy and Its Impact."

70. Weiss, "Reaching the Poor with Poverty Projects: What is the Evidence on Social Returns?," ADB Institute Discussion Paper.

71. Hal Hill, "Spatial Disparities in Developing East Asia: A Survey" *Asian-Pacific Economic Literature* (2002): 16, 1, 10-36.

72. Daimon, "The spatial dimension of welfare and poverty: lessons from a regional targeting programme in Indonesia."

73. Daimon, "The spatial dimension of welfare and poverty: lessons from a regional targeting programme in Indonesia."

74. John Weiss, "Poverty Targeting in Asia: Experiences from India, Indonesia, the Philippines, People's Republic of China and Thailand" (Tokyo: ADB Institute: 2005).

75. Chikako Yamauchi, "Effects of Grants for Productive Investment on Employment and Poverty: Lessons from Indonesia's Left-Behind Village Program," Australian National University Research School of Social Sciences Working Paper (Canberra: Research School of Social Sciences, 2005).

76. Daimon, "The spatial dimension of welfare and poverty: lessons from a regional targeting programme in Indonesia."

77. Yamauchi, "Effects of Grants for Productive Investment on Employment and Poverty: Lessons from Indonesia's Left-Behind Village Program."

78. Kimberley Niles, "Economic Adjustment and Targeted Social Spending: The Role of Political Institutions (Indonesia, Mexico and Ghana)." Prepared for World Development Report 2001 meetings, Castle Donington, 16-17 August (2001).

79. Niles, "Economic Adjustment and Targeted Social Spending: The Role of Political Institutions (Indonesia, Mexico and Ghana)" and Iyanatul Islam, "Dealing with Spatial Dimensions of Inequality in Indonesia: Towards a Social Accord." Paper for the Second Inequality and Pro-poor Growth Spring Conference on the theme of "How Important is Horizontal Inequality?," June 9-10 (Washington DC: World Bank: 2003).

80. Nicholas Minot and Bob Baulch, "Spatial Patterns of Poverty in Vietnam and Their Implications for Policy," *Food Policy* (2005): 30, 5-6, 461-475 and Nicholas Minot, Bob Baulch and Michael Epprecht, "Poverty and inequality in Vietnam: spatial patterns and geographic determinants" (Washington DC and Brighton: IFPRI and IDS, 2003).

81. Hill, "Spatial Disparities in Developing East Asia: A Survey."

82. Minot et al., 'Poverty and inequality in Vietnam: spatial patterns and geographic determinants'.

83. Ministry of Labor, Invalids and Social Affairs (MOLISA) and UN Development Programme (UNDP), "Taking Stock, Planning Ahead: Evaluation of the National Targeted Programme on Hunger Eradication and Poverty Reduction and Programme 135" (Hanoi: MOLISA and UNDP, 2004).

84. Tran Van Thuat and Ha Viet Quan, "A Case Study of the Program for Socio-economic Development of Communes Facing Extreme Difficulties in Ethnic Minority and Mountainous Areas of Vietnam (P135)." Workshop on Strengthening the Development Results and Impacts of the Paris Declaration Through Work on Gender Equality, Social Exclusion and Human Rights, London, 12-13 March (2008).

85. Ministry of Labor, Invalids and Social Affairs (MOLISA) and UN Development Programme (UNDP), "Taking Stock, Planning Ahead: Evaluation of the National Targeted Programme on Hunger Eradication and Poverty Reduction and Programme 135."

86. Tran Van Thuat and Ha Viet Quan, "A Case Study of the Program for Socio-economic Development of Communes Facing Extreme Difficulties in Ethnic Minority and Mountainous Areas of Vietnam (P135)."

87. Ministry of Labor, Invalids and Social Affairs (MOLISA) and UN Development Programme (UNDP), "Taking Stock, Planning Ahead: Evaluation of the National Targeted Programme on Hunger Eradication and Poverty Reduction and Programme 135."

88. Andres Rodríguez-Pose and Javier Sánchez-Reaza, "Economic Polarization Through Trade: Trade Liberalization and Regional Growth in Mexico," UNWIDER Discussion Paper 2003/60 (2003).

89. Sudhanshu Handa, Mari-Carmen Huerta, Raul Perez and Beatriz Straffon, "Poverty, Inequality and Spillover in Mexico's Education, Health and Nutrition Program," FCND Discussion Paper, 101 (Washington, DC: IFPRI, 2001).

90. Emmanuel Skoufias, Benjamin Davis and Sergio de la Vega, "Targeting the Poor in Mexico: An Evaluation of the Selection of Households in PROGRESA," *World Development* (2001): 29, 10, 1769-1784.

91. Natalia Caldés, David Coady and John Maluccio, "The Cost of Poverty Alleviation Transfer Programs: A Comparative Analysis of Three Programs in Latin America," *World Development* (2006): 34, 5, 818-837.

92. Ricardo Fuentes and Andres Montes, "Mexico: Country Case Study Towards the Millennium Development Goals at the Sub-National Level," Human Development Office Occasional Paper, HDOCPA-2003-07 (2003).

93. Priya Deshingkar, "Time to recognize the importance of internal migration for poverty reduction and development," ODI Blog, 15 December (London: ODI, 2006).

94. Priya Deshingkar, "Improved livelihoods in improved watersheds: can migration be mitigated," in *Watershed Management Challenges: Improving Productivity, Resources and Livelihoods* (Columbo: International Water Management Institute, 2003).

95. Priya Deshingkar, "Migration, remote rural areas and chronic poverty in India," *ODI/CPRC Working Paper Series*, ODI WP323, CPRC WP163 (London: ODI and Manchester: CPRC, University of Manchester, 2010).

96. Priya Deshingkar and Sven Grimm, "Voluntary Internal Migration: an update." Paper commissioned by the Urban and Rural Chang team and the Migration Team, Policy Division, DFID (2004).

97. Priya Deshingkar and Ed Anderson, "People on the move: new policy challenges for increasingly mobile populations," *Natural Resource Perspectives,* 92 (London: ODI, 2004).

98. Priya Deshingkar, "Seasonal migration: how rural is rural?," *ODI Opinion,* 52 (London, ODI: 2005) and Priya Deshingkar and Ed Anderson, "People on the move: new policy challenges for increasingly mobile populations," *Natural Resource Perspectives,* 92 (London: ODI, 2004).

99. Priya Deshingkar, "Migration, remote rural areas and chronic poverty in India."

100. Steve Wiggins and Priya Deshingkar, "Rural employment and migration: in search of decent work," *ODI Briefing Paper,* 27 (London: ODI, 2007).

101. Priya Deshingkar, "Migration, remote rural areas and chronic poverty in India."

102. Chronic Poverty Research Centre (CPRC), *Chronic Poverty Report 2004-05* and Andries du Toit and Sam Hickey, "Adverse Incorporation, Social Exclusion and Chronic Poverty," *CPRC Working Paper,* 81 (Manchester: CPRC, 2007).

103. Priya Deshingkar, "Migration, remote rural areas and chronic poverty in India."

104. Priya Deshingkar, "Migration, remote rural areas and chronic poverty in India." *ODI/CPRC Working Paper Series,* (London: ODI and Manchester, 2010)

105. Andres Rodríguez-Pose and Nicholas Gill (2004) "Is There a Global Link Between Regional Disparities and Devolution?," *Environment and Planning,* 2004: 36, 12, 2097-2117.

106. Amita Shah, "Patterns of poverty in remote rural areas: a case study of a forest-based region in Southern Orissa in India," *ODI/CPRC Working Paper Series,* ODI WP325, CPRC WP165 (London: ODI and Manchester: CPRC, University of Manchester, 2010).

## 4

ASSESSING THE POTENTIAL IMPACT ON POVERTY OF RISING
CEREALS PRICES: THE CASE OF MALI[1]

*George Joseph and Quentin Wodon*

The issue of the increase in food prices has received renewed attention in recent years as the increase in prices worldwide has had large negative impacts on households (e.g. Dessus et al. 2008; Ivanic and Martin 2008; World Bank 2008a and 2008b; IMF 2008; Wodon and Zaman 2010; Wodon et al. 2008). Multiple factors have contributed to this increase in prices, including climate change (which results in a higher frequency of droughts, as well as pressures to use food production for biofuels), policy decisions (such as limitations to exports in selected food grain producing countries) and global market trends (including higher energy prices, a depreciating dollar and increased food grain demand especially in rapidly developing middle income countries).

In Mali, prices for several commodities such as rice, millet and sorghum were in 2009 about 25 percent higher than they were a year before. This has led the authorities as well as development partners to consider a range of compensatory measures that could help offset part of the negative impact on the poor of this increase in prices. However, at least from a conceptual point of view, the net impact of an increase in food prices on the poor is not obvious. Indeed, when discussing the link between rice and other cereal prices and poverty, a key issue is to assess the double and opposite impact that a change in prices can have through producers (who benefit from an increase in prices) and consumers (who lose out when the price increases).

The techniques for the analysis of the short-term producer and consumer impacts of food commodity price changes are well developed in the literature. Early work in this area was conducted by Deaton (1989) using data from Thailand (see also Singh et al. 1986). Similar methods have been used in sub-Saharan Africa by Barrett, and by Dorosh (1996) for Madagascar and Budd (1993) for Cote d'Ivoire, among others. These are also the methods that we use in this chapter. Most of these studies have found that food price increases tend to lead to an increase in poverty because the consumption effects dominate the production effects as many countries are net importers of food, at least in sub-Saharan Africa.

There has also been a literature on assessing whether in the medium to long term, the increase in prices is compensated by an increase in wages, for those workers who contribute to the production of food crops (see for example

Ravallion 1990; Boyce and Ravallion 1991; Rashid 2002; Christaensen and
Demery 2006; and Ivanic and Martin 2007). The findings from these studies
suggest that wage offset compensates only in a limited way for the initial
increase in food prices. Finally, there has also been a substantial amount of
work looking at the impact of various policies to deal with food production and
prices. This can be illustrated with the case of rice. Indonesia is a country that
used to import substantial amounts of rice, but where restrictions were
progressively placed on imports in order to help local producers, with imports of
rice actually banned after 2004. Using a general equilibrium model, Warr
(2005) found that the ban on rice imports raised the price of domestically
produced rice, and that this led to an increase in poverty by almost one
percentage point (on the Indonesia story as well as for a more general discussion
on the experience of governments in Asia to stabilize the price of rice, see
Timmer and Dawe 2007). Another paper on Indonesia by Sumarto et al. 2005,
using panel data suggests that the practice of subsidizing rice as part of a social
safety net led to a reduction in the risk for a household to be poor. Papers on
Vietnam by Niimi et al. (2004) and Minot and Goletti (1998) suggest that the
liberalization of rice exports probably led to a reduction in poverty despite an
increase in the price of rice in the country, thanks essentially to increased rice
production.

In this chapter, our objective is to assess what could be the short-term
impact on poverty of the increase in the price of cereals in Mali (for a dynamic
analysis of the medium-term impact of higher rice prices in Mali, see Nouve and
Wodon 2008). The impact of a change in the price of rice is not ambiguous
because about half of the rice consumed in the country is imported. In the case
of wheat and bread as well, the impact is also not ambiguous since wheat is
imported and bread is produced from imported wheat. For these goods, an
increase in price will tend to result in higher poverty in the country as a whole
(even if some local producers will gain from this increase, in the case of rice).
For millet and sorghum, as well as for corn, by contrast, the impact on poverty is
less obvious as these are commodities that are produced for the most part in the
country for local consumption. Overall, when considering the various cereals
together, the impact of the price increase is likely to be an increase in poverty,
but whether this increase will be severe depends on a number of parameters,
including who consumes and produces what, and in what amounts. It is thus an
empirical question to assess what might be the impact on poverty of higher
cereals price in a country such as Mali.

For the sake of simplicity, we will use a number of assumptions to provide
estimates of the impact on poverty of higher food prices. First, we will assume
that the cost of an increase in food prices for a household translates into an
equivalent reduction of its consumption in real terms. This means that we do
not take into account the price elasticity of demand which may lead to
substitution effects and thereby help offset part of the negative effect of higher
prices for certain food items. Similarly, an increase for producers in the value of

their net sales of food translates into an increase of their consumption of equivalent size, and we again do not take into account the role that the price elasticity of supply may play here. As for food auto-consumed by producers (which represents a large share of total consumption), it is not taken into account in the simulations since changes in prices do not affect households when food is auto-consumed. Poverty measures obtained after the increase in prices are then compared to baseline poverty measures to assess impacts. This implicitly means that we do not take into account the potential spill-over effects of the increase in food prices for the food items included in the analysis on the prices for items not included.

A difficult question is whether increases in consumer prices do translate into increases in producer prices. At least two factors may dilute the impact of rising food prices on the incomes of farmers. First, production costs for farmers as well as transport costs are likely to be rising due to higher costs for oil-related products. Second, market intermediaries may be able in some cases to keep a large share of the increase in consumer prices for themselves without paying farmers much more for their crops. Because it is difficult to assess whether producers will benefit substantially from higher food prices, especially in the short term, we could consider our estimates obtained when considering only the impact on consumers as an upper bound of the impact of the rise in prices on poverty, and interpret the results obtained when factoring in a proportional increase in incomes for net sellers or producers as a lower bound of the impact.

The rest of the chapter is structured as follows. The next section presents basic data on cereals production and consumption in Mali based on an analysis of food consumption and production using the 2006 ELIM survey for a number of food categories. Then, we provide estimates of the overall impact of higher food prices on poverty. A brief conclusion follows.

# PATTERNS OF FOOD CONSUMPTION AND PRODUCTION

## *Food Consumption*

Cereals prices are the focus of this chapter, and within cereals, we focus further on rice, millet and sorghum, and corn as these goods represent a large share of total consumption and have also experienced increases in prices in recent months. Table 4.1 provides summary data on rice consumption. Table 4.2 provides similar data for the consumption of millet and sorghum, and in table 4.3 data are provided on the consumption of maize.

We see important differences in the weight of the various cereals in the overall consumption basket of the population, as well as differences between various types of households in their consumption patterns. Consider first rice. For those households who consume rice, total consumption is at about 153,000 FCFA per year. Since ninety percent of the population consumes rice, the average household consumption of rice in the population as whole is only slightly lower at about 146,000 FCFA (about $300 at the current exchange rate). Knowing that the poverty line is at 149,000 FCFA per person per year, and that the average size of a poor household in Mali is about 10.9 adults (as compared to 7.3 persons for non-poor households), the value of rice consumption among the poor on average is roughly equivalent to about one-tenth of what is needed for a typical household to not be poor, which is large.

The survey does not distinguish between imported and locally produced rice, but by comparing the income received from rice production (in table 4.5) with the consumption of rice, one can see that the average value of the consumption of rice is about two times higher than the average income received from rice. It is likely that consumers pay a mark up over the producer price (given the need to transport and market the locally produced rice), but it is also likely that some of the rice produced in Mali is exported to neighbouring countries. Therefore, one can assume that about half of the rice consumed in the country is locally produced, which is indeed the common perception in the country. Rice is consumed as frequently in rural as in urban areas, although rural consumption of rice relies more on auto-consumption. Rice consumption is slightly less frequent in the Sikasso area (this is the cotton producing part of the country, which is also one of the poorest areas), but even there close to nine persons out of ten consume rice.

In terms of the amounts consumed on average in the whole population, urban areas stand out as consuming 50 percent more rice than rural areas per household, despite the fact that rural households are larger. The lowest average amounts consumed are observed in areas where households are poorer, such as Sikasso. Clearly, rice is a good that is consumed in significantly higher amounts by the better off population as represented by the top two quintiles of consumption. In the top quintile, rice consumption per household is almost five times higher than in the bottom quintile. Although the data do not permit us to differentiate imported and locally produced rice, it is likely that imported rice is more consumed in urban areas.

Table 4.2 provides data for millet and sorghum consumption. The average consumption of millet and sorghum in the population as a whole is 136,000 FCFA per household per year, which is of the same order of magnitude as the consumption for rice. As for rice, more than nine out of ten households consume millet and sorghum. Production as recorded in the survey is higher than for rice (see table 4.6), but still well below the consumption level. Given that Mali does not import much millet and sorghum, it could be that the survey underestimates local production. In terms of patterns of consumption, there is a substantial difference with rice, given the fact that consumption of millet and sorghum is lower for the bottom and top quintile, and fairly stable between the second and fourth quintile, with in general fewer differences between quintiles than for rice. Thus consumption of millet and sorghum is not done more by the poor or the better off as the good is consumed in similar quantities by all types of households. Rural consumption is much higher than urban consumption, while the reverse was observed for rice. In terms of geographic patterns, consumption is lower in the areas of Kidal and Tombouctou than elsewhere.

Tables 4.3 and 4.4 provide consumption data for corn (maize) and wheat products. Consumption levels are substantially lower, at about 30,000 FCFA per household per year for corn, and 24,000 for wheat and related products such as bread. Slightly less than half of the population consumes corn, but the proportion reaches three-fourths for wheat products. Corn is consumed most heavily in the Sikasso area, while consumption of wheat products is highest in the Kidal area. While corn consumption is not affected much by the quintile of well-off of households, wheat products are as expected much more consumed by better off households, and this difference also reproduces itself when comparing urban and rural areas.

Together, the consumption of rice, millet and sorghum, corn and wheat products represents an outlay of about 320,000 FCFA per household per year. Using the same comparison as above, this represents about one-fourth of what is needed for a typical poor household not to be poor. An increase of 25 percent of cereals price would then represent about 5 percent of what a poor family needs in order not to be poor. Yet this would take into account only the impact of higher prices on the cost of food, and not the extra revenues that some households would probably get as producers of food.

Beyond statistical tables, it is useful to visualize the data so as to better understand differences in consumption patterns between various household groups as defined by their level of consumption, since this is ultimately what affects the impact on poverty of price changes. We focus here on rice, millet and sorghum, and corn because these are the cereals for which average consumption is highest (especially in the case of rice and millet and sorghum), and these are also the cereals for which we observe both consumption and production in the country (as discussed below in the case of production). For the graphical analysis we use simple on-parametric techniques to present kernel estimates for various variables (this follows previous similar work by a number of authors, as noted in the introduction). All figures presented in this section, as well as in following ones, share a common variable for the horizontal axis, namely the level of well-being of households according to the logarithm of their consumption per equivalent adult.

Figure 4.1 first provides the distribution density for the logarithm of consumption in urban and rural areas as well as at the national level. As urban households are richer, the urban density is to the right of the rural density. In rural areas, the mode of the density is about a value of 12, while in urban areas, the mode is around 13. As the distributions appear to have a normal shape, the modes are similar to the mean values. We present these figures to highlight the fact that for low values of log consumption (below 11) and for high values (above 14), the shares of the population in these areas are very low, so that in future graphs, the impact of what takes place at these extremes should be discounted by the fact that those impacts affect a very small share of the population. Said differently, what matters most is what is taking place roughly between values of 11 and 14 on the graphs, as this will drive the overall average effects. Note also that the poverty line would be a vertical line at a value of about 12 on the graph, with some differences depending on the region considered.

In figures 4.2a-c, we provide data on the consumption shares for rice, millet and sorghum, and corn among the population as a whole. For rice we have an inverted U shape, suggesting that rice consumption represents a higher share of total consumption for the population with intermediate levels of consumption near the middle of the distribution. This is because the poor tend to consume on average much less rice than better off households, while for those at the top of distribution, even if rice consumption is high, total consumption is even higher so that rice consumption as a share of total consumption is low. For millet and sorghum, the consumption share is much higher for the very poor and the poor than for other households. This echoes our earlier findings in terms of the consumption levels of households (albeit this was not discussed in terms of consumption shares) in tables 4.1 and 4.2. Thus, if one were to not include the producer side impact of higher prices on poverty, clearly, on the consumption side an increase in the price of millet and sorghum would likely hurt the poor more than an increase in the price of rice, both because the poor are comparatively larger users of millet and sorghum than the better off (while this is not the case for rice), and because for the poor at least the share of their consumption allocated to millet and sorghum is substantially higher than that allocated to rice. In figure 4.2c, the exercise is repeated for corn, with again higher consumption at the bottom of the distribution, but also with smaller shares of total consumption accounted for.

## Food Production

We now turn to the production side, where the data enable us to assess who produces rice, millet and sorghum, and corn (for wheat, the products are essentially imported).

As shown in table 4.5, a very large share of households produces rice (39.5 percent) for auto-consumption, for sales or both.   Rice production is concentrated in rural areas, and especially in Tombouctou, Segou, Gao, and Mopti. In terms of welfare status, the share of producers in the bottom four quintiles is similar at 40 percent to 50 percent, while it drops to 26 percent in the top quintile. The average income from rice in the population as a whole is substantial, at 206,000 FCFA for producers, which translates to an average income of about 81,500 FCFA per household in the population as a whole. While households in the bottom four quintiles have a similar probability of being rice producers, the amount produced among producers is as expected higher among richer households, with producers in the fourth quintile selling about two and a half times more rice than producers in the bottom quintile.

Table 4.4 provides the same data for millet and sorghum. The share of producers in the population is even higher, at close to 54 percent, with this time a much higher probability of production in the poorest three quintiles. Millet and sorghum production tends to be more concentrated in the areas that are not major producers of rice, although households in Tombouctou, Mopti and especially Segou have a high probability of being producers. On average, sales among producers are a bit lower to rice sales among rice producers, at about 173,000 FCFA per household. But because the proportion of millet and sorghum producers is larger, average income from millet and sorghum in the population as a whole (including for auto-consumption) is slightly higher for millet and sorghum than for rice. A big difference however between the two crops is that most of the rice produced in the country is sold, while most of the millet and sorghum produced is used by households for their auto-consumption.

## COMBINING CONSUMPTION AND PRODUCTION DATA TO ASSESS POVERTY IMPACTS

The impact on poverty of the change in food prices is the result of the combined impacts on the consumption and production sides. We first provide in this section estimates of the impact of changes in the price of rice, millet and sorghum, corn, wheat, as well as cereals as a whole. We measure likely impacts of food price increases on three poverty measures. The headcount index of poverty is the share of the population with a level of consumption per equivalent adult below the poverty line. The poverty gap takes into account the distance separating the poor from the poverty line (while giving a zero distance to the non-poor). The squared poverty gap takes into account the square of that distance (and thus inequality among the poor; for an introduction to the concepts of poverty measurement, see Coudouel et al. 2002).

We carry out the simulations in a very simple way. First, for rice, millet and sorghum, and corn producers, we measure the additional income or the loss in income obtained from the sale of the crops by households due to an increase or reduction in the price of the crops. We assume that this difference in income translates into an equivalent difference in the consumption per person of households used to measure poverty. We then compute again the poverty measures keeping the poverty line intact. For consumers, we do essentially the same thing, but considering also wheat products as well as cereals as a whole. That is, we estimate the increase or decrease in the cost of rice, millet and sorghum, corn and wheat products following a change in price, taking into account the actual spending of the household. In the case of a reduction in price, we then add to the consumption aggregate the reduction in the total cost of food for the household, since this reduction in cost means that the household can actually consume other goods (as if the household consumption had increased). In the case of an increase in the price of food, we subtract from the consumption aggregate the value of this increase, since the household will have to give up other consumption goods in order to be able to purchase the food it needs. For either an increase or a decrease in the price of food we then compute again poverty with the adjusted consumption level. What determines if a household is considered as a net producer or consumer is the level of the net sales of the consumer (negative for consumers, positive for producers; auto-consumption is not taken into account on either the producer or the consumer side).

This procedure is admittedly a rough approach, but it has the merit of being simple. The approach may slightly overestimate the impact on poverty of changes in prices because we do not take into account the price elasticity of rice, millet and sorghum, corn and wheat consumption, but this price elasticity is likely to be low in any case, due to the fact that all these products are important in the diet of the population and that the prices of the various food items seem to increase jointly at least in the medium term (so that it is not clear that households can offset the loss in purchasing power associated with the price increase by shifting to other foods). Also, the approach does not take into account any ripple effects of changes in the price of the various cereals on other parts of the economy. More sophisticated methods could be used to measure the "general equilibrium" effect of a change in the price of rice (such as using a Social Accounting Matrix, as done by Parra Osorio and Wodon 2008), but such simulations require a much larger number of assumptions, some of which are the subject of debate (especially when more complex computable general equilibrium models are used). The estimations given here thus provide "first round" likely poverty effects from lower or higher food prices paid to producing households or paid by consuming households, assuming that households don't change their consumption and production patterns for rice as well as for the other commodities after the change in their price.

Before providing the results, one more word of caution is required. As mentioned in the introduction, it remains an open question as if to the higher prices paid by consumers translate into higher prices paid to producers. If one has doubts as to producers will really benefit in the short run from higher prices, one could consider estimates based on consumption impacts only as an upper bound for the impact on poverty and estimates taking into account both consumer and producer impacts as a lower bound.

Key results from the simulations are provided in tables 4.8 to 4.10. Consider first table 4.8, which is based only on data on the consumption of food. At the time of the survey, the share of the population in poverty was 47.45 percent. If the price of rice increases by 25 percent, and if we look only at the impact on the consumer side, the headcount index would increase by about one and a half percentage point to 48.9 percent. The increase for millet and sorghum is smaller, at less than one point, due in part to the fact that much of production is auto-consumed. For corn, the increase is minimal, and the same can be said for wheat. For all cereals combined, the increase is 2.5 percentage points. This is large, and it would mean that some 290,000 persons would fall into poverty. The other poverty measures (poverty gap and squared poverty gap) follow a similar pattern, but with smaller increases in absolute terms since these measures are also smaller to start with. For information, we provide also the impacts for those households who consume the various foods, but in the case of rice, millet and sorghum and wheat, since most households are consumers, this does not much change the overall estimates.

If we now look at the impact of changes in producer prices in table 4.9, the impacts are reversed. The beneficial impact of the increase in rice prices is however smaller than the negative impact on the consumption size since perhaps half of the total consumption of rice is imported. With a 25 percent increase in prices, and if we look only at the impact on the producer side, national poverty measures would be reduced by about half a point (but there would be a larger reduction in poverty among rice producers of about four points). In the case of millet and sorghum, if the price for producers were to increase by 25 percent, the headcount index of poverty would decrease nationally by only a third of a point. For corn, the impact is one-tenth of a point. The combined impact of higher producer prices for rice, millet and sorghum, and corn is a reduction in the national estimate of the headcount index of slightly less than one point.

The total impact of changes in the price of the various cereals on poverty is obtained by taking both consumers and producers into account, and the results are given in table 4.10. If the price of rice increases by 25 percent, the headcount index of poverty increases in the population as a whole to 48.3 percent, while if the price of millet and sorghum increases by the same percentage, the headcount index of poverty increases to 47.9 percent. In the case of corn, poverty actually decreases when prices go up, because producers tend to be poorer than consumers. For cereals as a whole though, taking into account both consumer and producer impacts, the headcount index increases by 1.7 point to 49.2 percent with a 25 percent increase in prices.

Data on the poverty impacts separately for urban and rural areas are provided in the annex table 4.1, 4.2 and 4.3. The increase in the headcount index is much higher in urban areas (at close to three percentage terms) than in rural areas (at 1.2 percentage point). When one considers the poverty gap or squared poverty gap, the difference in impact on poverty between urban and rural areas is reduced, but there remains a gap, with urban areas affected more, as expected since urban dwellers are clearly net consumers of food.

As before, to understand the differences in results for rice, millet and sorghum and corn, it is useful to visualize the data. Figures 4.3a-c provides the shares of households who are net producers, net consumers, or neither for the various commodities (the last group of households does not consume nor does it produce rice, millet or sorghum, or corn). The sum of the three proportions is to one. The pattern is as expected very different for the three types of cereals.

In figure 4.3a for rice (which combines imported and domestically produced rice) the proportion of net consumers increases with the level of income (remember that most households are located in the middle of the graphs while the share of households located at the two extremes are low). Net producers are located for the most part among the poor, but even among the poor, there is a larger number of net consumers than net producers. This means that an increase in the price of rice is unambiguously going to increase poverty, with the impact on standards of living more generally being larger towards the middle of the distribution.

In figure 4.3b for millet and sorghum, and in figure 4.3c for corn, the pictures and messages are different. In the case of millet, there is a very large share of households in autarky especially at low levels of total consumption (this would be for the most part households who produce millet and sorghum for their own consumption). Among the rest of the population, there are more net producers than net consumers at comparatively lower levels of consumption, with a reversal at higher levels. For corn, the proportion of households in autarky is smaller and, below the poverty line, which is at a value of about 12 on the horizontal axis, we have a large number of net producers who should benefit from a price increase. Note also that these graphs give the proportion of households in various situations, but to assess poverty impact, we also need to have information on the actual quantities sold and purchased. By taking into account these quantities we can look at net incomes for the various crops at various levels of household consumption.

In figures 4.4a-c, we provide the data on the net income from sales of rice, millet and sorghum, and corn, with net income defined as the difference between sales and purchases of the good. As expected, for rice, net income is negative for almost all households, although for the very poor in rural areas, the magnitude of the negative net income is small (it is much larger in urban areas). For millet and sorghum in rural areas and at the national level, net income is positive for the ultra-poor (those with a log consumption value below 11), but it becomes negative for many among the poor (proxied by those with a log consumption between 11 and 12), and then the curves go into larger negative territory as consumption rises. In urban areas, net income is negative throughout. For corn, the picture is a bit different, as more households below a value of log income of 12 are net sellers. Thus, the graphical analysis confirms the results obtained with the poverty simulations. That is, for corn an increase in prices could very well be poverty reducing, while for millet and sorghum, even if some among the very poor benefit, the effect of higher prices is still likely to be an increase in poverty given the fact that many more households have consumption levels between values of 11 and 12 than below a value of 11.

Finally, figures 4.5a-c provide data on the net benefit ratio for rice, millet and sorghum, and corn, with this ratio defined as the net income from the commodity divided by the consumption level of the household. The figures are very similar to those for the net income, but they are scaled in such a way that the magnitude of net income effects is compared to the total consumption level of the households. The figures clearly show a negative impact again for rice in urban areas (but a gain in rural areas), and a small benefit for part of the distribution (the very poor) in the case of millet and sorghum, as well as corn.

## CONCLUSION

When assessing the potential impact of a change in the price of cereals on poverty, it is important to consider both the impact on producers (who tend to benefit from an increase in prices) and consumers (who tend to lose out when the price increases). If producers tend to be poor and if consumers live in urban areas and are better off, an increase in the price of cereals, despite its impact on the cost of food, may well be poverty reducing. In Mali, the main cereals that are sold for consumption (as opposed to auto-consumed) are rice as well as wheat, although there is also a substantial production and consumption of millet and sorghum, as well as corn. In the case of rice, the impact of an increase in price is not ambiguous at all since about half of the rice consumed in the country is imported. In the case of wheat and bread as well, an increase in prices is poverty increasing. For millet and sorghum, as well as for corn however, price increases could be potentially poverty reducing, at least if we assume that the higher price paid by consumers translates into a higher price received by producers. In the case of corn, we do find in our simulations a reduction in poverty with an increase in prices, while for millet and sorghum, because consumption levels are higher than production levels as recorded in the survey, we find that a price increase is poverty increasing.

Overall, we find that an increase in the price of the various cereals of 25 percent would lead to an increase in poverty which is substantial, since the share of the population in poverty would increase by 1.7 percentage point (this would represent close to 300,000 persons falling into poverty). If the increase in prices were at 50 percent, the increase in poverty would be substantially larger, at close to 3.5 points. If the increase in prices were to affect consumers only, without benefit for producers (for example if there is no trickle down of higher consumer prices to producers due to high intermediation costs), the increase in the headcount of poverty would be even higher, at 2.5 percentage points for a 25 percent price increase, and more than five percentage points for a 50 percent price increase.

Given that the national impact on poverty may not be substantial, the food price crisis could justify the implementation of compensatory measures to protect the most vulnerable households. These measures should probably not be in the form of broad import tax or value added tax cuts or food subsidies, as much of the proceeds from such measures would probably not reach the poor more than other household groups. Targeted interventions to reach poor households who are less likely to have the means to cope with price shocks would probably be more effective, as would interventions designed to increase rice production in the country (as documented in Nouve and Wodon 2008).

## NOTE

1. This work benefitted from funding from the World Bank. This chapter as well as broader work on the food and fuel price crisis in Mali was presented in Bamako in 2009, at a workshop organized jointly by the Ministry of Economics and Finance, the IMF and the World Bank. Support for conducting this work from Clara Ana Coutinho De Sousa is gratefully acknowledged. The opinions expressed in the chapter are those of the authors only, and need not represent those of the World Bank, its Executive Directors, or the countries they represent.

## BIBLIOGRAPHY

Boyce, K. James, and Martin Ravallion. "A Dynamic Econometric Model of Agricultural Wage Determination in Bangladesh," *Oxford Bulletin of Economics and Statistics* 53, no. 4 (1991): 361-76

Barrett, B. Christopher, and Paul A. Dorosh. "Farmers' Welfare and Changing Food Prices: Nonparametric Evidence from Rice in Madagascar," *American Journal of Agricultural Economics* 78, no. 3 (1996): 656-69.

Budd, J. W. "Changing Food Prices and Rural Welfare: A Non-Parametric Examination of the Cote d'Ivoire," *Economic Development and Cultural Change* 41, no. 3 (1993): 587-603.

Christiaensen, Luc and Lionel Demery. "Down to Earth: Agriculture and Poverty Reduction in Africa," Directions in Development. World Bank: Washington DC, 2007.

Coudouel, Aline, Jesko, Hentschel and Quentin Wodon. "Poverty Measurement and Analysis," in J. Klugman, editor, A Sourcebook for Poverty Reduction Strategies, Volume 1: Core Techniques and Cross-Cutting Issues, World Bank: Washington DC, 2002.

Deaton, Angus. "Rice Prices and Income Distribution in Thailand: A Non-Parametric Analysis," *The Economic Journal* 99, no. 395(1989):1-37.

Dessus, Sébastien, Santiago, Herrera and Rafael De Hoyos. "The Impact of Food Inflation on Urban Poverty and its Monetary Cost: Some Back of the Envelope Calculations," *Agricultural Economics* 39 (2008):417–29.

International Monetary Fund. "Food and Fuel Prices: Recent Developments, Macroeconomic Impact, and Policy Responses," mimeo, Washington D.C: IMF, 2008.

Ivanic, Maros, and Martin William. "Implications of Higher Global Food Prices for Poverty in Low-Income Countries," *Agricultural Economics* 39(2008):405–16.

Minot, Nicholas, and Francesco Goletti. "Export Liberalization and Household Welfare: The Case of Rice in Vietnam," *American Journal of Agricultural Economics* 80, no. 4 (1998): 738-49.

Niimi, Yoko, Puja Vasudeva-Dutta and Alan L. Winters. "Storm in a Rice Bowl: Rice Reform and Poverty in Vietnam in the 1990s," *Journal of the Asia Pacific Economy* 9, no. 2 (2004):170-90.

Nouve, Kofi, and Quentin Wodon. "Impact of Rising Rice Prices and Policy Responses in Mali: Simulations with a Dynamic CGE Model," mimeo, World Bank: Washington DC, 2008.

Ravallion, Martin. "Welfare changes of food price changes under induced wage responses: Theory and evidence for Bangladesh," *Oxford Economic Papers* 42 (1990): 574-85.

Ravallion, Martin, and Dominique van de Walle. "The impact on poverty of food pricing reforms: a welfare analysis for Indonesia," *Journal of Policy Modeling* 13, no. 2 (1991): 281-99.

Rashid, Shahidur. "Dynamics of Agricultural Wage and Rice Price in Bangladesh: a Reexamination," MSSD Discussion Paper No. 44, International Food Policy Research Institute: Washington DC, 2002.

Singh, Inderjit, Lyn, Squire and John Strauss. "Agricultural Household Models: Extensions and Applications." Baltimore: Johns Hopkins University Press, 1986.

Sumarto, Sudarno, Asep Suryahadi and Wenefrida Widyanti. "Assessing the Impact of Indonesian Social Safety Net Programmes on Household Welfare and Poverty Dynamics," *European Journal of Development Research* 17, no. 1 (2005): 155-77.

Timmer, C. Peter, and David Dawe. "Managing Food Price Instability in Asia: A Macro Food Security Perspective," *Asian Economic Journal* 21, no. 1(2007): 1-18.

Tsimpo, Clarence, and Quentin Wodon. "Rice Prices and Poverty in Liberia", mimeo, World Bank: Washington D.C, 2008a.

Tsimpo, Clarence, and Quentin Wodon." Impact sur la pauvreté de la hausse des prix alimentaires au Sénégal," mimeo, World Bank : Washington DC, 2008b.

Warr, P. "Food Policy and Poverty in Indonesia: A General Equilibrium Analysis," *Australian Journal of Agricultural and Resource Economics* 49, no. 4 (2005): 429-51.

Wodon, Quentin, Clarence Tsimpo, Prospere Backiny-Yetna, George Joseph, Frank Adoho and Harold Coulombe. "Impact of Higher Food Prices on Poverty in West and Central Africa," mimeo. World Bank, Washington DC, 2008.

Wodon, Quentin, and Hassan Zaman. "Higher Food Prices in Sub-Saharan Africa: Poverty Impact and Policy Responses," *World Bank Research Observer* 25, 2010: 157-76.

World Bank. "Addressing the Food Crisis: The Need for Rapid and Coordinated Action", Background paper for the Finance Ministers Meetings of the Group of Eight, Poverty Reduction and Economic Management Network: Washington DC, 2008a.

World Bank. "Guidance for Responses from the Human Development Sectors to Rising Food and Fuel Prices," Human Development Network: Washington DC, 2008b.

**Figure 4.1. Rural and urban welfare distributions (in logarithm)**

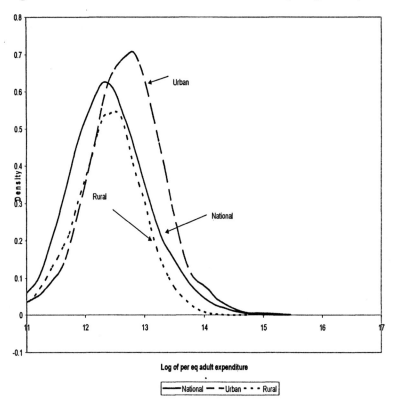

*Source:* Authors' estimation using ELIM 2006.

**Figure 4.2a. Budget share of rice expenditure**

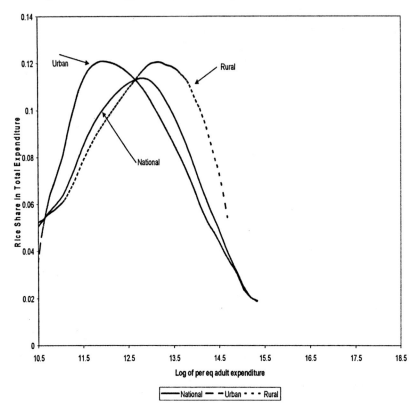

*Source:* Authors' estimation using ELIM, 2006.

**Figure 4.2b. Budget share of millet and sorghum expenditure**

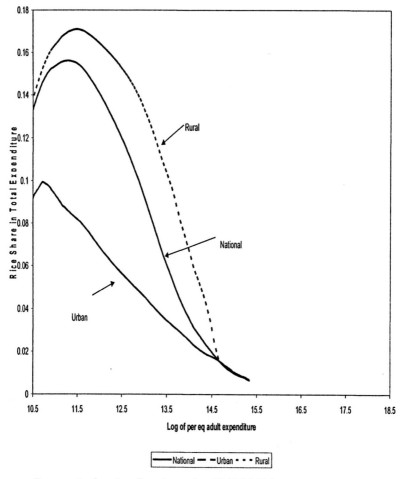

Source: Authors' estimation using ELIM 2006.

**Figure 4.2c. Budget share of corn expenditure**

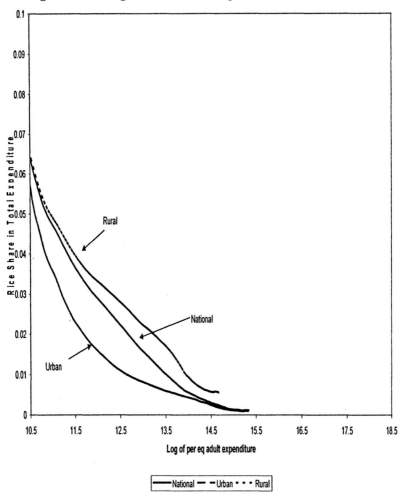

*Source*: Authors' estimation using ELIM 2006.

**Figure 4.3a. Net producers, net consumers, and autarky households for rice**

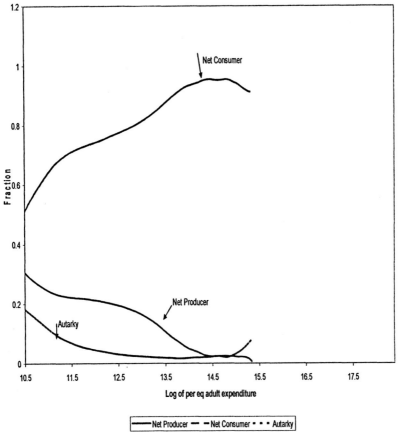

*Source*: Authors' estimation using ELIM 2006.

**Figure 4.3b. Net producers, net consumers, and autarky households for millet and sorghum**

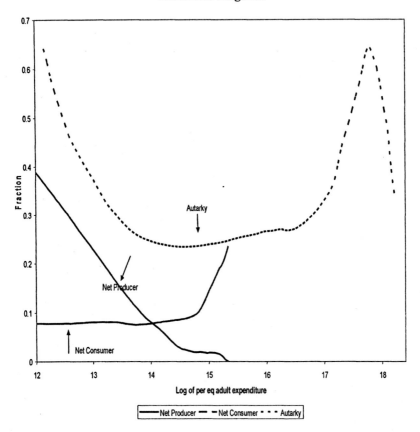

*Source*: Authors' estimation using ELIM 2006.

**Figure 4.3c. Net producers, net consumers, and autarky households for corn**

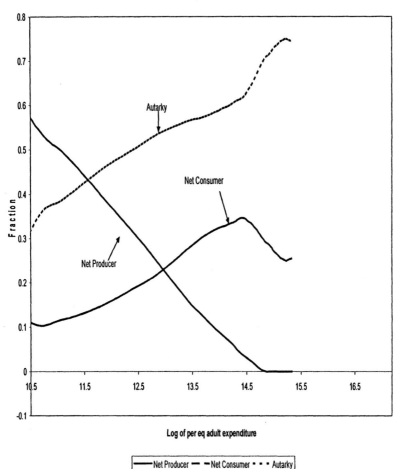

*Source*: Authors' estimation using ELIM 2006.

Joseph and Wodon

**Figure 4.4a. Income per capita (net sales) from rice production**

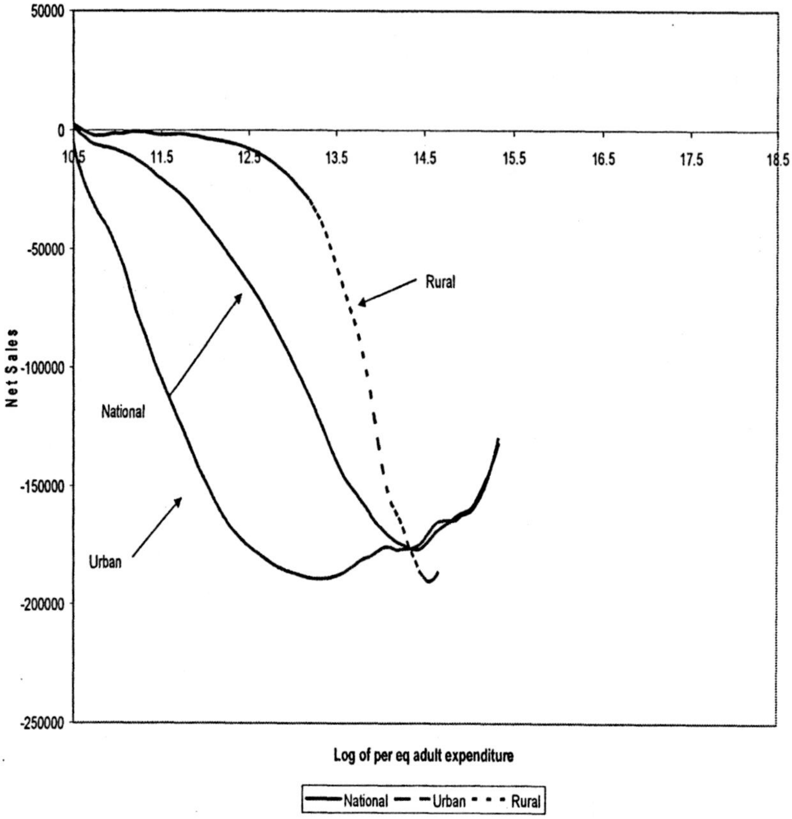

*Source:* Authors' estimation using ELIM 2006.

**Figure 4.4b. Income per capita (net sales) from millet and sorghum production**

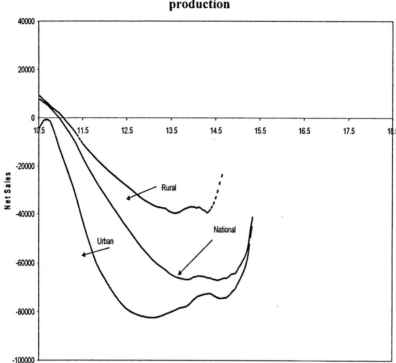

*Source*: Authors' estimation using ELIM 2006.

**Figure 4.4c. Income per capita (net sales) from corn production per equivalent adult**

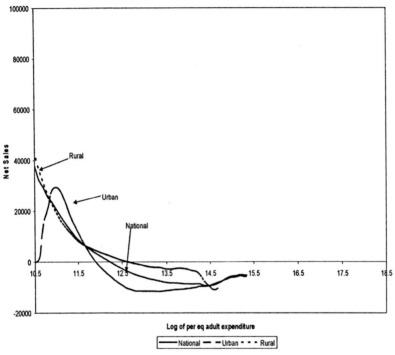

*Source*: Authors' estimation using ELIM 2006.

**Figure 4.5a. Net benefit ratio for rice**

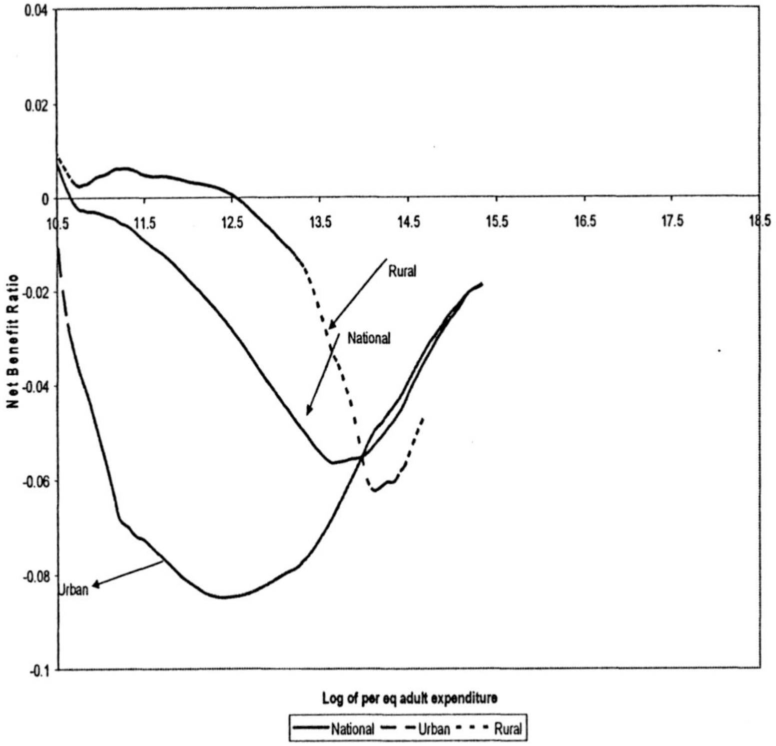

*Source*: Authors' estimation using ELIM 2006.

**Figure 4.5b. Net benefit ratio for millet and sorghum**

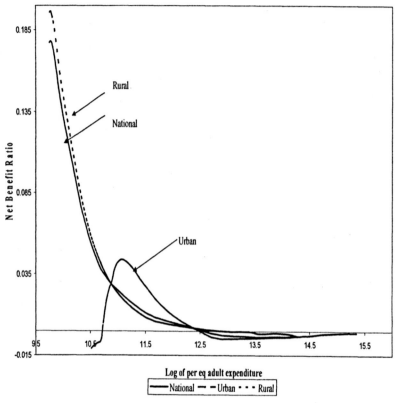

*Source:* Authors' estimation using ELIM 2006.

**Figure 4.5c. Net benefit ratio for corn**

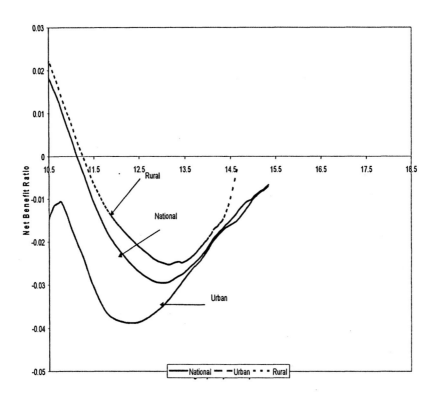

*Source*: Authors' estimation using ELIM 2006.

Table 4.1. Rice Consumption in Mali for Different Household Groups, 2006

| | Percentage of households consuming (%) | | | Average consumption for all households | | | Average consumption for households with positive consumption | | |
|---|---|---|---|---|---|---|---|---|---|
| | Purchase | Auto consumption | Total | Purchase | Auto consumption | Total | Purchase | Auto consumption | Total |
| **All** | 82.20 | 38.50 | 95.10 | 106,264.57 | 39,639.63 | 145,904.20 | 129,290.57 | 102,881.81 | 153,440.03 |
| **Residence area** | | | | | | | | | |
| Urban | 93.30 | 19.60 | 96.00 | 171,547.58 | 14,430.79 | 185,978.38 | 183,803.39 | 73,606.23 | 193,772.14 |
| Rural | 75.50 | 49.80 | 94.60 | 67,299.98 | 54,685.70 | 121,985.67 | 89,091.18 | 109,757.41 | 129,006.10 |
| **Region** | | | | | | | | | |
| Kayes | 88.90 | 24.50 | 95.40 | 109,893.48 | 9,957.25 | 119,850.73 | 123,573.18 | 40,570.47 | 125,605.61 |
| Koulikoro | 79.90 | 31.10 | 93.70 | 90,337.99 | 17,629.43 | 107,967.42 | 113,085.18 | 56,680.58 | 115,179.77 |
| Sikasso | 77.10 | 32.00 | 89.40 | 60,452.56 | 17,606.31 | 78,058.87 | 78,458.20 | 54,998.14 | 87,316.51 |
| Ségou | 72.40 | 52.60 | 96.70 | 76,444.23 | 48,060.66 | 124,504.90 | 105,593.58 | 91,372.99 | 128,719.45 |
| Mopti | 87.70 | 47.50 | 96.70 | 111,418.89 | 69,796.45 | 181,215.34 | 127,099.69 | 147,071.86 | 187,359.78 |
| Tombouctou | 78.10 | 70.80 | 97.80 | 88,144.03 | 122,662.75 | 210,806.78 | 112,911.94 | 173,185.98 | 215,465.30 |
| Gao | 81.70 | 51.70 | 97.90 | 159,453.73 | 85,852.87 | 245,306.61 | 195,236.78 | 166,030.19 | 250,564.86 |
| Kidal | 99.30 | 10.40 | 99.30 | 201,732.66 | 4,498.24 | 206,230.91 | 203,185.17 | 43,310.52 | 207,715.80 |
| Bamako | 94.30 | 15.10 | 95.90 | 201,619.48 | 9,575.45 | 211,194.94 | 213,827.44 | 63,301.80 | 220,238.63 |
| **Quintile** | | | | | | | | | |
| Q1 | 65.90 | 41.00 | 87.90 | 22,811.22 | 21,468.60 | 44,279.82 | 34,640.33 | 52,307.88 | 50,385.83 |
| Q2 | 77.70 | 43.20 | 94.90 | 53,176.76 | 40,056.87 | 93,233.63 | 68,471.18 | 92,808.03 | 98,267.83 |
| Q3 | 81.10 | 48.10 | 95.70 | 86,490.69 | 52,960.09 | 139,450.78 | 106,639.09 | 110,215.71 | 145,694.09 |
| Q4 | 83.20 | 41.00 | 96.60 | 117,981.55 | 48,836.13 | 166,817.67 | 141,728.09 | 119,088.93 | 172,773.52 |
| Q5 | 93.40 | 25.70 | 97.30 | 186,913.28 | 31,995.06 | 218,908.34 | 200,193.16 | 124,531.98 | 224,969.84 |

*Source*: Authors' estimation using ELIM 2006.

**Table 4.2. Millet and Sorghum Consumption in Mali for Different Household Groups, 2006**

| | Percentage of household consuming (%) | | | Average consumption for all households | | | Average positive consumption for all households | | |
|---|---|---|---|---|---|---|---|---|---|
| | Purchase | Auto consumption | Total | Purchase | Auto consumption | Total | Purchase | Auto consumption | Total |
| **All** | 65.80 | 52.90 | 91.00 | 57,707.51 | 78,094.34 | 135,801.86 | 87,677.81 | 147,703.45 | 149,153.08 |
| **Residence area** | | | | | | | | | |
| Urban | 79.40 | 26.90 | 88.40 | 68,675.63 | 21,316.04 | 89,991.67 | 86,505.74 | 79,230.87 | 101,854.02 |
| Rural | 57.70 | 68.40 | 92.70 | 51,161.11 | 111,982.84 | 163,143.95 | 88,640.02 | 163,784.73 | 176,072.59 |
| **Region** | | | | | | | | | |
| Kayes | 51.40 | 67.70 | 91.00 | 47,363.89 | 122,121.98 | 169,485.86 | 92,167.77 | 180,406.70 | 186,201.56 |
| Koulikoro | 58.60 | 73.70 | 94.80 | 62,123.73 | 134,332.16 | 196,455.89 | 106,092.76 | 182,160.38 | 207,288.59 |
| Sikasso | 34.50 | 64.30 | 79.80 | 14,373.50 | 91,588.62 | 105,962.12 | 41,669.61 | 142,516.41 | 132,858.56 |
| Ségou | 64.30 | 55.00 | 94.00 | 44,103.07 | 80,295.56 | 124,398.63 | 68,612.18 | 145,887.83 | 132,365.69 |
| Mopti | 87.60 | 51.70 | 97.70 | 94,695.56 | 74,001.27 | 168,696.83 | 108,125.43 | 143,248.41 | 172,620.14 |
| Tombouctou | 78.30 | 39.70 | 87.00 | 49,946.46 | 24,439.19 | 74,385.64 | 63,820.40 | 61,624.04 | 85,498.63 |
| Gao | 91.80 | 26.30 | 94.30 | 91,119.48 | 14,043.52 | 105,163.01 | 99,297.43 | 53,335.23 | 111,536.47 |
| Kidal | 38.90 | 5.70 | 42.20 | 26,701.72 | 1,416.54 | 28,118.26 | 68,625.52 | 24,992.43 | 66,618.44 |
| Bamako | 86.80 | 16.40 | 91.00 | 82,602.02 | 7,722.35 | 90,324.38 | 95,194.28 | 47,105.36 | 99,274.81 |
| **Quintile** | | | | | | | | | |
| Q1 | 42.60 | 73.40 | 87.60 | 25,911.89 | 91,306.84 | 117,218.73 | 60,764.43 | 124,468.88 | 133,836.50 |
| Q2 | 57.20 | 70.50 | 92.90 | 41,725.15 | 105,180.62 | 146,905.77 | 72,919.83 | 149,173.31 | 158,192.34 |
| Q3 | 61.20 | 64.80 | 93.00 | 58,529.32 | 111,379.59 | 169,908.91 | 95,575.94 | 172,013.97 | 182,710.06 |
| Q4 | 69.20 | 48.00 | 90.70 | 67,669.66 | 79,811.85 | 147,481.52 | 97,773.89 | 166,220.25 | 162,646.38 |
| Q5 | 83.60 | 26.90 | 90.70 | 75,461.63 | 29,760.69 | 105,222.32 | 90,282.78 | 110,452.22 | 116,073.06 |

*Source*: Authors' estimation using ELIM 2006.

**Table 4.3. Corn Consumption in Mali for Different Household Groups, 2006**

| | Percentage of households consuming (%) | | | Average consumption for all households | | | Average positive consumption for households | | |
|---|---|---|---|---|---|---|---|---|---|
| | Purchase | Auto consumption | Total | Purchase | Auto consumption | Total | Purchase | Auto consumption | Total |
| **All** | 21.20 | 33.00 | 48.10 | 8,322.46 | 21,979.48 | 30,301.94 | 39,273.56 | 66,654.80 | 63,018.39 |
| **Residence area** | | | | | | | | | |
| Urban | 34.10 | 13.20 | 42.00 | 14,183.10 | 9,014.41 | 23,197.52 | 41,650.15 | 68,407.95 | 55,256.50 |
| Rural | 13.50 | 44.80 | 51.70 | 4,824.49 | 29,717.77 | 34,542.27 | 35,699.30 | 66,346.95 | 66,778.31 |
| **Region** | | | | | | | | | |
| Kayes | 25.10 | 54.60 | 70.60 | 9,259.89 | 33,016.04 | 42,275.93 | 36,878.54 | 60,483.50 | 59,917.70 |
| Koulikoro | 13.30 | 50.30 | 58.00 | 3,552.65 | 27,992.64 | 31,545.29 | 26,787.73 | 55,597.68 | 54,364.83 |
| Sikasso | 37.10 | 71.20 | 90.70 | 22,855.75 | 81,110.92 | 103,966.67 | 61,663.67 | 113,899.93 | 114,569.77 |
| Ségou | 11.90 | 23.80 | 33.70 | 3,497.97 | 4,767.49 | 8,265.46 | 29,499.56 | 19,989.54 | 24,507.50 |
| Mopti | 13.30 | 14.50 | 24.40 | 3,439.66 | 3,786.25 | 7,225.91 | 25,925.79 | 26,149.35 | 29,658.65 |
| Tombouctou | 8.90 | 9.60 | 16.30 | 1,238.85 | 1,362.21 | 2,601.06 | 13,929.48 | 14,198.70 | 15,941.03 |
| Gao | 20.50 | 1.90 | 22.10 | 5,158.79 | 885.74 | 6,044.52 | 25,165.03 | 46,733.75 | 27,298.29 |
| Kidal | 9.40 | 0.70 | 9.40 | 2,459.44 | 50.33 | 2,509.76 | 26,233.99 | 7,040.00 | 26,770.81 |
| Bamako | 40.50 | 7.40 | 43.50 | 15,125.78 | 2,334.42 | 17,460.21 | 37,356.38 | 31,674.81 | 40,096.08 |
| **Quintile** | | | | | | | | | |
| Q1 | 9.30 | 49.10 | 53.50 | 2,706.66 | 24,508.26 | 27,214.92 | 29,084.81 | 49,951.54 | 50,837.66 |
| Q2 | 14.80 | 42.60 | 51.20 | 5,217.82 | 29,494.45 | 34,712.27 | 35,325.74 | 69,160.13 | 67,851.52 |
| Q3 | 17.90 | 40.90 | 51.40 | 8,161.21 | 29,720.05 | 37,881.26 | 45,611.14 | 72,713.82 | 73,751.90 |
| Q4 | 22.30 | 29.30 | 45.60 | 9,197.36 | 23,089.27 | 32,286.63 | 41,214.13 | 78,669.09 | 70,804.25 |
| Q5 | 32.70 | 16.10 | 43.10 | 12,575.49 | 9,666.79 | 22,242.28 | 38,415.70 | 60,090.62 | 51,596.29 |

*Source*: Authors' estimation using ELIM 2006.

# Table 4.4. Wheat and Bread Consumption in Mali for Different Household Groups, 2006

| | Percentage of households consuming (%) | | | Average consumption for all households | | | Average positive consumption for households | | |
|---|---|---|---|---|---|---|---|---|---|
| | Purchase | Auto consumption | Total | Purchase | Auto consumption | Total | Purchase | Auto consumption | Total |
| **All** | 72.90 | 6.40 | 74.00 | 23,114.22 | 921.54 | 24,035.76 | 31,695.91 | 14,334.69 | 32,493.31 |
| **Residence area** | | | | | | | | | |
| Urban | 82.90 | 4.80 | 83.60 | 36,533.92 | 1,194.34 | 37,728.26 | 44,053.31 | 24,972.86 | 45,137.62 |
| Rural | 67.00 | 7.40 | 68.20 | 15,104.58 | 758.72 | 15,863.30 | 22,560.10 | 10,237.35 | 23,248.55 |
| **Region** | | | | | | | | | |
| Kayes | 75.80 | 4.80 | 77.00 | 34,309.36 | 997.85 | 35,307.22 | 45,273.90 | 20,933.59 | 45,842.73 |
| Koulikoro | 76.30 | 9.90 | 77.40 | 18,550.55 | 1,008.97 | 19,559.51 | 24,322.94 | 10,225.70 | 25,270.15 |
| Sikasso | 81.30 | 6.20 | 82.00 | 18,345.41 | 443.23 | 18,788.64 | 22,558.31 | 7,194.62 | 22,907.07 |
| Ségou | 61.40 | 8.60 | 63.20 | 16,919.90 | 567.06 | 17,486.96 | 27,537.75 | 6,566.75 | 27,657.89 |
| Mopti | 72.40 | 3.00 | 72.90 | 12,482.85 | 159.57 | 12,642.42 | 17,235.26 | 5,256.15 | 17,330.43 |
| Tombouctou | 60.60 | 13.10 | 63.40 | 20,545.55 | 3,039.99 | 23,585.55 | 33,884.62 | 23,132.59 | 37,184.97 |
| Gao | 50.30 | 1.60 | 50.30 | 22,242.48 | 217.86 | 22,460.34 | 44,184.46 | 13,323.13 | 44,617.23 |
| Kidal | 93.90 | 5.40 | 96.20 | 118,776.22 | 553.87 | 119,330.08 | 126,556.77 | 10,213.64 | 124,016.15 |
| Bamako | 88.40 | 2.80 | 88.40 | 42,512.07 | 1,835.95 | 44,348.02 | 48,100.76 | 65,150.48 | 50,178.07 |
| **Quintile** | | | | | | | | | |
| Q1 | 58.00 | 4.80 | 58.60 | 7,404.80 | 440.82 | 7,845.61 | 12,768.96 | 9,173.68 | 13,398.24 |
| Q2 | 64.40 | 6.00 | 65.80 | 11,438.34 | 513.66 | 11,952.00 | 17,757.23 | 8,499.00 | 18,171.29 |
| Q3 | 66.50 | 8.20 | 68.00 | 16,233.94 | 634.08 | 16,868.01 | 24,401.63 | 7,688.97 | 24,823.89 |
| Q4 | 74.80 | 8.10 | 76.10 | 24,609.11 | 1,194.38 | 25,803.50 | 32,895.10 | 14,719.66 | 33,908.27 |
| Q5 | 88.90 | 4.90 | 89.50 | 42,123.95 | 1,401.56 | 43,525.51 | 47,395.67 | 28,784.88 | 48,622.97 |

*Source:* Authors' estimation using ELIM 2006.

## Table 4.5. Rice Income in Mali for Different Household Groups, 2006

| | Percentage of households consuming (%) | | | Average consumption for all households | | | Average positive consumption for households | | |
|---|---|---|---|---|---|---|---|---|---|
| | Purchase | Auto consumption | Total | Purchase | Auto consumption | Total | Purchase | Auto consumption | Total |
| **All** | 38.50 | 12.49 | 39.49 | 39,639.63 | 41,835.37 | 81,475.00 | 102,881.81 | 334,831.00 | 206,293.50 |
| **Residence area** | | | | | | | | | |
| Urban | 19.60 | 3.83 | 20.06 | 14,430.79 | 14,013.19 | 28,443.98 | 73,606.23 | 366,354.90 | 141,808.60 |
| Rural | 49.80 | 17.67 | 51.10 | 54,685.70 | 58,441.21 | 113,126.90 | 109,757.41 | 330,757.80 | 221,402.30 |
| **Region** | | | | | | | | | |
| Kayes | 24.50 | 2.49 | 24.90 | 9,957.25 | 1,676.04 | 11,633.29 | 40,570.47 | 67,219.67 | 46,722.14 |
| Koulikoro | 31.10 | 2.65 | 31.81 | 17,629.43 | 2,513.26 | 20,142.69 | 56,680.58 | 95,001.69 | 63,327.95 |
| Sikasso | 32.00 | 5.87 | 32.66 | 17,606.31 | 7,820.18 | 25,426.48 | 54,998.14 | 133,207.00 | 77,849.48 |
| Ségou | 52.60 | 29.53 | 53.52 | 48,060.66 | 174,259.90 | 222,320.60 | 91,372.99 | 590,135.80 | 415,367.50 |
| Mopti | 47.50 | 11.38 | 49.51 | 69,796.45 | 31,723.79 | 101,520.20 | 147,071.86 | 278,774.60 | 205,055.40 |
| Tombouctou | 70.80 | 38.94 | 73.44 | 122,662.75 | 34,265.89 | 156,928.60 | 173,185.98 | 87,992.68 | 213,676.90 |
| Gao | 51.70 | 22.02 | 52.92 | 85,852.87 | 18,117.87 | 103,970.70 | 166,030.19 | 82,296.15 | 196,451.50 |
| Kidal | 10.40 | 0.00 | 10.39 | 4,498.24 | 0.00 | 4,498.24 | 43,310.52 | 0.00 | 43,310.52 |
| Bamako | 15.10 | 0.23 | 15.13 | 9,575.45 | 1,045.17 | 10,620.62 | 63,301.80 | 450,000.00 | 70,211.27 |
| **Quintile** | | | | | | | | | |
| Q1 | 41.00 | 12.44 | 42.44 | 21,468.60 | 23,810.04 | 45,278.64 | 52,307.88 | 191,352.50 | 106,684.60 |
| Q2 | 43.20 | 13.38 | 44.14 | 40,056.87 | 38,529.16 | 78,586.02 | 92,808.03 | 288,066.60 | 178,025.20 |
| Q3 | 48.10 | 18.62 | 49.83 | 52,960.09 | 64,117.68 | 117,077.80 | 110,215.71 | 344,413.80 | 234,936.30 |
| Q4 | 41.00 | 14.92 | 41.58 | 48,836.13 | 60,887.89 | 109,724.00 | 119,088.93 | 408,077.90 | 263,888.00 |
| Q5 | 25.70 | 5.74 | 26.18 | 31,995.06 | 22,178.38 | 54,173.43 | 124,531.98 | 386,168.30 | 206,901.40 |

*Source:* Authors' estimation using ELIM 2006.

Table 4.6. Millet and Sorghum Income in Mali for Different Household Groups, 2006

| | Percentage of households consuming (%) | | | Average consumption for all households | | | Average positive consumption for households | | |
|---|---|---|---|---|---|---|---|---|---|
| | Purchase | Auto consumption | Total | Purchase | Auto consumption | Total | Purchase | Auto consumption | Total |
| **All** | 52.90 | 13.97 | 53.80 | 78,094.34 | 14,987.44 | 93,081.79 | 147,703.45 | 107,315.80 | 173,018.60 |
| **Residence area** | | | | | | | | | |
| Urban | 26.90 | 3.28 | 27.42 | 21,316.04 | 3,574.41 | 24,890.45 | 79,230.87 | 108,924.00 | 90,777.55 |
| Rural | 68.40 | 20.34 | 69.54 | 111,982.84 | 21,799.39 | 133,782.20 | 163,784.73 | 107,161.00 | 192,372.00 |
| **Region** | | | | | | | | | |
| Kayes | 67.70 | 20.57 | 68.14 | 122,121.98 | 24,916.44 | 147,038.40 | 180,406.70 | 121,154.60 | 215,799.40 |
| Koulikoro | 73.70 | 8.84 | 74.76 | 134,332.16 | 8,392.64 | 142,724.80 | 182,160.38 | 94,992.07 | 190,899.90 |
| Sikasso | 64.30 | 13.47 | 66.21 | 91,588.62 | 12,829.96 | 104,418.60 | 142,516.41 | 95,216.78 | 157,701.40 |
| Ségou | 55.00 | 23.48 | 56.27 | 80,295.56 | 30,793.69 | 111,089.30 | 145,887.83 | 131,140.40 | 197,424.10 |
| Mopti | 51.70 | 17.49 | 51.84 | 74,001.27 | 13,702.06 | 87,703.33 | 143,248.41 | 78,337.38 | 169,165.00 |
| Tombouctou | 39.70 | 12.92 | 41.48 | 24,439.19 | 11,937.62 | 36,376.80 | 61,624.04 | 92,371.10 | 87,696.58 |
| Gao | 26.30 | 5.04 | 27.08 | 14,043.52 | 3,226.20 | 17,269.72 | 53,335.23 | 63,953.33 | 63,767.09 |
| Kidal | 5.70 | 0.00 | 5.67 | 1,416.54 | 0.00 | 1,416.54 | 24,992.43 | 0.00 | 24,992.43 |
| Bamako | 16.40 | 0.59 | 16.51 | 7,722.35 | 909.88 | 8,632.23 | 47,105.36 | 154,417.50 | 52,286.93 |
| **Quintile** | | | | | | | | | |
| Q1 | 73.40 | 24.76 | 76.39 | 91,306.84 | 22,964.82 | 114,271.70 | 124,468.88 | 92,734.27 | 149,582.40 |
| Q2 | 70.50 | 20.83 | 71.35 | 105,180.62 | 17,906.77 | 123,087.40 | 149,173.31 | 85,954.71 | 172,511.20 |
| Q3 | 64.80 | 16.04 | 65.36 | 111,379.59 | 16,461.86 | 127,841.50 | 172,013.97 | 102,640.10 | 195,584.30 |
| Q4 | 48.00 | 12.41 | 48.51 | 79,811.85 | 15,621.36 | 95,433.21 | 166,220.25 | 125,917.10 | 196,730.40 |
| Q5 | 26.90 | 3.90 | 27.41 | 29,760.69 | 7,485.61 | 37,246.30 | 110,452.22 | 191,796.90 | 135,900.90 |

*Source:* Authors' estimation using ELIM 2006.

**Table 4.7. Corn Income in Mali for Different Household Groups, 2006**

| | Percentage of households consuming (%) | | | Average consumption for all households | | | Average positive consumption for households | | |
|---|---|---|---|---|---|---|---|---|---|
| | Purchase | Auto consumption | Total | Purchase | Auto consumption | Total | Purchase | Auto consumption | Total |
| **All** | 33.00 | 6.34 | 33.60 | 21,979.48 | 9,259.97 | 31,239.45 | 66,654.80 | 146,112.90 | 92,967.98 |
| **Residence area** | | | | | | | | | |
| Urban | 13.20 | 1.99 | 13.58 | 9,014.41 | 8,227.37 | 17,241.78 | 68,407.95 | 414,377.10 | 126,956.00 |
| Rural | 44.80 | 8.94 | 45.55 | 29,717.77 | 9,876.28 | 39,594.05 | 66,346.95 | 110,533.50 | 86,919.95 |
| **Region** | | | | | | | | | |
| Kayes | 54.60 | 10.74 | 55.52 | 33,016.04 | 8,182.98 | 41,199.02 | 60,483.50 | 76,166.44 | 74,205.51 |
| Koulikoro | 50.30 | 5.72 | 51.05 | 27,992.64 | 2,512.82 | 30,505.46 | 55,597.68 | 43,894.40 | 59,757.40 |
| Sikasso | 71.20 | 20.19 | 72.98 | 81,110.92 | 50,507.05 | 131,618.00 | 113,899.93 | 250,176.80 | 180,360.00 |
| Ségou | 23.80 | 4.31 | 24.30 | 4,767.49 | 1,833.74 | 6,601.23 | 19,989.54 | 42,571.72 | 27,163.02 |
| Mopti | 14.50 | 0.60 | 14.51 | 3,786.25 | 165.79 | 3,952.04 | 26,149.35 | 27,557.26 | 27,234.33 |
| Tombouctou | 9.60 | 2.29 | 10.47 | 1,362.21 | 1,015.81 | 2,378.02 | 14,198.70 | 44,431.14 | 22,716.80 |
| Gao | 1.90 | 0.00 | 1.90 | 885.74 | 0.00 | 885.74 | 46,733.75 | 0.00 | 46,733.75 |
| Kidal | 0.70 | 0.00 | 0.71 | 50.33 | 0.00 | 50.33 | 7,040.00 | 0.00 | 7,040.00 |
| Bamako | 7.40 | 1.45 | 7.37 | 2,334.42 | 963.08 | 3,297.50 | 31,674.81 | 66,305.63 | 44,742.41 |
| **Quintile** | | | | | | | | | |
| Q1 | 49.10 | 13.97 | 50.43 | 24,508.26 | 20,431.49 | 44,939.75 | 49,951.54 | 146,204.30 | 89,112.31 |
| Q2 | 42.60 | 9.42 | 43.39 | 29,494.45 | 23,861.67 | 53,356.13 | 69,160.13 | 253,358.80 | 122,970.70 |
| Q3 | 40.90 | 7.80 | 41.98 | 29,720.05 | 6,475.10 | 36,195.16 | 72,713.82 | 83,054.02 | 86,221.02 |
| Q4 | 29.30 | 3.85 | 29.51 | 23,089.27 | 2,927.30 | 26,016.57 | 78,669.09 | 76,080.76 | 88,156.43 |
| Q5 | 16.10 | 1.49 | 16.31 | 9,666.79 | 1,391.71 | 11,058.50 | 60,090.62 | 93,440.88 | 67,795.41 |

*Source*: Authors' estimation using ELIM 2006.

Table 4.8. Impact of a Change in the Price of Cereals on Consumers - National

| | -25% | -12.5% | No change | 12.5% | 25% | 50% | 100% |
|---|---|---|---|---|---|---|---|
| | **Rice** | | | | | | |
| **Poverty, population as a whole** | | | | | | | |
| Headcount index of poverty | 46.09 | 46.62 | **47.45** | 48.18 | 48.93 | 50.22 | 53.02 |
| Poverty gap | 16.28 | 16.46 | **16.66** | 16.87 | 17.10 | 17.62 | 18.82 |
| Squared poverty gap | 7.83 | 7.92 | **8.01** | 8.11 | 8.21 | 8.45 | 9.07 |
| **Poverty, rice consumers** | | | | | | | |
| Headcount index of poverty | 42.31 | 42.97 | **43.99** | 44.88 | 45.81 | 47.39 | 50.84 |
| Poverty gap | 13.82 | 14.05 | **14.29** | 14.55 | 14.84 | 15.47 | 16.95 |
| Squared poverty gap | 6.22 | 6.32 | **6.43** | 6.55 | 6.68 | 6.98 | 7.75 |
| | **Millet and Sorghum** | | | | | | |
| **Poverty, population as a whole** | | | | | | | |
| Headcount index of poverty | 46.56 | 46.91 | **47.45** | 47.99 | 48.22 | 49.36 | 51.23 |
| Poverty gap | 16.36 | 16.50 | **16.66** | 16.82 | 16.99 | 17.36 | 18.25 |
| Squared poverty gap | 7.86 | 7.93 | **8.01** | 8.09 | 8.18 | 8.37 | 8.85 |
| **Poverty, millet and sorghum consumers** | | | | | | | |
| Headcount index of poverty | 34.95 | 35.51 | **36.37** | 37.23 | 37.61 | 39.43 | 42.43 |
| Poverty gap | 10.39 | 10.63 | **10.87** | 11.13 | 11.41 | 12.00 | 13.42 |
| Squared poverty gap | 4.38 | 4.50 | **4.62** | 4.75 | 4.89 | 5.20 | 5.97 |
| | **Corn** | | | | | | |
| **Poverty, population as a whole** | | | | | | | |
| Headcount index of poverty | 47.41 | 47.45 | **47.45** | 47.51 | 47.51 | 47.58 | 47.90 |

**Table 4.8. Impact of a Change in the Price of Cereals on Consumers - National**

| | -25% | -12.5% | No change | 12.5% | 25% | 50% | 100% |
|---|---|---|---|---|---|---|---|
| Poverty gap | 16.61 | 16.63 | 16.66 | 16.68 | 16.70 | 16.75 | 16.86 |
| Squared poverty gap | 7.98 | 8.00 | 8.01 | 8.02 | 8.04 | 8.06 | 8.13 |
| **Poverty, corn consumers** | | | | | | | |
| Headcount index of poverty | 35.60 | 35.80 | 35.80 | 36.09 | 36.09 | 36.46 | 38.03 |
| Poverty gap | 11.28 | 11.40 | 11.52 | 11.63 | 11.75 | 12.00 | 12.53 |
| Squared poverty gap | 5.15 | 5.21 | 5.28 | 5.35 | 5.42 | 5.56 | 5.88 |
| **Wheat and bread** | | | | | | | |
| **Poverty, population as a whole** | | | | | | | |
| Headcount index of poverty | 47.03 | 47.29 | 47.45 | 47.53 | 47.61 | 47.76 | 48.33 |
| Poverty gap | 16.56 | 16.61 | 16.66 | 16.70 | 16.75 | 16.86 | 17.07 |
| Squared poverty gap | 7.96 | 7.98 | 8.01 | 8.03 | 8.06 | 8.12 | 8.23 |
| **Poverty, wheat and bread consumers** | | | | | | | |
| Headcount index of poverty | 43.02 | 43.37 | 43.59 | 43.69 | 43.80 | 44.00 | 44.78 |
| Poverty gap | 14.68 | 14.74 | 14.81 | 14.87 | 14.94 | 15.08 | 15.37 |
| Squared poverty gap | 6.95 | 6.99 | 7.02 | 7.06 | 7.09 | 7.17 | 7.32 |
| **All identified cereals** | | | | | | | |
| **Poverty, population as a whole** | | | | | | | |
| Headcount index of poverty | 44.85 | 46.05 | 47.45 | 48.84 | 50.09 | 52.79 | 58.41 |
| Poverty gap | 15.87 | 16.24 | 16.66 | 17.11 | 17.62 | 18.76 | 21.67 |
| Squared poverty gap | 7.62 | 7.80 | 8.01 | 8.23 | 8.48 | 9.05 | 10.68 |
| **Poverty, food consumers** | | | | | | | |
| Headcount index of poverty | 43.34 | 44.59 | 46.05 | 47.51 | 48.80 | 51.63 | 57.49 |
| Poverty gap | 15.01 | 15.40 | 15.83 | 16.31 | 16.84 | 18.03 | 21.06 |
| Squared poverty gap | 7.07 | 7.26 | 7.47 | 7.71 | 7.96 | 8.57 | 10.26 |

*Source:* Authors' estimation using ELIM 2006.

Table 4.9. Impact of a Change in the Price of Cereals on Producers - National

| | -25% | -12.5% | No change | 12.5% | 25% | 50% | 100% |
|---|---|---|---|---|---|---|---|
| | **Rice** | | | | | | |
| **Poverty, population as a whole** | | | | | | | |
| Headcount index of poverty | 48.25 | 48.02 | **47.45** | 47.04 | 46.94 | 46.42 | 46.02 |
| Poverty gap | 17.12 | 16.83 | **16.66** | 16.55 | 16.46 | 16.34 | 16.16 |
| Squared poverty gap | 8.33 | 8.10 | **8.01** | 7.96 | 7.91 | 7.85 | 7.75 |
| **Poverty, rice producers** | | | | | | | |
| Headcount index of poverty | 49.46 | 47.67 | **43.31** | 40.15 | 39.39 | 35.41 | 32.36 |
| Poverty gap | 19.15 | 16.95 | **15.59** | 14.75 | 14.13 | 13.17 | 11.78 |
| Squared poverty gap | 10.56 | 8.79 | **8.08** | 7.68 | 7.36 | 6.86 | 6.12 |
| | **Millet and Sorghum** | | | | | | |
| **Poverty, population as a whole** | | | | | | | |
| Headcount index of poverty | 47.78 | 47.65 | **47.45** | 47.45 | 47.15 | 46.92 | 46.20 |
| Poverty gap | 16.84 | 16.74 | **16.66** | 16.57 | 16.48 | 16.32 | 16.04 |
| Squared poverty gap | 8.13 | 8.07 | **8.01** | 7.95 | 7.90 | 7.80 | 7.64 |
| **Poverty, millet and sorghum producers** | | | | | | | |
| Headcount index of poverty | 69.92 | 69.11 | **67.90** | 67.90 | 66.04 | 64.64 | 60.18 |
| Poverty gap | 26.66 | 26.07 | **25.52** | 24.98 | 24.46 | 23.46 | 21.74 |
| Squared poverty gap | 13.77 | 13.37 | **13.00** | 12.65 | 12.32 | 11.73 | 10.76 |
| | **Corn** | | | | | | |
| **Poverty, population as a whole** | | | | | | | |
| Headcount index of poverty | 47.48 | 47.46 | **47.45** | 47.33 | 47.33 | 47.24 | 46.92 |
| Poverty gap | 16.81 | 16.73 | **16.66** | 16.58 | 16.52 | 16.41 | 16.23 |
| Squared poverty gap | 8.19 | 8.09 | **8.01** | 7.95 | 7.91 | 7.82 | 7.70 |

**Table 4.9. Impact of a Change in the Price of Cereals on Producers - National**

| | -25% | -12.5% | No change | 12.5% | 25% | 50% | 100% |
|---|---|---|---|---|---|---|---|
| **Poverty, corn producers** | | | | | | | |
| Headcount index of poverty | 80.13 | 79.95 | **79.83** | 78.27 | 78.27 | 77.09 | 72.96 |
| Poverty gap | 36.14 | 35.12 | **34.12** | 33.14 | 32.43 | 31.02 | 28.61 |
| Squared poverty gap | 21.22 | 19.94 | **18.94** | 18.21 | 17.62 | 16.58 | 14.98 |
| | *All identified cereals* | | | | | | |
| **Poverty, population as a whole** | | | | | | | |
| Headcount index of poverty | 48.50 | 48.25 | **47.45** | 46.92 | 46.44 | 45.60 | 44.35 |
| Poverty gap | 17.47 | 17.00 | **16.66** | 16.38 | 16.17 | 15.79 | 15.18 |
| Squared poverty gap | 8.64 | 8.24 | **8.01** | 7.84 | 7.71 | 7.48 | 7.14 |
| **Poverty, food producers** | | | | | | | |
| Headcount index of poverty | 61.69 | 60.83 | **58.07** | 56.22 | 54.58 | 51.67 | 47.35 |
| Poverty gap | 23.89 | 22.28 | **21.07** | 20.13 | 19.38 | 18.07 | 15.97 |
| Squared poverty gap | 12.71 | 11.32 | **10.51** | 9.94 | 9.47 | 8.68 | 7.51 |

*Source:* Authors' estimation using ELIM 2006.

Table 4.10. Impact of a Change in the Price of Cereals on Consumers and Producers - National

|  | -25% | -12.5% | No change | 12.5% | 25% | 50% | 100% |
|---|---|---|---|---|---|---|---|
| | Rice | | | | | | |
| **Poverty, population as a whole** | | | | | | | |
| Headcount index of poverty | 46.85 | 47.13 | **47.45** | 47.72 | 48.34 | 49.11 | 51.04 |
| Poverty gap | 16.73 | 16.63 | **16.66** | 16.75 | 16.90 | 17.27 | 18.23 |
| Squared poverty gap | 8.15 | 8.01 | **8.01** | 8.05 | 8.11 | 8.28 | 8.78 |
| **Poverty, rice consumers** | | | | | | | |
| Headcount index of poverty | 42.85 | 43.39 | **43.99** | 44.63 | 45.43 | 46.76 | 49.44 |
| Poverty gap | 14.03 | 14.12 | **14.29** | 14.50 | 14.75 | 15.32 | 16.66 |
| Squared poverty gap | 6.34 | 6.36 | **6.43** | 6.54 | 6.65 | 6.92 | 7.63 |
| **Poverty, rice producers** | | | | | | | |
| Headcount index of poverty | 48.13 | 46.57 | **43.31** | 40.71 | 40.20 | 38.09 | 35.16 |
| Poverty gap | 18.89 | 16.82 | **15.59** | 14.84 | 14.31 | 13.50 | 12.37 |
| Squared poverty gap | 10.42 | 8.74 | **8.08** | 7.72 | 7.44 | 6.99 | 6.36 |
| | Millet and sorghum | | | | | | |
| **Poverty, population as a whole** | | | | | | | |
| Headcount index of poverty | 46.87 | 47.16 | **47.45** | 47.98 | 47.91 | 48.78 | 50.12 |
| Poverty gap | 16.54 | 16.59 | **16.66** | 16.73 | 16.82 | 17.02 | 17.61 |
| Squared poverty gap | 7.98 | 7.99 | **8.01** | 8.03 | 8.07 | 8.16 | 8.47 |
| **Poverty, millet and sorghum** | | | | | | | |

Table 4.10. Impact of a Change in the Price of Cereals on Consumers and Producers - National

| | -25% | -12.5% | No change | 12.5% | 25% | 50% | 100% |
|---|---|---|---|---|---|---|---|
| **consumers** | | | | | | | |
| Headcount index of poverty | 35.09 | 35.70 | **36.37** | 37.21 | 37.51 | 39.18 | 41.87 |
| Poverty gap | 10.48 | 10.67 | **10.87** | 11.09 | 11.33 | 11.84 | 13.09 |
| Squared poverty gap | 4.43 | 4.52 | **4.62** | 4.73 | 4.85 | 5.11 | 5.79 |
| **Poverty, millet and sorghum producers** | | | | | | | |
| Headcount index of poverty | 68.92 | 68.78 | **67.90** | 67.90 | 66.23 | 65.07 | 62.37 |
| Poverty gap | 26.42 | 25.95 | **25.52** | 25.11 | 24.71 | 23.96 | 22.70 |
| Squared poverty gap | 13.64 | 13.30 | **13.00** | 12.71 | 12.45 | 11.99 | 11.28 |
| **Corn** | | | | | | | |
| **Poverty, population as a whole** | | | | | | | |
| Headcount index of poverty | 47.44 | 47.46 | **47.45** | 47.39 | 47.39 | 47.39 | 47.38 |
| Poverty gap | 16.77 | 16.71 | **16.66** | 16.60 | 16.57 | 16.51 | 16.42 |
| Squared poverty gap | 8.16 | 8.07 | **8.01** | 7.96 | 7.93 | 7.88 | 7.81 |
| **Poverty, corn consumers** | | | | | | | |
| Headcount index of poverty | 35.60 | 35.80 | **35.80** | 36.09 | 36.09 | 36.46 | 37.93 |
| Poverty gap | 11.33 | 11.42 | **11.52** | 11.61 | 11.71 | 11.92 | 12.37 |
| Squared poverty gap | 5.17 | 5.22 | **5.28** | 5.34 | 5.40 | 5.52 | 5.79 |
| **Poverty, corn producers** | | | | | | | |
| Headcount index of poverty | 80.13 | 79.95 | **79.83** | 78.27 | 78.27 | 77.35 | 73.26 |
| Poverty gap | 36.05 | 35.08 | **34.12** | 33.19 | 32.52 | 31.19 | 28.94 |
| Squared poverty gap | 21.17 | 19.92 | **18.94** | 18.23 | 17.66 | 16.66 | 15.12 |
| **All identified cereals** | | | | | | | |
| **Poverty, population as a whole** | | | | | | | |
| Headcount index of poverty | 45.96 | 46.73 | **47.45** | 48.32 | 49.19 | 50.89 | 55.13 |
| Poverty gap | 16.66 | 16.58 | **16.66** | 16.83 | 17.11 | 17.81 | 19.90 |
| Squared poverty gap | 8.23 | 8.03 | **8.01** | 8.06 | 8.17 | 8.48 | 9.65 |

Table 4.10. Impact of a Change in the Price of Cereals on Consumers and Producers - National

| | -25% | -12.5% | No change | 12.5% | 25% | 50% | 100% |
|---|---|---|---|---|---|---|---|
| **Poverty, cereals consumers** | | | | | | | |
| Headcount index of poverty | 44.44 | 45.26 | **46.05** | 47.04 | 47.96 | 49.83 | 54.40 |
| Poverty gap | 15.75 | 15.71 | **15.83** | 16.05 | 16.37 | 17.17 | 19.44 |
| Squared poverty gap | 7.63 | 7.47 | **7.47** | 7.56 | 7.70 | 8.08 | 9.36 |
| **Poverty, cereals producers** | | | | | | | |
| Headcount index of poverty | 59.45 | 58.89 | **58.07** | 57.61 | 57.51 | 56.60 | 55.99 |
| Poverty gap | 23.18 | 21.92 | **21.07** | 20.47 | 20.07 | 19.44 | 18.91 |
| Squared poverty gap | 12.32 | 11.14 | **10.51** | 10.10 | 9.80 | 9.34 | 8.92 |

*Source:* Authors' estimation using ELIM 2006.

**Annex table 4.1. Impact of a Change in the Price of Cereals on Consumers - Urban and Rural Areas**

| | -25% | -12.5% | No change | 12.5% | 25% | 50% | 100% |
|---|---|---|---|---|---|---|---|
| **Rice** | | | | | | | |
| **Poverty, Urban** | | | | | | | |
| Headcount index of poverty | 23.81 | 24.36 | **25.51** | 26.68 | 27.37 | 29.01 | 32.39 |
| Poverty gap | 7.36 | 7.55 | **7.76** | 7.99 | 8.26 | 8.84 | 10.26 |
| Squared poverty gap | 3.19 | 3.26 | **3.35** | 3.44 | 3.55 | 3.79 | 4.46 |
| **Poverty, Rural** | | | | | | | |
| Headcount index of poverty | 56.42 | 56.95 | **57.63** | 58.15 | 58.93 | 60.05 | 62.59 |
| Poverty gap | 20.41 | 20.59 | **20.78** | 20.98 | 21.20 | 21.68 | 22.79 |
| Squared poverty gap | 9.99 | 10.08 | **10.17** | 10.27 | 10.37 | 10.61 | 11.21 |
| **Millet and sorghum** | | | | | | | |
| **Poverty, Urban** | | | | | | | |
| Headcount index of poverty | 24.82 | 24.99 | **25.51** | 25.74 | 25.80 | 27.10 | 28.86 |
| Poverty gap | 7.59 | 7.67 | **7.76** | 7.85 | 7.94 | 8.15 | 8.63 |
| Squared poverty gap | 3.27 | 3.31 | **3.35** | 3.39 | 3.43 | 3.53 | 3.75 |
| **Poverty, Rural** | | | | | | | |
| Headcount index of poverty | 56.65 | 57.08 | **57.63** | 58.30 | 58.63 | 59.69 | 61.61 |
| Poverty gap | 20.43 | 20.60 | **20.78** | 20.98 | 21.19 | 21.64 | 22.71 |
| Squared poverty gap | 9.98 | 10.07 | **10.17** | 10.27 | 10.38 | 10.62 | 11.22 |
| **Corn** | | | | | | | |
| **Poverty, Urban** | | | | | | | |
| Headcount index of poverty | 25.43 | 25.51 | **25.51** | 25.64 | 25.64 | 25.76 | 26.51 |
| Poverty gap | 7.70 | 7.73 | **7.76** | 7.78 | 7.81 | 7.87 | 8.01 |
| Squared poverty gap | 3.31 | 3.33 | **3.35** | 3.37 | 3.39 | 3.42 | 3.51 |
| **Poverty, Rural** | | | | | | | |
| Headcount index of poverty | 57.61 | 57.63 | **57.63** | 57.65 | 57.65 | 57.71 | 57.81 |
| Poverty gap | 20.74 | 20.76 | **20.78** | 20.80 | 20.83 | 20.87 | 20.96 |

**Annex table 4.1. Impact of a Change in the Price of Cereals on Consumers - Urban and Rural Areas**

| | -25% | -12.5% | No change | 12.5% | 25% | 50% | 100% |
|---|---|---|---|---|---|---|---|
| Squared poverty gap | 10.15 | 10.16 | 10.17 | 10.18 | 10.19 | 10.22 | 10.27 |
| **Wheat and bread** | | | | | | | |
| **Poverty, Urban** | | | | | | | |
| Headcount index of poverty | 25.15 | 25.43 | 25.51 | 25.55 | 25.73 | 25.91 | 27.04 |
| Poverty gap | 7.67 | 7.71 | 7.76 | 7.80 | 7.85 | 7.94 | 8.15 |
| Squared poverty gap | 3.31 | 3.33 | 3.35 | 3.37 | 3.39 | 3.43 | 3.53 |
| **Poverty, Rural** | | | | | | | |
| Headcount index of poverty | 57.17 | 57.43 | 57.63 | 57.72 | 57.75 | 57.89 | 58.21 |
| Poverty gap | 20.68 | 20.73 | 20.78 | 20.83 | 20.89 | 20.99 | 21.21 |
| Squared poverty gap | 10.11 | 10.14 | 10.17 | 10.20 | 10.23 | 10.29 | 10.41 |
| **All identified cereals** | | | | | | | |
| **Poverty, Urban** | | | | | | | |
| Headcount index of poverty | 22.46 | 24.13 | 25.51 | 27.23 | 28.76 | 31.34 | 37.07 |
| Poverty gap | 7.09 | 7.40 | 7.76 | 8.17 | 8.65 | 9.75 | 12.66 |
| Squared poverty gap | 3.05 | 3.19 | 3.35 | 3.53 | 3.73 | 4.23 | 5.75 |
| **Poverty, Rural** | | | | | | | |
| Headcount index of poverty | 55.24 | 56.22 | 57.63 | 58.87 | 59.98 | 62.74 | 68.31 |
| Poverty gap | 19.94 | 20.35 | 20.78 | 21.26 | 21.78 | 22.94 | 25.85 |
| Squared poverty gap | 9.74 | 9.94 | 10.17 | 10.41 | 10.68 | 11.29 | 12.97 |

*Source*: Authors' estimation using ELIM 2006.

**Annex table 4.2. Impact of a Change in the Price of Cereals on Producers - Urban and Rural Areas**

| | -25% | -12.5% | No change | 12.5% | 25% | 50% | 100% |
|---|---|---|---|---|---|---|---|
| **Rice** | | | | | | | |
| **Poverty, Urban** | | | | | | | |
| Headcount index of poverty | 23.81 | 24.36 | **25.51** | 26.68 | 27.37 | 29.01 | 32.39 |
| Poverty gap | 7.36 | 7.55 | **7.76** | 7.99 | 8.26 | 8.84 | 10.26 |
| Squared poverty gap | 3.19 | 3.26 | **3.35** | 3.44 | 3.55 | 3.79 | 4.46 |
| **Poverty, Rural** | | | | | | | |
| Headcount index of poverty | 56.42 | 56.95 | **57.63** | 58.15 | 58.93 | 60.05 | 62.59 |
| Poverty gap | 20.41 | 20.59 | **20.78** | 20.98 | 21.20 | 21.68 | 22.79 |
| Squared poverty gap | 9.99 | 10.08 | **10.17** | 10.27 | 10.37 | 10.61 | 11.21 |
| **Millet and sorghum** | | | | | | | |
| **Poverty, Urban** | | | | | | | |
| Headcount index of poverty | 24.82 | 24.99 | **25.51** | 25.74 | 25.80 | 27.10 | 28.86 |
| Poverty gap | 7.59 | 7.67 | **7.76** | 7.85 | 7.94 | 8.15 | 8.63 |
| Squared poverty gap | 3.27 | 3.31 | **3.35** | 3.39 | 3.43 | 3.53 | 3.75 |
| **Poverty, Rural** | | | | | | | |
| Headcount index of poverty | 56.65 | 57.08 | **57.63** | 58.30 | 58.63 | 59.69 | 61.61 |
| Poverty gap | 20.43 | 20.60 | **20.78** | 20.98 | 21.19 | 21.64 | 22.71 |
| Squared poverty gap | 9.98 | 10.07 | **10.17** | 10.27 | 10.38 | 10.62 | 11.22 |
| **Corn** | | | | | | | |
| **Poverty, Urban** | | | | | | | |
| Headcount index of poverty | 25.43 | 25.51 | **25.51** | 25.64 | 25.64 | 25.76 | 26.51 |
| Poverty gap | 7.70 | 7.73 | **7.76** | 7.78 | 7.81 | 7.87 | 8.01 |
| Squared poverty gap | 3.31 | 3.33 | **3.35** | 3.37 | 3.39 | 3.42 | 3.51 |

**Annex table 4.2. Impact of a Change in the Price of Cereals on Producers - Urban and Rural Areas**

| | -25% | -12.5% | No change | 12.5% | 25% | 50% | 100% |
|---|---|---|---|---|---|---|---|
| **Poverty, Rural** | | | | | | | |
| Headcount index of poverty | 57.61 | 57.63 | **57.63** | 57.65 | 57.65 | 57.71 | 57.81 |
| Poverty gap | 20.74 | 20.76 | **20.78** | 20.80 | 20.83 | 20.87 | 20.96 |
| Squared poverty gap | 10.15 | 10.16 | **10.17** | 10.18 | 10.19 | 10.22 | 10.27 |
| *Wheat and bread* | | | | | | | |
| **Poverty, Urban** | | | | | | | |
| Headcount index of poverty | 25.15 | 25.43 | **25.51** | 25.55 | 25.73 | 25.91 | 27.04 |
| Poverty gap | 7.67 | 7.71 | **7.76** | 7.80 | 7.85 | 7.94 | 8.15 |
| Squared poverty gap | 3.31 | 3.33 | **3.35** | 3.37 | 3.39 | 3.43 | 3.53 |
| **Poverty, Rural** | | | | | | | |
| Headcount index of poverty | 57.17 | 57.43 | **57.63** | 57.72 | 57.75 | 57.89 | 58.21 |
| Poverty gap | 20.68 | 20.73 | **20.78** | 20.83 | 20.89 | 20.99 | 21.21 |
| Squared poverty gap | 10.11 | 10.14 | **10.17** | 10.20 | 10.23 | 10.29 | 10.41 |
| *All identified cereals* | | | | | | | |
| **Poverty, Urban** | | | | | | | |
| Headcount index of poverty | 22.46 | 24.13 | **25.51** | 27.23 | 28.76 | 31.34 | 37.07 |
| Poverty gap | 7.09 | 7.40 | **7.76** | 8.17 | 8.65 | 9.75 | 12.66 |
| Squared poverty gap | 3.05 | 3.19 | **3.35** | 3.53 | 3.73 | 4.23 | 5.75 |
| **Poverty, Rural** | | | | | | | |
| Headcount index of poverty | 55.24 | 56.22 | **57.63** | 58.87 | 59.98 | 62.74 | 68.31 |
| Poverty gap | 19.94 | 20.35 | **20.78** | 21.26 | 21.78 | 22.94 | 25.85 |
| Squared poverty gap | 9.74 | 9.94 | **10.17** | 10.41 | 10.68 | 11.29 | 12.97 |

*Source:* Authors' estimation using ELIM 2006.

**Annex table 4.3. Impact of a Change in the Price of Cereals on consumers and Producers - Urban and Rural Areas**

| | -25% | -12.5% | No change | 12.5% | 25% | 50% | 100% |
|---|---|---|---|---|---|---|---|
| **Rice** | | | | | | | |
| **Poverty, urban** | | | | | | | |
| Headcount index of poverty | 23.95 | 24.47 | **25.51** | 26.63 | 27.27 | 28.60 | 31.81 |
| Poverty gap | 7.50 | 7.61 | **7.76** | 7.98 | 8.24 | 8.81 | 10.17 |
| Squared poverty gap | 3.32 | 3.30 | **3.35** | 3.44 | 3.54 | 3.78 | 4.44 |
| **Poverty, rural** | | | | | | | |
| Headcount index of poverty | 57.48 | 57.65 | **57.63** | 57.51 | 58.11 | 58.62 | 59.96 |
| Poverty gap | 21.02 | 20.82 | **20.78** | 20.82 | 20.92 | 21.20 | 21.97 |
| Squared poverty gap | 10.39 | 10.19 | **10.17** | 10.19 | 10.24 | 10.37 | 10.80 |
| **Millet and sorghum** | | | | | | | |
| **Poverty, urban** | | | | | | | |
| Headcount index of poverty | 24.89 | 25.19 | **25.51** | 25.71 | 25.72 | 27.03 | 28.72 |
| Poverty gap | 7.62 | 7.68 | **7.76** | 7.83 | 7.91 | 8.09 | 8.52 |
| Squared poverty gap | 3.30 | 3.32 | **3.35** | 3.38 | 3.41 | 3.49 | 3.67 |
| **Poverty, rural** | | | | | | | |
| Headcount index of poverty | 57.07 | 57.35 | **57.63** | 58.30 | 58.20 | 58.87 | 60.04 |
| Poverty gap | 20.68 | 20.73 | **20.78** | 20.86 | 20.95 | 21.17 | 21.83 |
| Squared poverty gap | 10.16 | 10.16 | **10.17** | 10.19 | 10.23 | 10.33 | 10.69 |
| **Corn** | | | | | | | |
| **Poverty, urban** | | | | | | | |
| Headcount index of poverty | 25.43 | 25.51 | **25.51** | 25.40 | 25.40 | 25.52 | 26.05 |

**Annex table 4.3. Impact of a Change in the Price of Cereals on consumers and Producers - Urban and Rural Areas**

| | -25% | -12.5% | No change | 12.5% | 25% | 50% | 100% |
|---|---|---|---|---|---|---|---|
| Poverty gap | 7.86 | 7.81 | 7.76 | 7.71 | 7.73 | 7.77 | 7.88 |
| Squared poverty gap | 3.54 | 3.41 | 3.35 | 3.34 | 3.35 | 3.38 | 3.45 |
| **Poverty, rural** | | | | | | | |
| Headcount index of poverty | 57.64 | 57.64 | 57.63 | 57.59 | 57.59 | 57.54 | 57.28 |
| Poverty gap | 20.90 | 20.84 | 20.78 | 20.73 | 20.67 | 20.56 | 20.39 |
| Squared poverty gap | 10.30 | 10.23 | 10.17 | 10.11 | 10.06 | 9.97 | 9.84 |
| **All identified cereals** | | | | | | | |
| **Poverty, urban** | | | | | | | |
| Headcount index of poverty | 22.83 | 24.23 | 25.51 | 27.01 | 28.39 | 30.69 | 36.17 |
| Poverty gap | 7.41 | 7.55 | 7.76 | 8.07 | 8.51 | 9.54 | 12.28 |
| Squared poverty gap | 3.43 | 3.32 | 3.35 | 3.48 | 3.67 | 4.12 | 5.56 |
| **Poverty, rural** | | | | | | | |
| Headcount index of poverty | 56.70 | 57.16 | 57.63 | 58.21 | 58.84 | 60.26 | 63.93 |
| Poverty gap | 20.95 | 20.77 | 20.78 | 20.90 | 21.10 | 21.65 | 23.43 |
| Squared poverty gap | 10.46 | 10.22 | 10.17 | 10.19 | 10.25 | 10.51 | 11.54 |

*Source:* Authors' estimation using ELIM 2006.

# 5

## ASSESSING THE POTENTIAL IMPACT OF HIGHER FERTILIZER USE ON POVERTY AMONG FARM HOUSEHOLDS: ILLUSTRATION FOR RWANDA

*Enrique Hennings and Quentin Wodon*

The use of modern inputs such as fertilizers has been identified as a key tool for increasing farm productivity in sub-Saharan Africa. This is also the case in Rwanda where fertilizers have been given a high priority under the government's Crop Intensification Program. Using nationally representative household survey data and information from fertilizer trials, we estimate what the impact of higher fertilizer use could be on poverty reduction. Estimates from a farm production function combined with poverty data suggest that the reduction in poverty associated with various levels of fertilizer use could indeed be substantial.

## INTRODUCTION

As poverty remains primarily a rural phenomenon in most sub-saharan African countries, interventions to reduce it must focus in large part on ways to increase agricultural productivity. This is especially the case nowadays given that climate change is threatening the sustainability of traditional agricultural practices in some areas, and that higher food prices around the world are putting pressure on national governments to increase local food production. In this context, policies to expand the use of modern inputs, including fertilizers, among African farmers have received renewed attention (e.g., Minde et al. 2008; Minot and Benson 2009).

In Rwanda, a key objective of the Government over the last decade has been to transform the agricultural sector from a semi-subsistence system into a commercially-oriented and intensive production system. This objective was already stated in the country's first poverty reduction strategy, and it has been reaffirmed since (i.e., Republic of Rwanda 2009). The use of modern inputs such as fertilizers has been identified as a key tool for increasing labor and land productivity, and thereby raising farm incomes, which in turn could be spent in the rural non-farm sector in order to generate faster economic growth (i.e., Republic of Rwanda 2007; see also Abt Associates 2002; Desai 2002). Beyond fertilizers and other inputs, overall public funding for agriculture increased from 4.2 percent of the budget in 2008 to 6.6 percent in the 2010–11 budget (World Bank 2011).

Improving agricultural productivity in Rwanda is not an easy task, however. Population growth has led to changes in the traditional agricultural system as well as in the ability of the country to remain self-sufficient with respect to input uses and food consumption. The average farm has become smaller and more fragmented, due to a reduction in land availability. Moreover, there has been an expansion in the agricultural frontier, but this has pushed agricultural production to marginal lands or lands of steep slopes, covered by woods, with the risk of a rapid deterioration of the soil. More intensive farming not accompanied by the use of additional soil nutrients may have contributed to decreasing productivity in some areas. Already some fifteen years ago, Clay et al. (1995) provided evidence that erosion could reduce farm yields, thus threatening productivity and food security.

Besides soil conservation, other efforts must be made to improve yields and productivity. The use of manure could help increase productivity and maintain soil fertility, but its production is limited due to a decline in livestock and insufficient to achieve the increase in production necessary for poverty reduction and food security. This is one of the reasons why it has been suggested that traditional methods for preserving soil quality and increasing productivity should be accompanied by measures to promote the use of agricultural inputs such as pesticides, herbicides, and fertilizers.

The Crop Intensification Program launched by the Government in 2007 aimed to increase crop productivity in a sustainable way through increased use of modern inputs, and especially seeds and fertilizers. Efforts were also made to modernize agricultural technologies and reduce land fragmentation so that larger parcels could be cultivated. As noted in World Bank (2011), the program was instrumental in increasing the use of improved seeds for key crops such as maize, wheat and Irish potato, and doubling the average use of fertilizers to 16 kg per ha in 2010 from a base of 8.5 kilogram in 2006.

At the farm level, in order to recommend a level of use of pesticides, herbicides, and fertilizers by farmers, one must know what type of inputs to use and in what proportion. For fertilizers, profitability is typically assessed by estimating a ratio between the value of additional production that may be attributable to fertilizer treatment, and the cost of such treatment. The usual rule to classify a treatment as profitable is that the value/cost ratio be more than or equal to two. It is thus not enough that the costs of the investment be covered: for a treatment to be considered profitable, the returns must be much higher than the investments since access to financial resources is limited and the production and price risks may be considerable for farmers. It is suggested in the literature that whenever the value/cost ratio is over three, fertilizer promotion will succeed in the short-term while for a value cost ratio lower than three but larger or equal to two, success is expected after a longer period of time, and must be accompanied by technical assistance and monitoring.

In Rwanda, studies conducted as early as in the 1980s and 1990s suggested that fertilizer use could indeed be profitable, which led the Government to propose plans to increase the use of fertilizer, as explained above. Those plans aimed among others to identify the potential use of fertilizers by agro-ecological zones; promoting the increase of fertilizer supply; reinforcing farmers' technical capacity to use fertilizers through extension programs, as well as farmers' interest in and access to fertilizers; and encouraging a private sector fertilizer market by increasing their profitability. In order to update the analyses on fertilizers' profitability for the different crops and zones, a range of studies and demonstration projects were launched. In addition, and more recently, the Government created a voucher program in partnership with micro-finance organizations under the Crop Intensification Program, whereby maize and wheat farmers benefitted from subsidies for their fertilizer use.

In this chapter, we do not aim to tackle many of the complex policy issues related to the promotion of higher fertilizer use and other modern inputs in Rwanda today (see World Bank 2011 for a discussion). Instead, we focus on assessing what the impact of such higher fertilizer use might be on poverty reduction. The contribution of our work consists in making the link between estimates of the profitability of fertilizer use with the expected consumption level of households, and thereby poverty among farmers.

To do so, we use a survey which combines detailed information on farm production with an extensive consumption module that allows for poverty measurement. The basic idea consists in estimating a farm production function and using independent information from demonstration projects on optimal fertilizer use in order to simulate with that farm production function the gains from higher fertilizer use and thereby the reduction in poverty. The next section provides the estimates of the production function and the third section reports the results on the estimated likely impact of higher recommended fertilizer use on measures of poverty among farm households. A conclusion follows.

# FARM PRODUCTION FUNCTION

The data used for this study are from the nationally representative Household Living Condition Survey implemented in 2000–2001 (*Enquête Intégrale sur les conditions de vie des ménages au Rwanda 2000-2001*; see Ministry of Finance, 2002, for a description). The sample includes 6,420 households, most of which live in rural areas. Table 5.1 provides summary statistics on expenditure on fertilizers by quintiles of consumption per equivalent adult. Households in the top quintile have as expected a higher level of expenditure on fertilizers, with households in the bottom quintile spending approximately 20 times less on chemical fertilizers than households in the top quintile. Similar differences are seen for other inputs such as organic fertilizers, insecticides and herbicides. These differences suggest that modern inputs are not used by households in the lower quintiles. This could be due to high input prices, low supply, or even the lack of awareness of the potential benefits of using modern inputs in agricultural production. Another possible explanation could be that the valley lands tend to be more productive than the steep slopes where usually low-income farmers are located. These areas suffer from acidity that reduces the ability to cultivate, resulting in a lower consumption of fertilizers. Due to data limitations (we know the use of various inputs by households, but not for each of the crops cultivated by households) the production function is estimated at the household Level.[1]

The dependent variable is the total value of the crop production of the household. The independent variables include labor, land, capital, input costs, land fragmentation, land tenure and geographic location. The descriptive statistics for the variables used in the regression are shown in Table 5.2.

The value of output is defined as the aggregate value of production, which is the sum of each crop's production multiplied by its unit price as reported by the farmers in the survey; in cases of missing data on unit prices, mean unit prices at the Primary Sampling Unit level were used instead. Labor is divided into family labor (expressed in adult equivalents per-hectare[3]) and paid labor (aggregate value of wages paid for hired labor). The land variable is expressed in cultivated hectares. The capital variable is the market value of farm possessions (agricultural equipment) such as plows, tractors, picks and other basic farm tools. The 'fertilizer variable is the sum of farm expenditure for organic and inorganic fertilizer and pesticides. The fragmentation variable reflects the number of parcels per-hectare. The share of the land that is owned by the farmer is also used, as is the level of education of the household head expressed in year of education (as a proxy for the skills of the farmer). The different regions in the survey are represented through dummy variables, with Kigali Ngali as the reference.

A translog functional form is used, with interaction terms between several variables. In generic notation, if we denote expenditure on fertilizers by $F_i$ and the other independent variables by the vector $X_i$ (which includes a constant) we estimate:

$$\text{Log } O_i = \beta'X_i + \gamma F_i + \varepsilon_i \tag{1}$$

The results of the estimation are presented in table 5.3. The model is able to explain one-fourth of the variation $(R^2)$ between households in the value of production, which is good for this type of analysis. The parameter estimates are similar to previous studies conducted by Michigan State University in 1990 using different data. One difference is that the sign of the fragmentation variable here is positive and statistically significant while it was negative and not significant in that previous study. Our results are also similar to those of Yamauchi (2002).

Family labor, paid labor, the extension of cultivated land, the fragmentation level, expenditure on fertilizer, capital, the share of owned land, and the education of the household head all have positive and statistically significant impacts on output, as predicted by economic theory. The location variables are often statistically significant, suggesting the need to control for location. The majority of the interaction variables are not significant except for the interaction between capital and fertilizer use, as well as the interaction between fertilizer and fragmentation. The interaction between capital and fertilizer use is negative, suggesting that the use of agricultural tools can improve productivity through the use of manual weed control while substituting the use of herbicides. The interaction between fragmentation and capital is also negative, perhaps because the use of capital such as machinery is more difficult in small plots than in larger ones, reducing the economies of scale and affecting negatively the value of production. The interaction between fertilizer use and fragmentation is positive, perhaps because the use of fertilizer can be better controlled in smaller parcels when the use of machinery is limited, as is the case of Rwanda (this last interaction effect is not statistically significant).

# POTENTIAL IMPACT ON POVERTY OF HIGHER FERTILIZER USE

The next step in the estimation consists in computing the expected profit made by each household from using higher levels of fertilizer, and incorporating the additional profit in the consumption aggregate of the household in order to estimate the reduction in poverty that could be achieved with higher fertilizer use. We restrict our analysis to the subsample of farmers in the surveys.

Consider a household that increases its use of fertilizer from its current level $F_i$ to a higher level $F_i + \Delta F_i$ with a unit cost of acquiring fertilizers at $C_F$. The original level of consumption per equivalent adult of the household is denoted by $y_i$, and the household size (in terms of equivalent adults) is $h_i$. Ranking households by increasing level of consumption, denoting the number of poor households by $q$, and denoting the poverty line by $z$, the estimate of the FGT (Foster et al. 1984) poverty measures (in expected terms) after an increase in the use of fertilizer by households will be:

$$P\alpha = \frac{1}{n}\sum_{i=1}^{q}\left[\frac{z - [y_i + (\Delta O_i - (\Delta F_i)C_F)/h_i]}{z}\right]^{\alpha}$$

With $\Delta O_i = \exp(\beta' X_i + \gamma(F_i + \Delta F_i)) - \exp(\beta' X_i + \gamma F_i)$          (2)

In equation 2, using appropriate weights in the household survey, the headcount index (the share of the population in poverty) is obtained for a value of $\alpha$ equal to zero. The poverty gap (which takes into account the distance separating the poor from the poverty line multiplied by the headcount index) is obtained when $\alpha$ is equal to one. The squared poverty gap, which places a higher weight on the poorest farmers in the sample, is obtained with $\alpha$ equal to two.

The increase in fertilizer use for the simulations is set using different methods. First, we simply rely on the recommendations for fertilizer use by crop provided in Kelly and Murekezi (2000). That is, for each farm household, using the information on the various crops produced, we compute an optimal value of fertilizer use that corresponds to recommendations from trials. These trials relied on fertilizer demonstrations and assessed the importance of combining organic and inorganic fertilizers together with other factors such as lime to make use of acid soils.

Key findings from the trials (FSRP/FAO–SFI study) indicate that profitable fertilizer treatments for climbing beans involved use of urea ranging from 10 to 40 kg/ha and use of DAP (diammonium phosphate) from 80 to 120 kg/ha.

Results suggested that if levels of K were adequate, the use of NPK (nitrogen phosphorus potassium) plus urea in lieu of DAP would not be not profitable; similarly the use of only manure is not profitable in some of the zones. For maize, the profitable fertilizer treatments included DAP from 60 to 90 kg/ha and urea from 80 to 110 kg/ha, as well as combining urea with up to 300 kg/ha of NPK, noting that maize requires high moisture levels and that irrigation may be necessary in zones that present rain deficits.

Recommendations for rice fertilization included use of DAP, DAP combined with urea and also the NPK-urea combination. The suggestions for the use of urea and DAP ranged from 50 to 100 kg/ha, and for NPK they ranged from 140 to 250 kg/ha. The studies also suggested that there is a large margin between paddy and processed rice, suggesting that farmers may be able to increase their rents from rice by turning to selling a more processed product. Sorghum fertilizer promotion is recommended in several zones also, with suggestions for combinations of DAP (76 to 110 kg/ha) and urea (70–80 kg/ha). As sorghum is more tolerant to the lack of moisture than maize, maize could be cultivated in swamps using additional irrigation if necessary in zones where the use of fertilizers shows to be profitable both for maize and sorghum, while sorghum could be cultivated on the hills where moisture is much lower.

For potatoes, profitable fertilizer treatments identified included the combination of DAP and urea (90–110 kg/ha and 30–100 kg/ha), as well as the use of NPK by itself (300 kg/ha). Kelly and Murekezi (2000) suggest that the promotion of potato production should not be conducted in areas that may favor the spread of bacterial diseases, unless a specific program to control them is implemented. Soybean's most profitable fertilization is usually the use of DAP and urea (60–80 kg/ha and 10–20 kg/ha). Somewhat less profitable is the use of NPK, although there are recommendations for its use from 120 to 230 kg/ha; if possible, the most profitable fertilizer combination should however be used.

For sweet potato, the DAP-urea combination again results in higher profitability. For this crop, the value-cost ratio is much higher than two, which could allow farmers to increase their production per hectare significantly and lead them to substitute crops. Finally, the promotion of fertilizer treatments is not recommended for three crops: wheat, peas, and cassava. The low market prices for wheat, the high cost of liming, and the limits imposed by cassava's production cycle are some of the reasons that prevent fertilization from being profitable.

All these recommendations are used when simulating the impact on farm profits of higher fertilizer use and computing the expected reduction in poverty from such decisions by farm households. Results are provided in Table 5.4.

The results suggest that if farmers were able to follow these recommendations, the increase in fertilizer use would be dramatic. While in 2001, actual expenditure on fertilizer was very low at approximately 600 RWF/ha, the optimal values suggest average spending of 17,175 RWF/ha. This would, according to our regression estimates, double revenues, generating a value–cost ratio under the optimal allocation of 5.33, versus the current estimate of 2.33 (the traditional cut-off point for profitability is a value of two, as mentioned earlier). These results are in line with those of Kelly and Murekezy (2000; see also Kelly and Nyirimana 2002).

To analyze the effects on poverty, we rely on the poverty lines estimated by the Ministry of Finance (2002), namely 45,000 RWF per year per equivalent adult for the extreme poor and 64,000 RWF for the poor. The baseline estimates before the increase in fertilizer use suggest that in the sample of producers, 38.15 percent of the population is in extreme poverty, and 57.31 percent in poverty. Under the optimal fertilizer use scenario, the headcount indices of extreme poverty and poverty would drop respectively to 27.79 percent and 45.08 percent, which is a very large reduction.

In practice however, it is unlikely that households would invest in the use of fertilizer to the levels suggested by the trials. For many households, this level of expenditure is simply not affordable, and too risky. In order to be more realistic, we computed expected levels of output and the associated costs with different levels of fertilizer use that correspond to shares of household total consumption. First we let households spend up to 2 percent of their consumption on fertilizers, with the caveat that if the optimal level of fertilizer use by any given household is less than 2 percent of consumption, then the household will limit its investment to the optimal use. The same experiment is repeated for various thresholds in terms of consumption of households, namely 5 percent, 10 percent and 20 percent of household consumption. Throughout, the results are similar, in that higher fertilizer use could lead to an important reduction in poverty. However, it is important to note that the reduction in poverty estimated with a smaller level of fertilizer use than the recommended value (equivalent to, say, 10 or 20 percent of the consumption level of households) would already go a long way in delivering the benefits in terms of higher output value at a limited additional cost for farmers.

## CONCLUSION

In order to improve food security, the government of Rwanda has long identified three main objectives: increase farm productivity and profitability, combat soil degradation, and diversify rural household incomes. Among the strategies that could be used to achieve these objectives, increasing the use of organic matter, fertilizers and lime, as well as soil conservation investments are high on the agenda. The government has seen its role as that of a facilitator for the development of a competitive market for the distribution and use of fertilizers. This limited role is nevertheless complex. Beyond the need to promote fertilizer imports, farmers must be enabled to access credit, adequate transportation infrastructure and distributional channels must be maintained, and technical assistance/demonstration programs for the use of fertilizers on the crop/zone combinations for which profitability has been proven already must be expanded so farmers may acquire the necessary information and skills to use fertilizers and thereby increase agricultural production and profitability. Ensuring secure land tenure and reducing land fragmentation are also important to encourage farmers to invest more in agriculture. Similarly, political stability is essential to promoting productivity investments. Apart from inorganic fertilizer, an increase in livestock (through research of livestock diseases to reduce mortality rates) and use of animal manure would also help to improve soil quality.

The results provided in this chapter deal with only a small part of the agricultural growth agenda in Rwanda, but they are encouraging. Increasing the use of fertilizers has been given a high priority in Rwanda under the Crop Intensification Program, and substantial progress has been achieved. Using data from the 2000–2001 EICV, and on the basis of the estimation of a farm production function as well as results from fertilizer trials, our results suggest that higher fertilizer use could indeed contribute substantially to poverty reduction among farm households.

Our econometric results are of course indicative only. In practice and for policy, further research remains necessary to improve productivity increases resulting from the use of fertilizers and guarantee the profitability of programs that promote their usage. The responses of each crop and region to a specific fertilizer need to be tested regularly so that the success of fertilizer programs is most likely. Weather and soil conditions should always be taken into account when fertilizer recommendations are being made. Success stories from the use of fertilizers should be documented and shared with the community so that the know-how becomes a part not only of the research community but also of the farmers involved. Potential problems from high fertilizer use must also be recognized, and guarded against. Finally, research on technology improvements for intensification of intercropping and mixed cropping techniques are also an important tool for increasing productivity and improving the quality of products.

NOTES

1. A majority of farmers produce several crops at the same time (crop mixing). Unfortunately, even though data are available on the allocation of land to specific crops, the use of various types of labor and fertilizers are available only at the household level.

2. The estimation of the number of adult equivalents was based on coefficients of 1 for adults between 16 and 60 years of age, and 0.25 for children (aged 6 to 15) and for individuals over 60 years of age.

BIBLIOGRAPHY

Abt Associates. "Fertilizer Use and Marketing Policy Workshop Proceedings." Agricultural Policy Development Project Research Report No. 10. Bethesda, MD: Abt Associates, 2002.

Clay, Daniel, Fidele Byiringiro, Jaakko Kangasniemi, Thomas Reardon, Bosco Sibomana, Laurence Uwamariya and Douglin Tardif-Douglin. "Promoting Food Security in Rwanda Through Sustainable Agricultural Productivity: Meeting the Challenges of Population Pressure, Land Degradation, and Poverty." MSU International Development Paper No. 17. Michigan State University. Michigan: Department of Agricultural Economics, 1995.

Clay, Daniel, Valerie Kelly, Edson Mpyisi and Thomas Reardon. "Input Use and Conservation Investments among Farm Households in Rwanda: Patterns and Determinants." FSRP Working Paper, Food Security Research Project. Kigali, Rwanda: Ministry of Agriculture, Animal Resources, and Forestry, 2001.

Desai, Gunvant M. "Key Issues in Achieving Sustainable Rapid Growth of Fertilizer Use in Rwanda." Agricultural Policy Development Project Research Report No. 16. Bethesda, MD: Abt Associates, 2002.

Kelly, Valerie and Joseph Nyirimana. "Learning from Doing: Using Analysis of Fertilizer Demonstration Plots to Improve Programs for Stimulating Fertilizer Demand in Rwanda." Food Security Research Project Report, MINAGRI, Rwanda: Food Security Project, 2002.

Kelly, Valerie and Anastese Murekezi. "Réponse et rentabilité des engrais au Rwanda." Synthèse des résultats des études du MINAGRI menées par le Food Security Research Project et l'Initiative sur la fertilité des Sols de la FAO. Mimeo. Rwanda, 2000.

Ministry of Finance. "A Profile of Poverty in Rwanda." Kigali, 2002.

Minot, Nicholas and Todd Benson. "Fertilizer Subsidies: Are Vouchers the Answer?" IFPRI Issue Brief 60. Washington, DC: International Food Policy Research Institute, 2009.

Minde, Issac,  T. S. Jayne, Eric Crawford, Joshua Ariga and Jones Govereh. "Promoting Fertilizer Use in Africa: Current Issues and Empirical Evidence from Malawi, Zambia, and Kenya." Food Security Collaborative Working Paper, Michigan State University: Department of Agricultural, Food, and Resource Economics, 2008.

Mukamana, Josepha. "The Chemical Fertilizer Market in Rwanda." Agricultural Policy Development Project Research Report No. 18. Bethesda, MD: Abt Associates, 2002.

Republic of Rwanda. "Strategy for Developing Fertilizer Distribution Systems in Rwanda". Kigali: Ministry of Agriculture and Animal Resources, 2007.

Republic of Rwanda. "Strategic Plan for the Transformation of Agriculture in Rwanda," Phase II (PSTA II)–Final Report. Kigali: Ministry of Agriculture and Animal Resources, 2009.

World Bank. "Rwanda Economic Update: Seeds for Higher Growth." Washington, DC: The World Bank, 2011.

Yamauchi C. "The Use of Fertilizers and its Impact on Agricultural Production in Rwanda." Mimeo. Washington, DC: The World Bank, 2002.

**Table 5.1. Average Private Expenditure on Fertilizers in Rural Areas, by Quintile, 2001**

|              | Poorest (Q1) | Q2    | Q3     | Q4     | Richest (Q5) |
|--------------|--------------|-------|--------|--------|--------------|
| Chemical     | 60.58        | 68.16 | 206.55 | 279.77 | 1299.77      |
| Organic      | 32.39        | 22.56 | 35.83  | 71.86  | 315.07       |
| Insecticides | 42.77        | 85.88 | 221.31 | 309.41 | 706.35       |
| Herbicides   | 0.70         | 6.55  | 22.95  | 65.93  | 81.54        |

*Source:* Authors' estimation using EICV 2000-2001 survey.

**Table 5.2. Descriptive Statistics for Variables Used in Production Function Estimation, 2001**

| Variable | Mean | Std. Dev | Min | Max |
|---|---|---|---|---|
| Log value of production | 1999 | 2.12 | 0.00 | 17.57 |
| Log family labor (adjusted household size) | 8.19 | 0.52 | 5.54 | 9.57 |
| Log hired labor (at cost) | 2.04 | 3.68 | 0.00 | 13.38 |
| Log land cultivated | -0.97 | 1.49 | -6.91 | 2.91 |
| Fragmentation index | 29.88 | 77.67 | 0.14 | 1000.00 |
| Log fertilizer use | 1.11 | 2.66 | 0.00 | 12.40 |
| Log capital | 6.53 | 2.04 | 0.00 | 14.41 |
| Share of land owned | 0.89 | 0.24 | 0.00 | 1.00 |
| Log years of education of household head | 1.48 | 0.59 | 0.00 | 2.83 |
| Log fertilizer *Log labor costs | 3.16 | 15.19 | 0.00 | 162.69 |
| Log fertilizer * Log land cultivated | -0.48 | 3.55 | -44.77 | 32.08 |
| Log fertilizer *capital | 4.13 | 15.27 | 0.00 | 120.13 |
| Log Fertilizer*fragmentation | 35.62 | 228.41 | 0.00 | 4605.17 |
| Fragmentation * Log capital | 201.34 | 531.51 | 0.00 | 8655.41 |
| Butare region | 0.08 | 0.28 | 0.00 | 1.00 |
| Byumba region | 0.09 | 0.29 | 0.00 | 1.00 |
| Cyangugu region | 0.08 | 0.26 | 0.00 | 1.00 |
| Gikongoro region | 0.06 | 0.24 | 0.00 | 1.00 |
| Gisenyi region | 0.10 | 0.29 | 0.00 | 1.00 |
| Gitarama region | 0.11 | 0.31 | 0.00 | 1.00 |
| Kibungo region | 0.08 | 0.27 | 0.00 | 1.00 |
| Kibuye region | 0.06 | 0.23 | 0.00 | 1.00 |
| Kigali (capital) | 0.11 | 0.32 | 0.00 | 1.00 |
| Ruhengeri region | 0.12 | 0.32 | 0.00 | 1.00 |
| Umutara region | 0.04 | 0.19 | 0.00 | 1.00 |

*Source:* Authors' estimation using EICV 2000-2001 survey.

**Table 5.3. Farm Production Function Estimates, 2001**

| Variable | Coefficient | Standard Error |
|---|---|---|
| Log family labor (adjusted household size) | 0.280*** | 0.072 |
| Log hired labor (at cost) | 0.073*** | 0.010 |
| Log land cultivated | 0.270*** | 0.034 |
| Fragmentation index | 0.008*** | 0.003 |
| Log fertilizer use | 0.310*** | 0.072 |
| Log capital | 0.258*** | 0.041 |
| Share of land owned | 0.649*** | 0.139 |
| Log years of education of household head | 0.163*** | 0.059 |
| Log fertilizer * Log labor costs | -0.003 | 0.003 |
| Log fertilizer * Log land cultivated | 0.014 | 0.012 |
| Log fertilizer* Log capital | -0.027*** | 0.009 |
| Log fertilizer * fragmentation | 0.0002 | 0.000 |
| Fragmentation * Log capital | -0.0007** | 0.000 |
| Butare region | 1.394*** | 0.304 |
| Byumba region | 1.820*** | 0.306 |
| Cyangugu region | 0.758*** | 0.309 |
| Gikongoro region | 1.126*** | 0.311 |
| Gisenyi region | 1.068*** | 0.309 |
| Gitarama region | 0.700** | 0.302 |
| Kibungo region | 2.004*** | 0.307 |
| Kibuye region | 0.794** | 0.319 |
| Kigali (capital) | -0.439 | 0.306 |
| Ruhengeri region | 1.435*** | 0.301 |
| Umutara region | 1.220*** | 0.328 |
| Constant | 3.957*** | 0.709 |
| Adjusted R² | 0.247 | |

*Source:* Authors' estimation using EICV 2000-2001.
*Note:* ***Significant at 1 percent level.
       **Significant at 5 percent level.
        *Significant at 10 percent level.

**Table 5.4. Profitability Analysis for Higher Fertilizer Use, 2001**

| Variable | Value (RwF, except ratios) |
|---|---|
| Optimal revenue with fertilizer | 98375.72 |
| Predicted revenue | 41486.02 |
| Change in predicted revenue | 55734.54 |
| Cost of optimal fertilizer | 17775.41 |
| Observed spending on fertilizer | 599.83 |
| Change in spending on fertilizer | 17175.58 |
| Optimal profit | 65130.55 |
| Predicted profit | 32236.22 |
| Change in profit | 32259.23 |
| Value/cost ratio under optimal fertilizer use | 5.53 |
| Predicted value/cost ratio based on observed spending | 2.33 |

*Source:* Authors' estimation using EICV 2000–2001.

**Table 5.5. Potential Impact on Poverty of Higher Fertilizer Use in Rwanda**

| | All producer | Rural producers | Extreme poor, rural |
|---|---|---|---|
| **Extreme poverty headcount** | 0.3815 | 0.3920 | 1.0000 |
| 2% use | 0.3322 | 0.3414 | 0.8709 |
| 5% use | 0.3210 | 0.3300 | 0.8417 |
| 10% use | 0.3179 | 0.3267 | 0.8333 |
| 20% use | 0.3207 | 0.3294 | 0.8337 |
| Optimal use | 0.2779 | 0.2860 | 0.7154 |
| **Poverty headcount** | 0.5731 | 0.5871 | 1.0000 |
| 2% use | 0.5323 | 0.5454 | 0.9958 |
| 5% use | 0.5237 | 0.5366 | 0.9903 |
| 10% use | 0.5224 | 0.5353 | 0.9899 |
| 20% use | 0.5229 | 0.5355 | 0.9894 |
| Optimal use | 0.4508 | 0.4624 | 0.8247 |
| **Extreme poverty gap** | 0.1245 | 0.1283 | 0.3273 |
| 2% use | 0.0993 | 0.1024 | 0.2613 |
| 5% use | 0.0955 | 0.0985 | 0.2514 |
| 10% use | 0.0935 | 0.0964 | 0.2459 |
| 20% use | 0.0929 | 0.0958 | 0.2444 |
| Optimal use | 0.1001 | 0.1030 | 0.2668 |
| **Poverty gap** | 0.2304 | 0.2367 | 0.5270 |
| 2% use | 0.2004 | 0.2059 | 0.4724 |
| 5% use | 0.1957 | 0.2011 | 0.4626 |
| 10% use | 0.1931 | 0.1984 | 0.4565 |
| 20% use | 0.1928 | 0.1981 | 0.4537 |
| Optimal use | 0.2003 | 0.2055 | 0.4721 |
| **Extreme squared poverty gap** | 0.0569 | 0.0587 | 0.1499 |
| 2% use | 0.0428 | 0.0442 | 0.1126 |
| 5% use | 0.0408 | 0.0421 | 0.1074 |
| 10% use | 0.0397 | 0.0410 | 0.1045 |
| 20% use | 0.0393 | 0.0406 | 0.1035 |
| Optimal use | 0.0447 | 0.0460 | 0.1195 |
| **Squared poverty gap** | 0.1197 | 0.1232 | 0.2989 |
| 2% use | 0.0985 | 0.1014 | 0.2492 |
| 5% use | 0.0953 | 0.0981 | 0.2413 |
| 10% use | 0.0935 | 0.0962 | 0.2366 |
| 20% use | 0.0978 | 0.1006 | 0.2436 |
| Optimal use | 0.0995 | 0.1022 | 0.2533 |

*Source:* Authors' estimation using EICV 2000-2001.

# 6

## POVERTY REDUCTION CHALLENGES IN SOUTH ASIA

### *Arup Mitra*

A very large percentage (roughly 40 percent) of the world's poor live in South Asia. Hence, poverty reduction strategies are of crucial significance. The poverty challenges in South Asia include equitable sharing of growth and development among a highly dispersed population, reducing the incidence of consumption poverty and improving the nutritional levels both in the rural and urban areas, bridging gender and rural-urban inequalities, creating productive employment opportunities, augmenting productivity both in the farm and non-farm sectors, increasing public health coverage to reduce morbidity and mortality and improving educational attainment and imparting skills and market-based knowledge to the population.[1]

In this chapter, we analyze various analytical and empirical aspects concerning South Asia's poverty reduction challenges. The relationship between economic growth and poverty is complex. How the structural elements at the country level can be modified to derive greater benefits of growth, particularly in the context of globalization, is a basic question. One has to go beyond the standard growth-oriented approach, emphasizing the importance of "growth plus various interventions" approach. The latter is indeed essential in the context of South Asia, which has witnessed faster growth in certain parts and sluggish growth with persistent poverty in certain other parts. The standard poverty–growth elasticity of –2 has not been experienced by several countries and actual poverty rates in some of the countries are far higher than the predicted magnitude, possibly indicating that the changes in income distribution are weakening the expected impact of growth, and/or the income distribution around the poverty line might have been changing with an increased differentiation amongst the poor. In all, the general trade liberalization through the reduction of import tariffs, and in particular, trade reform in a specific market; infrastructure investment and in particular, roads investment; population policy; poverty targeting based on cash transfers and in particular, the role of micro-finance; and food for work programmes of poverty targeting are important to assess the benefits in terms of positive changes in well-being levels.

In a computable general equilibrium framework Khan (2006) after carrying out certain modifications to suit the South Asian context shows that trade liberalization can lead to reduction in poverty both at the national level and at the level of various household groups though the extent of reduction is limited. Since the model allows reverse migration from urban to rural areas and since beyond a certain point further reduction in tariff levels becomes attenuated for most groups, the impact in terms of poverty reductions is limited. Trade liberalization with labor market flexibility reveals a slightly faster decline in the head count measure of poverty though the increment is rather small. With flexibility in the formal sector labor market, rural poor seem to gain more compared to the situation of inflexibility. However, it may not be justified to argue for flexibility in the labor market since the model does not allow one to capture the multi-dimensional aspect of poverty. A mere improvement in the headcount measure of consumption/income poverty does not reflect on the health, education and other important dimensions of well-being. Whether these aspects actually improve or rather deteriorate are more fundamental questions, which need to be thoroughly investigated. The author argues for enhancing the capabilities of the poor and also perusal of other growth enhancing reforms since trade reforms do not bear significant effects in terms of poverty reduction.

Jalilian and Weiss (2006) assess the effect of infrastructure on total factor productivity growth and poverty both. Based on the principal component analysis different infrastructure variables have been combined to construct the infrastructure quantity and quality index. After controlling for income growth, initial inequality and inflation, the impact of infrastructure on poverty is assessed. Though there is no significant relationship directly between poverty and infrastructure index in this equation, after physical infrastructure index is interacted with human capital the role of infrastructure in reducing poverty becomes evident. Using the household survey data Warr points out distinctly that around one-sixth of the reduction in rural poverty incidence can be attributed to road improvements. Similarly access to all-weather roads is seen to reduce rural poverty by seven percentage points.

Orbeta (2006) highlights the adverse effect of population factor in explaining poverty. With a rise in the number of children, family savings, labor force participation and earnings for mothers decline implying that poverty shoots up. In the light of the findings the author argues for an active population policy, suggesting that family planning facilities for those who need them should be an integral part of poverty reduction efforts. However, the idea of willingness to accept family planning facilities is a bit unrealistic because most of the low-income households view large family size or more children as an asset which would yield income in the near future. Awareness programmes and education are some of the important strategies in dealing with this situation of poverty accentuated by a high fertility rate.

Arif (2006) assesses the effectiveness of two important anti-poverty programmes in Pakistan. Given the rise in the poverty ratio and the worsening of income distribution during the nineties the new legislation insists on incurring an expenditure of 4.5 percent of GDP in the "pro-poor" sector. *Zakat* in this context is a programme targeted at the poor and the destitute. This is a charge, which is deducted at source by financial institutions and disbursed to the needy on the recommendation of the local communities. On the other hand, the Lady Health Worker Programme aims at providing basic health services to the rural and urban poor through a designated female health worker. However, as the author notes, both these programmes have certain shortfalls. The former includes poor governance and the latter involves a heavy burden per health worker. Though these two programmes are quite unique to Pakistan and a large number of beneficiaries under the *Zakat* programme are actually poor, they provide respite only in the short run. Hence, many more initiatives need to be taken in terms of education and infrastructure investment, which can be beneficial only in the long run. Montgomery (2006) rather considers a more interesting issue, that is, the role of the commercially based micro-finance sector in the context of poverty reduction. The author tends to refute the view that micro-finance may not be a suitable instrument for reaching the "core poor." Based on the survey results of 3,000 households in both rural and urban areas of Pakistan findings suggest that the beneficial effects of micro-finance are stronger in the case of the poorest borrowers.

Srivastava (2006) evaluates the effectiveness of a food-for-work programme for the poor in the state of Uttar Pradesh, India. He considers the conventional definition of poverty to identify the poor and compares them with those who have been identified as poor by the villagers themselves. Surprisingly the overlaps are quite limited. The poor identified by villagers seem to possess fewer assets than those identified by the conventional definition of poverty. The author argues that the effectiveness of the food-for-work programme in terms of reaching the core poor is quite weak if the conventional definition of poverty in terms of expenditure is considered to identify the poor.

Though some of the country-specific case studies and their policy implications can be generalized across space, the structural factors, institutions and the investment patterns vary considerably across nations. Hence, a mere extension of some of the measures which have been successful in certain regions need not reap similar returns elsewhere. In a "growth plus" approach to dealing with poverty issues, more case studies, relating to participatory approach and suggesting involvement of the poor in taking initiatives and managing the programmes, need to be brought in.

The present chapter aims at bringing out the role of various factors and sectors in reducing poverty in the South Asian context. It is structured as follows. The first section focuses on population growth and poverty, the second section deals with the composition of growth that is taking place in the South Asian context and the third section refers to growth–poverty links. The fourth section examines the role of industry in reducing poverty and the fifth section highlights the new changes that are taking place within the services sector and assesses if such changes are conducive to poverty reduction. The sixth section refers to the nature of relationship between agriculture and the rural non-farm sector and the latter's role in reducing poverty. The final section examines urban poverty as a spill-over of rural poverty and argues for urban and rural anti-poverty programmes separately as poverty in one space cannot be reduced in terms of the other. Finally section makes concluding remarks.

## POPULATION GROWTH AND POVERTY

There is a general consensus that persistent widespread poverty has influences on and is in turn influenced by demographic parameters such as population growth, structure and distribution. Macroeconomic and sectoral policies have, however, rarely paid due attention to population considerations. An explicit integration of population into economic and development strategies would both speed up the pace of sustainable development and poverty alleviation and contribute to the achievement of population objectives and an improved quality of life. Development strategies must realistically reflect the short-, medium- and long-term implications of, and consequences for, population dynamics as well as patterns of production and consumption. In this chapter we take the position that the mismatches between macroeconomic changes and demographic changes can lead to adverse outcomes in terms of low productivity employment, sluggish growth in wages and deterioration in the overall well-being of the population.

A large population size and a rapid population growth both deteriorate the standard of living. Even when production takes place on a large scale and expands further with a positive growth, a large population base and/or a fast growth rate can reduce the income in per capita terms, reducing households' ability to improve their quality of life. Poor quality of population tends to retard productivity and shoot up the population growth rate. A high incidence of poverty manifests itself in a high fertility rate, as every addition to the household is perceived by the poor as a potential earning member. Given the fact that an early entry into the job market takes place in low-income households, this may raise the household income; however, it would not necessarily result in a rise in per capita income or consumption.

Despite recent declines in birth rates in many countries, large increases in population size are inevitable. Due to the youthful age structures the coming years will still bring substantial increases in population and labor supplies in absolute terms. Population movements, rapid growth of cities and the unbalanced regional distribution of population would continue. In the face of all this, countries in transition face major development obstacles some of which are related to persistence of trade balances, the slow-down in the world economy, the persistence of the debt servicing problem, and the need for technologies and external assistance. The achievement of sustainable development and poverty eradication would not be possible, therefore, without any major support from macroeconomic policies designed to provide an appropriate international economic environment, good governance, effective national policies and effective national institutions (United Nations 1994).

Widespread poverty is often accompanied by unemployment, malnutrition, illiteracy, low status of women, exposure to environmental risks, and limited access to social and health services. All these factors contribute to high levels of fertility, morbidity and mortality and to low economic productivity and population size, which raises poverty in turn. Though in some of the studies population size is not seen to influence poverty independently of per capita agricultural output, population growth does exercise an independent effect on the level of poverty.[2]

The agricultural situation in many parts of rural South Asia is one of economic stagnation resulting from population growth and its subsequent effect on land-holding, manifested in terms of increased landlessness, without being accompanied by any commensurate increase in work opportunities in either agricultural or non-agricultural activities. This implies that population growth can adversely influence the distribution of income for any given level of per capita output.

Other than the fact that a rapid population growth distributes the benefits among a large number of individuals, it also exerts tremendous pressure on the existing infrastructure. Resources required for a better quality of life in terms of human capital formation and health, are also too large and could be unavailable. Providing basic requirements to a large population base would have to be at the cost of other developmental expenditure, which could add to productivity, and thus national income.

In fact a large population size is both a cause and consequence of poverty, and reduces the economic growth too. Though economic development is expected to reduce population growth in the long run and bring improvement in the quality of population, development itself is dependent upon a reduction in heavy population pressures, lest it be increasingly difficult to remove the shortages of capital, food, foreign exchange, and skills that limit the rate of present as well as future development.

A large relative size of population located in the economically active age groups of 15 to 59, raises the supplies of labor. The limited role of the industrial sector to absorb the unskilled and semi-skilled work force on a large scale leads to a residual absorption of labor in the low-productivity informal sector and results in large-scale poverty. In other words the mismatch between supplies and demand for labor gets accentuated resulting in huge expansion in the class of working poor. On the other hand, with a rise in life expectancy this implies that old members (60 and above) from the low-income households may have to struggle in the informal sector in search of a livelihood, facing increased competition from the workers in the younger age groups.[3]

# STRUCTURAL COMPOSITION OF ECONOMIC GROWTH

Despite a gradual decline in its relative contribution to GDP over the decades agriculture still accounts for a substantial proportion of GDP in different SAARC countries. Though the share of industry increased over time, its relative size in comparison to that of the tertiary sector has been on the low side except in Bhutan. In the recent years the tertiary or service sector dominates agriculture and industry in most of the countries except Bhutan and Nepal. This is quite unique in relation to the experience of the developed nations though in comparison to the developing countries it is not uncommon. The increasing role of the government, simultaneous growth of several tertiary activities, along with manufacturing and, more importantly, the expansion in physical and financial infrastructure and the emergence of new activities accompanied by information technology and business process outsourcing services are the main factors behind such pattern of growth. In Maldives, tourism acted as the main engine of growth with a sizeable share in GDP. However, the rate of growth of per capita GDP has been quite sluggish in many of these countries. This is mainly because a large percentage of the work force is engaged in low-productivity activities. The proportion of the work force employed in agriculture is still very high, particularly in Bangladesh, India, Nepal, Pakistan and Sri Lanka. On the other hand, the percentage of workers engaged in manufacturing/industry dwindles at a considerably low level except in Maldives and Pakistan. All this tends to suggest that there has been a mismatch between the change in the structure of value added and the change in the structure of work force, with low-productivity activities being located largely in agriculture and tertiary sectors. The majority of South Asia's labor force is engaged in the informal sector, working as casual laborers and self-employed workers. The share of informal employment in India exceeds 90 percent, in Sri Lanka it is estimated to account for about two-thirds of employment and in other South Asian countries it is large as well (World Bank 2004).

Though the open unemployment rates are not necessarily high the relative size of unorganized or informal sector employment in both rural and urban areas is dominant (Mitra 1994). Among the poor the open unemployment rates are not too high because they cannot afford to remain unemployed for long. The working poor are mostly engaged in low-productivity activities with meager earnings. The composition of rural unorganized sector activities shows that a large majority would fall into the agricultural sector. Though in some of the regions the demand-induced component of the rural non-farm sector has gained momentum, in several regions it is a manifestation of supply-push phenomenon. The urban informal sector comprises both manufacturing and tertiary activities though the latter would constitute about 70 percent of the total informal sector activities conducted in the urban areas (Mitra 2001). Earnings in the informal sector are meager partly because supplies of labor exceed demand and partly because products manufactured in the informal sector are of poor quality and have a limited market. Some of the activities are purely of a residual type and this self-employment is characterized by low productivity.

The overlaps between poverty and informal sector employment and low-quality housing are expected. A large percentage of the population in SAARC countries live in slums with inadequate safe drinking water, sanitation and sewerage and electricity. The lack of basic amenities and productive employment opportunities make them prone to several fatal and contagious diseases. The lack of education and vocational training restricts their upward income mobility. In terms of the education index, adult literacy rate (age 15 and above) and the overall human development index, most of the South Asian countries (except Maldives and Sri Lanka) are characterized by low magnitudes, with implications in terms of large-scale absorption in the informal sector for sources of livelihood. The poor quality of teaching, unqualified teachers, irrelevant curriculum, lack of textbooks and other learning materials are chronic in South Asia (RIS 2002).

In East Asia the diffusion of primary education was possibly the single most important factor accounting for the reduction in poverty and income inequality. The East Asian countries in general allocated a much larger proportion of their public investment for agriculture and rural development than most other developing countries at comparable stages of their development.

This, together with universal primary education made growth broad-based and labor-intensive with skill intensity, resulting in higher growth and improved income distribution (Hashim 1998). Investment in physical and human capital with special emphasis on developing human resources and effective participation in international markets, leading to expanding employment at higher productivity, contributed to both reduction in poverty and enhancement of growth.

Inequality and poverty are highly inter-related. Higher inequality results in higher poverty and vice-versa. However, it is also possible to have a high incidence of absolute poverty despite modest or low inequality, and similarly with a very a high level of relative inequality absolute poverty could still be nominal. As income or wealth data are not often reliable, expenditure inequality is taken to be a proxy. But expenditure inequality could be a gross underestimate of the actual inequality (income or wealth) prevailing in any economy partly because the expenditure statements of the top income groups do not reflect their actual purchasing power if a large part of the income is saved. Given the level of inequality if growth takes place, it is expected to reduce poverty, and given the level of per capita income a rise in inequality results in a higher level of poverty.

## ECONOMIC GROWTH AND POVERTY

Whether economic growth is sufficient to reduce poverty is, as mentioned above, an important question from a policy point of view. Economic growth in some of the South Asian countries has picked up sizeably in the past two decades or so, and hence the possible connection between growth and poverty has generated new interest. From the available literature we note that the poor may benefit from economic growth only indirectly and, hence, the proportional benefits of growth going to the poor could be always less than those accruing to the non-poor. In other words, the positive effects of growth on the poor tend to get offset by the adverse effects of rising inequality emerging in the process of economic growth in the initial stages. However, if economic growth is accompanied by a decline in inequality, the poor benefit more than the non-poor—the situation is described in the literature as pro-poor growth (see Kakwani 2000). Even when inequality rises, observed poverty may still decline if the growth effect dominates the inequality effect, that is, the extent of fall in poverty due to growth is larger than the rise in poverty due to a rise in inequality.

In the Indian context, with wide regional variations in terms of socio-economic development, economic reforms have been initiated at different levels across states. Most of the economic reforms have been pursued in the industrial sector, the spread and growth of which show considerable regional variations. Availability of infrastructure, which is a strong determinant of industrial productivity and competitiveness on the one hand and occupation, mobility and earnings of the population on the other, also varies significantly across state. As an outcome of this economic growth is expected to have wide regional variation, and further the changing income distribution in the process, growth would also be different across states. Hence, it would be interesting to delineate the effects of growth and inequality separately on poverty in different states. And this could be done for the pre- and reform periods so as to identify the changes that can be attributed strongly to economic reforms.

Keeping in view these outcomes in the process of economic growth, Bhanumurthy and Mitra (2010), based on the methodology of Mazumdar and Son (2002), made an attempt to decompose the change in poverty over two time points in terms of pure growth effect (holding inequality constant), inequality effect (holding growth constant) and population shift effect. Such an analysis enables a critical analysis of the policy issues and offers a profound understanding of the reform process. However, it may be noted that in doing such an exercise only the expenditure inequality has been considered, which is a gross underestimate of income/asset inequality. Similarly the growth effect is envisaged in terms of mean effect—the mean of consumption expenditure per capita—which is again a gross under-estimate of per capita income. As data on income and its distribution are not available, these crude proxies are followed and are based on the National Sample Survey data on expenditure per capita and its size distribution. In assessing the population shift effect the percentages of population residing in urban and rural areas are considered, which in addition to rural–urban migration includes also the differentials in rural–urban natural growth of population, and the rise (fall) in the level of urbanization (percentage of rural population) due to reclassification of areas. Hence, it may not be justified to perceive the population shift effect purely in terms of rural-urban migration, which can again comprise inter-state as well as intra-state streams, not deciphered in the study.  While interpreting the results, these limitations, therefore, need to be kept in view.

As economic reforms are likely to bring in higher growth, the growth or mean effect is expected to have gone up in the reform period. Bhanumurthy and Mitra (2010) noted that the growth/mean effect dominated over the inequality effect as well as the population shift effect in most of the states during the pre-reform period, which brought in a decline in the observed poverty ratio.

In many states, the decline in the observed poverty incidence in the rural and urban areas is mainly attributed to the growth effect.

Though the inequality effect was positive in sign, suggesting that in the process of growth, inequality rose and accentuated poverty, it could not neutralize the beneficial effects of growth on poverty. Due to rising levels of urbanization (percentage of population residing in the urban areas) though urban poverty increased, the decline in the percentage of rural population ushered in more than a proportionate fall in rural poverty in several states. Thus in these states, the population shift effect helped the all-area combined poverty to decline during the pre-reform period.

In the reform period, the growth effect continued to dominate over the other two effects, and it also accounted for much of the decline in the incidence of poverty in most of the states. In urban India, although the growth effect continued to be dominant in the reform period, it is interesting to note that the adverse effect of inequality fell in this period compared to the eighties and became almost negligible. Secondly, despite a rise in the adverse effects of inequality in overall rural India in the reform period relative to the eighties, in a large number of states the inequality effect turned out to be beneficial in the sense that, like the growth effect, it too helped poverty to decline. This is possibly because growth became pro-poor in these states by generating employment opportunities. As regards the population shift effect, it continued to be beneficial at the all-India level. In other words, the fall in the incidence of rural poverty due to a decline in the percentage of rural population continued to be more than the rise in the incidence of urban poverty caused by the rise in the level of urbanization in the reform period. However, in terms of magnitude the population shifts effect for all areas fell marginally from −0.3 in the pre-reform period to −0.21 in the reform period.

On the whole, the beneficial effect of growth on poverty increased in magnitude in the reform period relative to the pre-reform period in several states both in the rural and urban areas. The adverse effect of inequality corresponding to all areas (rural–urban combined) fell in several states, though at the all-India level it went up from 1.07 percent to 1.63 percent.

It has been widely noted that economic growth varies considerably across space; even within India, states have recorded different growth rates, which do not seem to have any tendency towards convergence in the long run (Sachs, Bajpai and Ramiah 2002). Such divergence across states is noted not only in terms of growth in the state domestic product but also in terms of growth in average per capita consumption expenditure (Deaton and Dreze 2002). Among several factors that influence economic growth, industrial performance has been described as an engine of growth implying that equalization of industrial productivity can bring in equalization of economic growth across space. In stepping up the economic growth we often refer to industrialization, though from argumentative point of view any sector can result in higher levels of growth. But since agriculture, particularly in a developing economy context, is severely constrained by social, institutional and economic factors it is less likely to accelerate the economic growth. Besides, as per Kuznet's (1966) perception of modern economic growth, it is agriculture that loses its share both in terms of value added and work force in the process of economic growth. On the other hand, the tertiary sector value added cannot be interpreted similarly to the value added originating from the commodity producing sector since the factor (labor) income and value added are not distinguishable in the case of the tertiary sector. Besides, the concept of value added is not clear relating to various tertiary activities (Bhattacharya and Mitra 1991, 1997). Hence, it is the industry-led growth that has been the prime focus, and factors relating to manufacturing productivity are considered crucial for economic progress, enabling the low-income regions to catch up with their high-income counterparts within a finite time horizon.

## INDUSTRALIZATION AND POVERTY

The role of industry in generating high-productivity employment opportunities and enhancing the standard of living of the population is quite important. The unskilled and semi-skilled labor released from the agriculture sector is likely to get absorbed productively in the organized industrial sector (Hoselitz 1957). Large-scale production in the industrial sector hastens the growth of per capita income, and simultaneously generates productive employment opportunities, justifying its role "as the engine of growth". On the other hand, the lack of industrial growth and/or sluggish absorption of labor in this sector due to adoption of capital intensive technology and labor market rigidities can lead to the growth of low-productivity informal sector activities and poverty (Davis and Golden 1954; Mitra 1994).

The concept of growth with productive employment generation has important implications in terms of industrialization and the technology adopted in the industrial sector. The large spread of the industrial sector, and adoption of labor-intensive technology can create demand, which may absorb a large fraction of labor available to the non-agricultural sector. Even when the high productivity organized industry cannot absorb labor directly, ancillarisation, sub-contracting and outsourcing can also create employment outside the organized industry due to the complementary relationship between the organized and unorganized (or formal and informal) sectors (Papola 1981).

Though the relative size of the informal sector is almost equally high in both the situations of sluggish industrialization and rapid industrialization, the latter situation envisages the growth of productive activities even within the informal sector.

The other connection can be perceived in terms of the inter-sectoral wage linkages. The organized industrial wage rate, which may be high because linked to high levels of technology and productivity, can have a positive impact on wages in other activities including those in the urban informal sector. The wages across activities may be inter-connected for a number of reasons. As rapid industrialization takes place it generates several complementary activities and thus the demand for labor increases. The backward and forward linkages between activities help productivity growth in industry to spill over to other activities, thus raising the labor earnings in the rest of the economy (Shaw 1990). All this would explain how the direct and indirect effects of industrialization may have a beneficial effect on the standard of living, thus reducing poverty.

In assessing the role of industry in reducing poverty Mitra (2007), therefore, considered capital–employment ratio, productivity and wages per worker. Each of these variables is taken separately as the mechanism of their influence on poverty is somewhat similar. Higher wages benefit the workers who are directly employed in the organized industry and through inter-sectoral wage linkages they help poor, employed elsewhere, experience upward mobility. Similarly, higher productivity levels would mean higher wages to workers, and hence, lower levels of poverty. A rise in capital–employment ratio is likely to raise labor productivity and thus wages which in turn reduce poverty.

It may be argued that this specification assessing the effect of industry on poverty is an excluded variable model as many relevant variables like growth index, other development indicators and variables representing human capital formation are not included. But since the thrust of the argument lies in industry as the engine of growth, industry is taken to encompass the effect of those excluded variables. Also, one would like to assess how much of the total variation can be explained in terms of industry only. Hence, it is a deliberate choice to include the industry-specific variable only.

The gross value added per employee is significant in the case of the following industries: food and food products (20–21), beverages (22), cotton textiles (23), wool, silk (24), textile products (26), paper and paper products (28), leather and leather products (29), basic chemicals (30), non-metallic mineral products (32), basic metals (33), metal products (34), machinery and equipment (35–36), transport equipment (37), other manufacturing industries (38), and the aggregate ASI. The sign is negative, which supports the hypothesis that improvement in labor productivity in the organized industry reduces poverty. This may come through both a rise in the wage rate in the organised industry and its positive effect on the wage rate in the informal sector, where the poor are largely engaged. Capital-employee ratio is also negatively associated with poverty in a large number of industries: food and food products (20–21), beverages (22), cotton textiles (23), wool, silk (24), jute textile (25), textile products (26), paper and paper products (28), leather and leather products (29), basic chemicals (30), rubber, plastic etc. (31), non-metallic mineral products (32), basic metals (33), metal products (34), machinery and equipment (35–36), other manufacturing industries (38), and the aggregate ASI. All this indicates that higher levels of capital per head raises productivity and thus reduces poverty. Man-days per worker turn out to be significant with a negative sign for the following industry groups: food and food products (20–21), cotton textiles (23), wood and wood products (27), paper and paper products (28), basic chemicals (30), basic metals (33), machinery and equipment (35–36) and the aggregate ASI. Wages per worker also takes a negative sign and this is statistically significant except for the industry groups cotton textiles (23), jute textile (25), leather and leather product (29) and basic metals (33), where it is significant, with a wrong sign in the classical regression equation and insignificant in the fixed effect model.

In the regression of the change in the poverty incidence over time on the rate of growth of workers, the latter turns out to be significant only in a few cases: food and food products (20–21), beverages (22), wool, silk etc. (24), leather and leather products (29), non-metallic mineral products (32), and the aggregate organized manufacturing sector. The rate of growth of wages turns out to be significant only for food and food products (20–21), beverages (22), wool, silk etc. (24), paper and paper products (28), leather and leather products (29), non-metallic mineral products (32) and the aggregate ASI. Productivity growth is seen to cause larger declines in poverty only in two industry groups: beverages (22) and wool, silk etc. (24). On the whole, in terms of inter-temporal changes the role of industry in reducing poverty has been quite negligible, implying that, in reality, organized industry's performance has not been pro-poor. However, the significance of industry-specific variables in the regression of poverty level offers certain insights from long-term point of view. Considering the cross-sectional dimension of the data set the significance of the industry-specific variables suggests what role industry can play in the long run.

And from this point of view the beneficial effects of industry on poverty are noteworthy. Though industry as such does not account for any significant proportion of the work force, the indirect effects of industrialization can benefit the poor and thus industry-led growth can be considered to be pro-poor.

## SERVICE SECTOR AND POVERTY

The interpretations regarding the growth of the tertiary sector have undergone vast changes over the past four decades. In the work of Kuznets (1966), based on the historical experience of the nations the tertiary sector is seen to expand in relative terms only when development matures with a considerable rise in per capita income following from rapid industrialization. However, in the context of the developing countries the phenomenon of a relatively large tertiary sector has been widespread (Gemmell 1986). It is easy to rationalize this pattern of growth in the context of the developed countries because, following the rapid progress of industrialization, the demand for several services grows faster which in turn reduces the share of the secondary sector in the total product of the economy. But in the case of the developing countries the dominance of the tertiary sector before the secondary sector's relative size could increase to a reasonably high level did invite concerns, at least in the past.

Bhattacharya and Mitra (1990) further argued that a wide disparity arising between the growth of income from the services and commodity-producing sector tends to result in inflation and/or higher imports leading to adverse balance of trade. Mitra (1992) observed that the rapid growth of income from the tertiary sector did not necessarily reduce poverty, as it did not create new employment opportunities. In other words, the rise in the income from the tertiary sector was seen to have resulted from a rise in income of those who were already engaged in the high-productivity segment rather than the addition of income of the new entrants to the job market.

In sharp contrast to this view, factors like the increasing role of the government in implementing the objectives of growth, employment generation, and poverty reduction; expansion of defence and public administration; the historical role of the urban middle class in wholesale trade and distribution and demonstration effects in developing countries creating demand patterns similar to those of high-income countries have been highlighted to offer a rationale to the expansion of the tertiary sector (Panchamukhi et al. 1986). As the elasticity of service consumption with respect to total consumption is higher than unity even in countries with very low per capita consumption (Sabolo 1975), the rapid growth of the tertiary sector has been further rationalized in terms of a strong demand base existing in the economy. Sub-sectors like transport, communication and banking do contribute significantly to the overall economic growth. Especially the role of information technology (IT) and business process outsourcing services (BPOS) in enhancing the economic growth has been noticed widely in the post-reform period in India (World Bank 2004; Gemmel 1986).

Equally important as the effects of the sectors on growth are the effects on the well-being of the population, especially the poor. Ravallion and Datt (1996) noted that the changing composition of growth in favor of the tertiary sector has been important for poverty reduction in India as it has generated employment and simultaneously enhanced real income. The effects of the sectoral growth on poverty are strongly linked to human development. The skill requirement to participate in the new activities varies across different sub-sectors. Ravallion and Datt (2002) show that the elasticity of poverty to non-agricultural growth depends on the initial conditions of human development, i.e. the education and the health status of the population. Similarly, Thorbecke and Hong-Sang (1996) show in the case of Indonesia that the low skills needed for agriculture and some services make these sectors more pro-poor than the usually skill-intensive manufacturing.

Mitra and Schmid (2008) noted that the tertiary sector has been expanding rapidly since the eighties. With the opening up of the economies several new activities have emerged and grown. Several tertiary sector activities like trade; hotel and restaurants; transport storage and communications; and community, social and personal services are also seen to influence the aggregate growth. Considerable linkages are seen to exist among sub-sectors within both secondary as well as tertiary activities. In fact, within tertiary activities several sub-sectors—particularly those falling into the domain of infrastructure—impact on both the aggregate growth rate and industrialization. The tertiary sector, both in terms of share in GSDP as well as the growth rate, is associated with a reduction in poverty. Within the tertiary sector, particularly transport (in terms of percentage share) tends to reduce poverty. The growth rate in value added in banking is also associated with larger declines in poverty between two time points. Hence, these activities within the tertiary sector not only enhance economic growth but also tend to reduce poverty suggesting the possibilities of concomitant growth in employment. The results from factor analysis also confirm that several activities within the tertiary sector (relative size of transport and public administration) are positively associated with other development indicators like life expectancy. Growth in transport and real estate also reduces poverty. Growth in both manufacturing and some of the sub-sectors in tertiary activities raises the labor force participation rate, possibly because of expansion in employment.

On the whole, results are indicative of an association between the composition of growth and poverty reduction. A large tertiary sector, which has emerged in the Indian context much before the share of industry could dominate the value added composition, is not all that superfluous, as it was often thought to be. It has the capacity to enhance economic growth, which contributes to poverty reduction.

## RURAL NON-FARM SECTOR AND POVERTY

In South Asia poverty is considered to be basically a rural problem. Therefore, a significant gain in rural poverty reduction in this sub-region is said to be crucial to reach the international poverty reduction target. Based on the analysis and experience of the International Fund for Agricultural Development (IFAD), Thapa (2004) argues that to be successful, poverty reduction policies in South Asia must focus on the less-favoured rural areas and on the most disadvantaged sections of the rural poor (mainly women, the landless and indigenous peoples). In order to overcome disadvantages arising from remoteness, the lack of social services, insecure and unproductive jobs, and discrimination as women or ethnic minorities, the rural poor need legally secure access to productive assets (mainly land, forests and water); sustainable or regenerating agricultural technology; access to markets; opportunities to participate in decentralized resource management; and access to financial services.

What role the rural non-farm sector plays in the process of development of a country has been a matter of serious question. Its pattern of expansion has been so diverse across countries that it is difficult to explain in terms of any single uncontested framework. Apparently each region has a different story to unfold. In the literature three important stages of the rural non-farm sector transformation have been identified. Africa and South Asia are possibly in the first stage, in which there is a production or expenditure linkage with agriculture, and not much rural–urban links. A tendency towards a greater mix of situations is seen in Latin America, where the non-farm sector includes activities based on linkages with agriculture as well as others that are separate (e.g. tourism, mining and service sector activities). On the other hand, East Asia is in the third stage, where urban-rural links are stronger as manifested in terms of more advanced forms of business linkages, such as sub-contracting arrangements and labor commuting.

Unfortunately, the rural non-farm sector in several parts of South Asia is characterized by low productivity and it does not seem to provide a sustainable livelihood. On the other hand, the components which have grown in response to demand do not tend to derive the growth stimulus from the rural sector itself. In an attempt to reduce the diseconomies of scale and to take advantage of the positive externalities available within the large cities, the urban activities are spilling over rapidly to the rural hinterland of the big cities. There is thus a tremendous change that is taking place in the land use pattern, away from agriculture towards commercial activities. Besides, these urban activities relocated in the rural areas are less likely to provide job opportunities to the rural job seekers as the skill mismatches are phenomenal.

From the policy point of view the productivity growth in the rural non-farm sector deserves a special mention. The poor income potential of agriculture has become a serious concern in a country where large numbers of citizens are farmers and more often small farmers. Several studies have noticed a relatively lower incidence of out-migration from the rural areas that showed improved performance of the non-farm sector. This in turn reduces the excessive pressure on urban infrastructure. Hence, if the urban areas—rather than a selected few cities—which draw much of the migrants from the rural areas, have to match the world class, the development disparities across space have to be reduced. It is not just providing urban services in the rural areas—more importantly, productive employment opportunities have to be created in a big way. A viable non-farm sector would be an important answer.

Usually large farmers access productive avenues in the non-farm sector. However, education for rural households (as has been the case in the Philippines), has a major impact on non–farm employment. Besides, rural infrastructure is immensely important. Usually those who have access to capital in the rural areas experience a steady flow of income from the non-farm activities. In this context micro-finance can be critically important: while keeping interest rates at affordable levels is essential, the lenders' risk in the rural setting needs to be recognized as well. A balance between the two requires the regulators' attention while not derailing the process, as feared in the recent Andhra Pradesh incidents.

The implications of the nexus between traditional agriculture and its supplement in the form of the non-farm sector and particularly the concern for food security loom large in the process. Indeed, to the extent that there is a backward linkage and such activities can be located on the supply chains of agro-goods the non-farm sector is likely to improve agricultural prices through the demand side. However there is a possibility of labor shortages arising at critical stages of farm activities, leading to wage increases. The net outcome is not totally clear. While farm mechanization and a search for efficient methods may be a response so that crop production is not hurt, adverse effects on production are possible when extension and research do not receive adequate attention. In this context the impact of NREGA in India on farm wages, food prices and production can deserve investigation as a starting point. On the contrary, to the extent that the withdrawn labor is truly surplus in the "Lewisian" sense or non-farm activities are scheduled in line with farm timetables, the apprehension may be misplaced.

Probably of greater significance is the demand for land which is known as a scarce resource in rural India. While not as demanding as agriculture, non-farm activities also require space. This calls for rational planning of land use based on fertility, water endowments and logistical requirements. Land distribution schemes for the poor and landless and ways to improve efficiency in resource use in agriculture are also considered to be effective in this direction.

A number of studies in the past have focused on determinants of the non-farm sector growth. A positive relationship between agricultural growth productivity and share of non-agricultural employment is taken to substantiate the hypothesis of agriculture-led growth (Unni 1991). In favour of this hypothesis Unni (1991) also cited the positive impact of land concentration, rural incomes, and cropping pattern (inclined towards the non-food crops) on the proportion of male work force engaged in non-agricultural activities. An increase in agricultural productivity can raise non-agricultural employment either by raising the demand for non-agricultural products and services or through a residual absorption of labor displaced from agriculture because of mechnisation, into non-agricultural activities. From a simple correlation between the two variables it may not be therefore possible to conclude that a demand linkage of rural industries exists. However, in the context of rural industries in particular, a positive association between their performance and agricultural productivity (or the growth rate of agricultural output) is seen to be a reflection of the positive impact of rising purchasing power and investible resources generated by the agricultural sector (Papola 1987). On the other hand, demand and production linkage between agriculture and non-agriculture is said to be weak because large farmers tend to demand goods, which are produced in urban areas (Kumar 1993). The HYV technology is also urban based. Rather a strong association between unemployment and non-agricultural employment has been noted by Kumar (1993).

Shukla (1991) observed that the aggregate non-farm employment, particularly the manufacturing employment, in the rural sector varies positively with urbanization. Industrial dispersal in the rural areas around the periphery of the big cities—which is quite limited in nature—may be attributed to the diseconomies of the agglomeration, measures adopted for controlling environmental pollution, scarcity and high price of urban lands, problems of labor organization in large urban centers and so on. Subsequently, these villages, as Kundu (1992) argues, produce commodities and services quite similar to those produced in the urban localities and tend to get integrated into the national market. However, he maintains that only in agriculturally prosperous districts are non-agricultural activities in rural and urban areas found to be highly inter-related. Since land and labor productivity are not strongly related to non-agricultural activities in the rural or urban areas of these districts, rural-urban linkages are said to derive their strength from the development dynamics in the region and they need not necessarily stem from labor or land productivity (Kundu 1992).

Acharya and Mitra (2000) noted that unorganized manufacturing employment in the rural areas is negatively influenced by both urbanization and agricultural value added per rural population. In other words, with a rise in urbanization, manufacturing tends to shift from their rural location to urban areas, which does not necessarily mean physical transfer, rather it could be an outcome of reclassification of areas. Improvements in agricultural production are, however, likely to generate demand for manufacturing goods and therefore raise employment in this sector; given the argument of shifting of activities towards "agglomerations" or a reclassification of areas, this effect probably did not show up. On the flip side it can also be argued that a rise in agricultural incomes perhaps generates demand for urban-based non-agricultural goods rather than those manufactured in rural settings. Both rural poverty and agricultural labor are found to influence employment in rural manufacturing positively, lending support to the "residual sector" hypothesis. However, the literacy coefficient bears a positive sign, favoring the view that education facilitates occupational diversification.

In the case of wholesale and retail trade the coefficients of urbanization and agricultural labor show negative and positive signs, respectively. The latter is indicative of a residual sector growth in trading, which is prompted by the fact that entry barriers in terms of skill etc., do not exist in this sector. The total non-farm sector employment, however, showed a positive association with infrastructure and a negative correlation with rural poverty.

It is difficult to suggest any clear-cut demarcation between agriculture and non-agriculture activities in the rural areas. Particularly among the subsidiary status workers some of the self-employed and casual workers can be found in both the sectors. Even some of the principal status workers in agriculture or non-agriculture have a second subsidiary activity in agriculture. The work pursued on the basis of principal status does not necessarily yield a high income thus compelling some to augment their earnings by working in the capacity of subsidiary status in the agriculture sector as such possibilities exist to a larger extent in this sector than the non-agriculture sector. In terms of wages and incidence of poverty also it has not been possible to suggest that all rural non-farm workers are invariably better off than their agricultural counterparts. The decomposition exercise is again indicative of certain urban-based activities spilling over to the nearby rural areas due to space constraints and rising diseconomies rather than revealing the process of rural transformation originating from factors that determine rural development (Mitra 2002).

As Thapa (2004) points out the rural poor in South Asia are characterized by a number of general economic, demographic and social features, the most common one being landlessness or limited access to land and other productive resources. Poor rural households are characterized by larger families, with higher dependency ratios, lower educational attainment and higher underemployment. They lack basic amenities like piped water supply, sanitation and electricity and their access to credit, inputs and technology is severely limited. Lack of information about markets, lack of business and negotiating experience, lack of a collective organization and informal networks that they use to collect information pertaining to the job market result in sluggish or lack of upward mobility in socio-economic terms. In addition low levels of social, financial and physical infrastructure raise their vulnerability to natural crises and reduce their productivity levels substantially.

## URBAN POVERTY: A RURAL SPILL-OVER?

Population mobility across space is an outcome of economic growth too. The spatial composition of growth reflected in terms of rural–urban development disparity motivates people to shift to areas with better employment prospects. As total poverty is a weighted average of rural and urban specific poverty ratios, the net effect of population mobility on poverty depends on changes in its rural and urban components. Since economic reforms are more urban based, the spatial composition of growth is expected to change, resulting in a migration of population from rural to urban areas. The decline in the incidence of poverty (rural-urban combined) depends on whether urban employment opportunities are large enough to absorb the increasing supplies of labor from the rural areas. A large number of empirical studies exist to suggest that rural migrants have been able to escape poverty though they could not graduate to the urban formal sector (Banerjee 1986; Mitra 1994; Papola 1981). Even when the incidence of urban poverty rises due to rural–urban migration, the decline in the combined poverty ratio may be evident with a fall in the rural poverty incidence occurring in response to out-migration. This is precisely because the weight of urban poverty is much less in the combined poverty compared to the rural poverty. There have been many important studies which referred to the "spill-over" effect of rural poverty while commenting on urban poverty. Dandekar and Rath (1971) observe: "The character of urban poverty is the consequence of the continuous migration of the rural poor into the urban areas in search of a livelihood, their failure to find adequate means to support themselves there and the resulting growth of pavement and slum life in the cities." The proponents of the 'over urbanization' argument (Hoselitz: 1953, 1957) in their attempts to explain the low industry–urban ratio in the developing countries, perceived the process of labor movement from the rural areas and their subsequent absorption in the low-productivity urban informal sector.

The Harris-Todaro (1970) framework in its more formal attempt to explain the rural-to-urban migration mainly in terms of rural-urban expected earning differentials implies that assetless and jobless ones in the rural areas will be attracted more to the urban areas. Their failure to be absorbed in the high-productivity sector demonstrates that they effectively transfer their poverty from the rural to the urban areas. It is in this context that this chapter attempts an' empirical verification of the 'over–urbanization' thesis conceptualizing urban poverty mainly in terms of the inflow of rural poor into the urban areas in search of jobs and their subsequent residual absorption in low productivity activities. However, the critics of the "over-urbanization" argument view it as "sweeping the urban ills under the rural carpet" as the underlying excess-supply–limited-demand paradigm is too simplistic to tackle this complex issue.

As far as the association between rural and urban poverty is concerned the empirical results demonstrate a positive link between them (Mitra 1992). Rural poverty has a tendency to increase the urban in-migration for employment which, in turn, expands the relative size of urban informal sector employment. The residual absorption of labor in the low-productivity informal sector reduces the consumption expenditure per capita and thus inflates the ratio of urban poverty. Since industrial employment affects the share of informal sector employment negatively and the latter responds positively to migration from the rural areas, it is quite likely that among the migrant workers a large majority are engaged in the informal sector. The above findings tend to lend support to the essence of the "over-urbanization" thesis. However, the elasticity of urban poverty with respect to rural poverty as estimated at mean values from the reduced form coefficient matrix is only 0.05 and 0.07 as estimated in different variants of the model. Further, recalling the modest rural-to-urban migration rates in the Indian context, it is not correct to interpret urban poverty as purely a spill-over of rural poverty. In fact, many of the urban poor have been located in the urban areas for a considerable period of time—the recent arrival of rural-to-urban migrants just adds to the existing magnitude of urban poverty. Therefore, viewing urban poverty entirely as a reflection of rural poverty can be quite misleading. In fact, in an earlier work (Mitra 1988), we noted that even in the major metropolitan cities all the poor were not rural migrants. However, it may be equally erroneous to ignore all the linkages existing between rural and urban poverty and abandon the "over-urbanization" stream of argument altogether. In fact, if the size of the informal sector employment is already large due to a high natural growth of population in the urban areas, the percentage of migrant workers to total workers in the informal sector can turn out to be nominal. To recapitulate, Mitra (1992) indicates that there exists a statistically significant and positive relationship between urban and rural poverty. This is consistent with the spillover hypothesis. However, the elasticity of urban poverty with respect to rural poverty turns out to be low in magnitude. Thus for the reduction of urban poverty one has to realize the importance of urban employment programmes.

## WHERE TO GO?

Poverty in the South Asian context is a persistent phenomenon and the success of poverty reduction in South Asia is indeed crucial for achieving the Millennium Development Goal of halving poverty by the year 2015 (Thapa 2004). Several lines of policy directions emerge from the foregoing discussion though in reality only a limited few have been pursued. Several studies have brought out systematically the limitations of the existing programmes.

Many multilateral and bilateral donors have accorded high priority to poverty reduction in their development assistance programmes, particularly in response to the Millenium Development Goals (MDGs). For example, the World Bank proposes to attack poverty in three distinct ways: promoting opportunities, facilitating empowerment, and enhancing security (see Thapa 2004). The Asian Development Bank (ADB) aims at pursuing economic growth, human development and sound environmental management. UNDP focuses on human capital formation in order to empower the poor and reduce inequality along the gender line as well (see Thapa 2004). The ILO recommends "decent work" for the poor, and emphasizes the importance of social security floor for those who are engaged in low-productivity activities. While pursuing policies relating to labor market deregulation and introducing various mechanisms of labor market flexibility, the national governments must introduce a social security net for the poor.

For rural poverty reduction, the access of the poor to productive resources (land, water, forests), technology, financial services, and markets have to improve and strategies need to be adopted to strengthen their capacity and their organizations. Also rural poverty reduction efforts should focus on the less-favored areas (dry-land for example) and on socially marginalized groups including women, child labor and elderly populations engaged in physical and labor intensive activities. On the other hand, urban poverty programmes have to be adopted separately and exclusively instead of interpreting the urban poor as a spillover from the rural areas. The urban poor who are largely engaged in the low productivity informal sector, comprising heterogeneous activities, are more difficult to deal with. While the provision of basic amenities like safe drinking water, sanitation and land tenure and health facilities are important for improving their well-being, attention also needs to be given to skill formation and training, quality education and mechanisms for disseminating information on job market and the availability of special schemes for the urban poor.

In addition, it is essential to implement productivity enhancement schemes (for the informal sector workers) in terms of technology up-gradation, credit and marketing assistance and information relating to new products for which demand is expanding. In the short run, the employment guarantee programmes need to be adopted on a large scale for the urban poor. Often they use informal networks to access jobs which land them in activities characterized by excess supplies of labor, resulting in low wages (Mitra 2004). Thus there is hardly any possibility for them to experience upward mobility. Intervention in terms of dissemination of job market information can reduce mismatches between demand for and supply of labor.

How the labor intermediary's share in the earnings of the contract labor can be kept within reasonable limits is another matter of concern. How informalization and growth strategies can be integrated with issues relating to human rights, in which areas the social protection floor can be strengthened and in what ways state delivery of services to the poor can be made efficient are pertinent questions.

Another important challenge to poverty reduction in South Asia includes non-income components. It is not just income/consumption poverty that is worrisome, rather various social, cultural, demographic, health and education specific variables determine well-being levels. Instances exist to suggest low consumption poverty which is however accompanied by poor well-being levels. Each of these areas requires active and effective policy intervention.

## NOTES

1. http://www.unescap.org/pdd/publications/RuralPoverty/ChIV.pdf
2. Walle (1985). In explaining the variations in poverty, even after accounting for the impact of agricultural output growth and prices on poverty, the time trend has been found to be significant, suggesting the effect of other variables like population.
3. Several other implications are also important. As with a high fertility rate, the percentage of population in the young age brackets increases with ever-greater pressure on education and training, reproductive and child healthcare facilities and an overall increase in the dependency ratio, a decline in the mortality rate ushers in a rise in the number of old people. The needs of young olds and the olds are varied. Pressures on healthcare facilities are enormous. Particularly in low-income households, the rising number of old members would mean increasing dependency and more pressure on the public healthcare system.

## BIBLIOGRAPHY

Acharya, Sarathi, and Arup Mitra. "The Potential of Rural Industries and Trade to Provide Decent Work Conditions: A Data Reconnaissance in India." SAAT Working Papers, New Delhi: SAAT-ILO, 2000.

Adelmann, Irma, and Cyntia Taft Morris. "Economic Growth and Social Equity in Developing Countries." Stanford: Stanford University Press, 1973.

Aghion, Philippe, Caroli Eve and Cecilia Garcia-Penalosa. "Inequality and Economic Growth: The Perspective of the New Growth Theories." *Journal of Economic Literature* 37, no. 4 (1999):1615-1660.

Aghion, Philippe, and Peter Howitt. "Endogenous Growth Theory." Cambridge: MIT Press, 1999.

Ahluwalia, M. S. "Rural Poverty and Agricultural Performance in India." *Journal of Development Studies* 14, no. 3 (1978): 298-23.

Alesina, Alberto, and Dani Rodrik. "Distributive Policies and Economic Growth," *Quarterly Journal of Economics* 109, no. 2 (1994).

Arif, G. M. "Poverty Targeting in Pakistan: The Case of *Zakat* and the Lady Health Worker Programme," in *Poverty Strategies in Asia: A Growth Plus Approach*, ed. John Weiss and Haider A. Khan. UK: A Joint Publication of the Asian Development Bank Institute and Edward Elgar Publishing, 2006.

Asian Development Bank (ADB). *Fighting Poverty in Asia and the Pacific: The Poverty Reduction Strategy.* Manila: ADB, 1999.

Banerjee, Biswajit. *Rural to Urban Migration and the Urban Labour Market.* New Delhi: Himalaya Publishing House, 1986.

Bhalla, G. S. *Trends in Poverty, Wages and Employment in Rural India in Empowering Rural Labour in India*, ed. Radha Krishna R. and Alakh N. Sharma. New Delhi: Empowering Rural Labour, Institute for Human Development, 1998.

Bhalla, A. S. "The Role of Services in Employment Expansion." *International Employment Review* 101, no. 5 (1970): 519- 39.

Bhanumurthy, N. R., and Arup Mitra. "Globalisation, Growth and Poverty in India," ed. Machiko Nissanke and Erik Thbecke. *The Poor under Globalization in Asia, Latin America and Africa.* New York: UNU-Wider, Oxford University Press, 2010.

Bhattacharya, B. B., and Arup Mitra. "Excess Growth of Tertiary Sector in Indian Economy, Issues and Implications." *Economic and Political Weekly*, November 3 (1990): 2445-50.

———. "Changing Composition of Employment in Tertiary Sector: A Cross-Country Analysis." *Economic and Political Weekly*, 15 March, 1997.

———."Agriculture-Industry Growth Rates: Widening Disparity: An Explanation." *Economic and Political Weekly*, Aug. 26, 1989.

Chen, Shaohua, Gaurav Datt, and Martin Ravallion. "Is Poverty Increasing in the Developing World?" *Review of Income and Wealth. December* 40, 4 (1994): 359-76.

Chen, Shaohua, and Martin Ravallion. "How Did the World's Poorest Fare in the 1990s?" World Bank Policy Research Working Paper No. 2409. Washington, DC: World Bank, 2000.

Datt, Gaurav, and Martin Ravallion. "Farm Productivity and Rural Poverty in India." *The Journal of Development Studies* 34, 4 (1998): 62-85.

———. "Growth and Redistribution Components of Changes in Poverty Measure: A Decomposition with Application to Brazil and India in the 1980s." *Journal of Development Economics* 38 (1992): 275-95.

———. "Is India's Economic Growth Leaving the Poor Behind?" *Journal of Economic Perspectives* 16, 2 (2002): 89-108.

Davis, K. and H. Hertz. Golden. "Urbanization and the Development of Pre-Industrial Areas," *Economic Development and Cultural Change* 3, no.1 (1954): 6-26.

Deaton, A., and Jean Dreze. "Poverty and Inequality in India: A Re-Examination." *Economic and Political Weekly*, September 7, 2002.

Dollar, David, and Aart Kraay. "Growth Is Good for the Poor." *Journal of Economic Growth, September* 7, 3 (2002): 195- 25.

Drèze, Jean, and Mamta Murthi. "Fertility, Education, and Development: Evidence from India." *Population and Development Review* 27, no.1 (2001): 33-63.

Fields, Gary S. *Distribution and Development: A New Look at the Developing World*. New York: Russel Sage Foundation, 2001.

Foster, J., J. Greer, and E. Thorbecke. "A Class of Decomposable Poverty Measures." *Econometrica* 52, no. 3 (1984): 761-66.

Gemmel, N. "*Structural Change and Economic Development: The Role of Service Sector.*" Hampshire: Macmillan Press, 1986.

Goldar, Bishwanath. "Trade Liberalisation and Real Wages in Organized Manufacturing Industries in India," ed. A. Karnik and L. G. Burange. *Economic Policies and the Emerging Scenario: Challenges to Government and Industry*. Mumbai: Himalaya Publishing House, 2004.

———. "Employment Growth in Organized Manufacturing in India," *Economic and Political Weekly* 35, no. 14 (2000): 1191-1195.

Harris, J. R., and M. P. Todaro. "Migration, unemployment and development: A two-sector analysis," *American Economic Review* 61, no.1 (1970): 6126-141.

Hashim, S. R. "Foreword" in ed. A. Chandra, H. Mund, T. Sharan and C. P. Thakur. *Labour, Employment and Human Development in South Asia*. Delhi: B.R. Pub Corp, 1998.

Hoselitz, B. F. "Urbanisation and Economic Growth in Asia," *Economic Development and Cultural Change* 6, no. 1 (1957).

———. "Generative and Parasitic Cities." *Economic Development and Cultural Change* 3, (1955).

———. "Urbanization and Economic Growth in Asia." *Economic Development and Cultural Change,* vol. 5 (1957).

———."The Role of Cities in the Economic Growth of Underdeveloped Countries," *Journal of Political Economy,* Vol. 6l (1953).

International Labor Office (ILO). "The Dilemma of the Informal Sector," Report of the Director General (Part –I), 78th Session. Geneva, 1991.

————. "Labour and Social Trends in Asia and the Pacific 2006: Progress Towards Decent Work." Geneva, 2006.

Jalilian, H., and J. Weiss. "Infrastructure and Poverty: Cross-country Evidence," in. *Poverty Strategies in Asia: A Growth Plus Approach*, ed. by John Weiss and Haider A. Khan. UK: A Joint Publication of the Asian Development Bank Institute and Edward Elgar Publishing, 2006.

Kakwani, Nanak. "Growth and Poverty Reduction: An Empirical Analysis."*Asian Development Review* 18, no. 2 (2000): 74-84.

Khan, H. A. "Macro–modeling of Poverty and the Dual–Dual Model," in *Poverty Strategies in Asia: A Growth Plus Approach*, ed. by John Weiss and Haider A. Khan. UK: A Joint Publication of the Asian Development Bank Institute and Edward Elgar Publishing, 2006.

Kumar, Alok. "Rural Non–Farm Employment: A Static and Dynamic Study of Inter–State Variations," *Indian Journal of Labour Economics* 36, no. 3 (1993).

Kundu, Amitabh. *Urban Development and Urban Research in India*. New Delhi: Khama Publishers, 1992.

Kundu, Amitabh, Niranjan, Sarangi, and Bal, Paritosh Dash. Rural Non–farm Employment: An Analysis of Rural Urban Interdependencies," Working Paper 196. UK, London: Overseas Development Institute, 2003.

Kuznets, S. *Modern Economic Growth: Rate, Structure and Spread*. New Haven: Yale University Press, 1966.

Mazumdar, D., and Hyun, Hwa. Son. *Vulnerable Groups and the Labour Market Micro Data from Delhi Slums*. Delhi: Manohar Publications, 2002.

Mitra, Arup. "Growth and Poverty: The Urban Legend." *Economic and Political Weekly*, 28 March 1992.

————. "Rural Non Farm Sector: Issues and Facts." *Manpower Journal* 38 (2-3): 2002.

————. "Occupational Choices, Networks and Transfers: An Exegesis Based on Micro Data from Delhi Slums." *Economic and Political Weekly* 39, no. 06, 2003.

————. "Spread of Slums: The Rural Spill-Over?" *Demography India* 17, no. 1 1988.

————. "Duality, Employment Structure and Poverty Incidence: The Slum Perspective." *Indian Economic Review* 25, no. 1, 1990.

————. "Rural Non-Farm Employment, Poverty and Women." *Indian Journal of Labour Economics* 36, no. 3, 1993.

————. *Urbanization, Slums, Informal Sector Employment and Poverty: An Exploratory Study*. Delhi: B.R. Publishing Corporation, 1994.

————. "Employment in the Informal Sector," in ed. A. Kundu and A. N. Sharma. *Informal Sector in India: Perspectives and Policy*. Delhi: IHD, IAMR, 2001a.

————."The Urban Labour Market in India: An Overview." *International Journal of Employment Studies* 9, no. 2, (2001b).

————."Informal Sector, Networks and Intra-City Variations in Activities: Findings from Delhi Slums." *Review of Urban and Regional Development Studies* 16, no. 2, July 2004.

————. "Industry and Poverty: Evidence from Indian States". *Indian Journal of Labour Economics*, April-June, 2007.

Mitra, Arup, Aristomene Varoudakis, and Marie Ange Veganzones Varoudakis "Productivity and Technical Efficiency in Indian States' Manufacturing: The Role of Infrastructure." *Economic Development and Cultural Change* 50, no. 2, 2002: 395-426.

Mitra, Arup, and J. P. Schmidt. "Growth and Poverty in India: Emerging Dimensions of the Tertiary Sector." *The Service Industries Journal* 28, no. 8, October 2008.

Montgomery, H. "Serving the Poorest of the Poor: The Poverty Impact of the Khushhali Bank's Microfinance Lending in Pakistan," in *Poverty Strategies in Asia: A Growth Plus Approach*, ed. by John Weiss and Haider A. Khan. UK: A Joint Publication of the Asian Development Bank Institute and Edward Elgar Publishing, 2006.

Orbeta Jr A.C. "Poverty, Vulnerability and Family Size: Evidence from the Philippines", in *Poverty Strategies in Asia: A Growth Plus Approach*, ed. by John Weiss and Haider A. Khan. UK: A Joint Publication of the Asian Development Bank Institute and Edward Elgar Publishing, 2006.

Palanivel, T. "Rising Food and Fuel Prices in Asia and the Pacific: Causes, Impacts and Policy Responses." *UNDP Regional centre in Asia and the Pacific*. Sri Lanka: Mimeo, UNDP Regional Centre in Colombo, 2008.

Panchamukhi, V. R., R. G. Nambiar, and R. Metha. *Structural Change and Economic Growth in Developing Countries*. New Delhi, 1986.

Papola, T. S. *Urban Informal Sector in a Developing Economy*. Vikas Publishing House, 1981.

————. "Rural Non-Farm Employment: An Assessment of Recent Trends," in R. Islam (ed.) *Rural Industrialisation and Employment in Asia*. New Delhi: ILO/ARTEP, 1987.

Ravallion, Martin. "Growth and Poverty: Evidence for Developing Countries in the 1980s." *Economics Letters* 48, 3-4, (1995): 411-17.

Ravallion, Martin, and Gaurav, Datt. "How Important to India's Poor Is the Sectoral Composition of Economic Growth?" *World Bank Economic Review* 10, no.1 (1996): 1-25.

———. "Why Has Economic Growth Been More Pro-poor in Some States of India Than Others?" *Journal of Development Economics* 68, no. 2 (2002): 381-400.

SAARC. "SAARC Regional Poverty Profile." Kathmandu: SAARC Secretariat, 2003.

Sachs, J. D., N. Bajpai, and A. Ramiah. "Understanding Regional Economic Growth in India." CID Working Paper No. 88. Cambridge, MA: Center for International Development at Harvard University, 2002.

Sabolo, Y. *Indian Service Industrie.* Geneva: ILO, 1975.

Shaw, A. "Linkages of Large Scale, Small Scale and Informal Sector Industries: A Study of Thane-Belapur." *Economic and Political Weekly*, February 17-24, 1990.

———. "Peri-Urban Interface of Indian Cities: Growth, Governance and Local Initiatives." *Economic and Political Weekly*, January 8, 2005.

Shukla, V. "Rural Non-Farm Activity: A Regional Model and Its Empirical Application to Maharashtra." *Economic and Political Weekly*, November 9, 1991.

Srivastava, P. "The Role of Community Preferences in Targeting the Rural Poor: Evidence from Uttar Pradesh," in *Poverty Strategies in Asia: A Growth Plus Approach*, ed. by John Weiss and Haider A. Khan. UK: A Joint Publication of the Asian Development Bank Institute and Edward Elgar Publishing, 2006.

Thapa, G. "Rural Poverty Reduction Strategy for South Asia," ASARC Working Paper 2004-06. Rome: International Fund for Agricultural Development, 2004.

Thorbecke, Erik, and Hong Sang Jung. "A Multiplier Decomposition Method to Analyze Poverty Alleviation." *Journal of Development Economic*, 48, no. 2 (1996): 279-300.

United Nations. "Programme of Action Adopted at the International Conference on Population and Development." *Population and Development,* vol. 1, 5-13 September 1994. Cairo: United Nations Publications, 1995.

Unni, J. "Non–Agricultural Employment, Livelihoods and Poverty in Rural India," Working Paper No. 88. Ahmedabad, Gujarat: Institute of Development Research, 1997.

———. "Regional Variations in Rural Non-Agricultural Employment: An Exploratory Analysis," *Economic and Political Weekly* 26, no. 3, 1991.

Walle, D. V. "Population Growth and Poverty: Another Look at the Indian Time Series Data," *Journal of Development Studies* 21, no. 3, 1985.

Warr, Peter. "Roads and Poverty Reduction in Lao PDR", in *Poverty Strategies in Asia: A Growth Plus Approach*, ed. by John Weiss and Haider A. Khan. UK: A Joint Publication of the Asian Development Bank Institute and Edward Elgar Publishing, 2006.

————."Poverty Reduction and Economic Growth: The Asian Experience." *Asian Development Review* 18, no. 2 (200): 131–47.

Willimason, J. "Migration and Urbanization," in ed. H. Chenery and T. N. Srinivasan. *Handbook of Development Economics* vol. 1, Elsevier Science Publishers B.V., 1988.

World Bank. *Sustaining India's Services Revolution: Access to Foreign Markets, Domestic Reform and International Negotiations.* India: South Asia Region, 2004.

————. *Country Programs and Policies to address Rising Food Prices Processes.* Washington DC: World Bank, 2008.

## 7

# OFFICIAL DEVELOPMENT ASSISTANCE: DOES IT REDUCE POVERTY?

*John Weiss*

Official Development Assistance (henceforth Aid) comes in various forms—project finance, technical assistance, general budget or balance of payments support, emergency assistance and more recently debt relief. In many countries its value in absolute terms is relatively modest and in some is considerably less than private flows such as foreign direct investment, portfolio investment and remittances. In the past it has been allocated between recipient countries on a variety of grounds including not just need, as reflected in average income or poverty levels, but also strategic political considerations (although this has lessened since the end of the Cold War), past colonial ties and commercial links. An analysis of data for 2008 from the World Development Indicators reveals no correlation between a country's income per capita and its aid receipts per capita, although Aid plays a much greater role in the poorest countries with a fairly close negative correlation between income per capita and the share of Aid in GDP.[1] There are various biases in Aid allocation with small countries receiving relatively large amounts of Aid per capita and sub-Saharan Africa and Eastern Europe and Central Asia benefiting relative to South Asia (see table 7.1). In sub-Saharan Africa there were fifteen countries in which Aid was more than 10 percent of GDP in 2008.

Table 7.2 summarizes uses of Aid under the broad headings applied by the Development Assistance Committee (DAC) of the OECD. It distinguishes between bilateral national donors (DAC members), EU institutions and the multilateral banks (the World Bank and the Regional Development Banks). Broadly speaking multilaterals now focus heavily on economic infrastructure—principally transport and energy—with bilateral donors providing relatively more in the way of general programme support, humanitarian relief and food Aid.

Given that since the late 1990s the international development community has defined the primary objective of Aid as bringing down global poverty—and highlighted this as the first of the Millennium Development Goals (MDGs)—it is a highly legitimate question to ask what has been achieved by Aid in this context. Superficially it might appear relatively little. Aid has been criticized for variously being too small to make a difference, for encouraging weak governance, for being delivered slowly with high transactions cost and for allowing recipient governments to avoid taking hard decisions on the mobilization of domestic resources. In terms of global trends there is a consensus that the first MDG—of halving extreme poverty defined in terms of a headcount ratio (the share of the poor in the total population) at a US$1.25 poverty line at 2005 prices—will be achieved. However this is predominantly due to reductions in poverty in China and India, both of which have received relatively little Aid and where it is implausible to suggest Aid played any role in their recent rapid growth.[2] The most recent updating of global poverty projections suggests that by 2015 the first MDG will be missed in sub-Saharan Africa as a whole and in parts of Eastern Europe, Central Asia and Western Asia, where the incidence of poverty has worsened not improved since 1990 (see table 7.3).

**Table 7.1. Aid Allocation 2008**

| Region | Aid per capita[a] (current US$) | Net Aid/GNI (%) | GNI per capita[a] (Atlas method current US$) | Correlation coefficient[b] |
|---|---|---|---|---|
| Sub-Saharan Africa | 75.2 | 10.1 | 1,126 | -0.41 |
| South Asia[c] | 35.2 | 3.0 | 1,082 | na |
| Eastern Europe and Central Asia | 82.7 | 2.9 | 6,793 | -0.60 |

*Source*: World Development Indicators, Washington DC: World Bank, 2001.
*Note*: Calculated from World Development Indicators, 2001.
[a] Simple average for the regions
[b] Correlation between Net Aid/ GNI per capita
[c] Excluding Afghanistan

**Table 7.2. Aid Uses by Type of Donor 2009 (% distribution)**

| Category | DAC members | EU institutions | World Bank | Regional Development Banks |
|---|---|---|---|---|
| **Social and administrative infrastructure:** | **42.7** | **33.7** | **36.7** | **36.1** |
| Education | 8.8 | 8.0 | 6.1 | 2.2 |
| Health | 4.6 | 3.2 | 4.7 | 0.8 |
| Water supply and sanitation | 6.2 | 3.3 | 7.4 | 6.9 |
| Government and civil society | 12.5 | 13.1 | 8.0 | 16.9 |
| **Economic infrastructure:** | **14.9** | **10.1** | **51.4** | **48.6** |
| Transport | 7.4 | 8.5 | 18.4 | 15.5 |
| Energy | 3.6 | 1.1 | 15.3 | 18.1 |
| **Production:** | **6.4** | **15.6** | **11.7** | **6.3** |
| Agriculture | 4.7 | 11.5 | 5.9 | 5.0 |
| **Multisector** | **8.8** | **13.9** | **0.2** | **8.4** |
| **Programme** | **5.0** | **11.7** | **0** | **0.4** |
| **Humanitarian** | **8.7** | **10.0** | **0** | **0.1** |
| **Other** | **13.5** | **5.0** | **0** | **0.1** |
| **Total** | **100** | **100** | **100** | **100** |

*Source*: DAC database

**Table 7.3. MDG One[a] and Regional Poverty**

| Region | Headcount[b] 1990 (%) | Headcount[b] 2005 (%) | Target Headcount[b] 2015 (%) |
|---|---|---|---|
| Sub Saharan Africa | 58 | 51 | 29 |
| South Asia | 49 | 39 | 24.5 |
| Eastern Europe and Central Asia | 6 | 19 | 3 |
| South East Asia | 39 | 19 | 19.5 |
| East Asia | 60 | 16 | 30 |
| Latin America and the Caribbean | 11 | 8 | 5.5 |
| West Asia | 2 | 6 | 1 |
| North Africa | 5 | 3 | 1.5 |
| All developing regions | 46 | 27 | 23 |

*Source*: United Nations, The Millennium Development Goals Report, 2010
[a]Halve extreme poverty 1990–2015
[b]Share of population with expenditure below $1.25/day at 2005 prices.

Nonetheless the story is complex and any balanced assessment of the contribution of Aid to poverty reduction must consider all the other factors at work in recipient countries and judge whether on balance Aid has made a difference. Thus the key question is if poverty reduction has been disappointingly slow in much of sub–Saharan Africa and in some countries elsewhere, would it have been even slower in the absence of Aid? Also given the diversity in types of aid and mechanisms of its delivery what have we learnt from recent experience on what works and what does not in terms reducing poverty? This chapter addresses these issues by dividing the discussion between what we know about the macro impact of aid on growth and thus on poverty and the micro or project level, where experience with different types of poverty–focused intervention are considered.[3] Particular attention is given to various specific forms of Aid delivery, such as microfinance, which in the view of some in the development community has been one of the key innovations in the drive for poverty reduction.

## HOW CAN WE INTERPRET THE MACRO STORY?

Since the late 1990s when poverty reduction became the explicit target of the Aid community (although in practice interventions may have been informed by a variety of motives including political and commercial) ideas on how this might be achieved have been dominated by a paradigm based around three inter-related pillars—economic growth, good governance and social development. Aid was to support each of these and each in turn would reinforce the others. Thus growth would create the jobs and resources to bring down poverty, whilst social development through expenditure on primary health care and education would improve non-monetary indicators of welfare and governance reforms would both support growth and ensure that its benefits were distributed in a way that helped the poor (Weiss 2008). Growth was the centrepiece of the strategy, since extensive empirical work has shown a clear association between economic growth and reductions poverty.

The change in income of the poor over a given time period can be decomposed into a growth effect (holding distribution constant), a distribution effect (holding income constant) and an interaction term allowing both income and distribution to change.[4] Since distribution tends to change relatively slowly empirical applications of a decomposition approach have found that the growth effect usually dominates the others. In addition, income distribution has worsened in many countries in recent decades, even where there is no pro-poor distributional shift Aid interventions by protecting the poor and reducing any worsening of the income distribution may still have a poverty reducing effect.

Cross-country regressions typically have found the poverty elasticity with respect to growth of between −1 and −2, so that on average a 1 percent growth would cause a fall in measured poverty of between 1 and 2 percent, depending upon the analysis. The strength of the growth–poverty relationship however varies between countries with the level of inequality at the start of a period of growth and the characteristics of growth (for example in terms of job creation or its sectoral composition). Inequality, in particular, always weakens the impact of growth on poverty.[5]

In poor countries there is little dispute that major reductions in poverty will require a period of relatively high and sustained growth, hence a key test for the impact of Aid on poverty is how far it has contributed to higher growth. The standard theory on which early justifications for Aid were based is the two-gap model where Aid finances the larger of the savings or balance of payments gaps. Growth in this analysis is generated by the level of investment and its efficiency. Hence,

$$g = (I/Y)/v \qquad\qquad (1)$$

Where, g is the growth rate, I is investment, Y is GDP and v is the capital output ratio (or the measure of investment efficiency).

$I$ will be the sum of domestic savings $(S)$ and Aid $(A)$,

$$\text{So, } g = ((S/Y) + (A/Y))/v \qquad\qquad (2)$$

Therefore, an increase in the aid share in GDP, holding everything else constant, leads to higher investment and thus higher growth. Similarly if there is a target g and domestic savings are too low to finance the investment needed to achieve the target (so there is a gap between necessary $I$ and planned or ex ante $S$) this can be filled by Aid.

This simple model has been overtaken by more refined growth models where there is only a temporary relationship between investment and growth (in the neoclassical version) or a non-linear one with a changing investment efficiency (the endogenous growth version) (Easterley 2003;31–32). However it serves to highlight some of the key criticisms leveled against Aid in terms of why it might have only a weak impact on growth.

First, if it may be that ultimately Aid finances consumption not investment, so $S/Y$ falls as $A/Y$ rises, leaving g changed only marginally or not at all. Ultimately Aid resources are fungible and may be diverted to consumption through misappropriation or by allowing governments to avoid difficult decisions on domestic resource mobilization. The pillar of governance reform noted above in principle was to avoid this problem, but this type of institutional change takes time and critics argue that in the past such problems were relatively common. Secondly, Aid may reduce the efficiency of investment in an economy, thus raising $v$ and reducing the impact of a given $I$ on growth. This could arise where, due to fungibility, Aid funds ultimately finance prestige projects with negative economic returns or where the policy advice or conditionality attached to the receipt of funds turns out to be misguided.[6] Thirdly, in economies where Aid is very low relative to domestic savings any rise in $A/Y$ will be too small to have a major impact on g. Hence whilst in principle we might expect the inflow of foreign savings through Aid to have a positive effect on growth and thus on poverty reduction this might be relatively weak and difficult to detect empirically.

Considerable effort has gone into examining the Aid–growth link with much of this literature based on regressions that attempt to establish an average relationship across countries. This approach always runs the risk of omitting key variables that explain the circumstances of individual countries and faces the difficulty that Aid allocation between countries is non-random, so that countries may receive higher levels of Aid because they are very poor and growing slowly (suggesting a spurious negative relationship between Aid and growth) or may be rewarded for good performance with more Aid (suggesting a spurious positive relationship).

The basic approach is to estimate an equation across countries of the form

$$gi = \alpha + \beta_1 Xi + \beta_2 Ai + \varepsilon i \tag{3}$$

Where, $g$ is growth, $i$ is country $i$, $X$ is a set of country characteristics (including income per capita at the start of the growth period and key potential "fundamental" causes of growth such as measures of governance, openness and geography), $A$ is measure of Aid, $\alpha$ is a constant and $\varepsilon$ is the error term.

Early empirical work applying a version of (3) often found no impact of Aid on growth (so $\beta_2$ was insignificant) creating the "macro-micro paradox" referred to earlier. Additions to the literature in the late 1990s extended this approach in several important ways—by allowing for diminishing returns in any Aid effect due to a limit on absorptive capacity (by including a separate term for Aid squared), by distinguishing between different types of Aid and including only the flows likely to have the greatest short-run growth effect and most critically by allowing for a differential Aid impact varying with the policy stance of recipient governments (through an interaction term between a measure of Aid and a measure of policy). With these amendments the test equation becomes

$$g = \alpha + \beta_1 Xi + \beta_2 Pi + \beta_3 Ai + \beta_4 (Ai)^2 + \beta_5 (Ai^*Pi) + \varepsilon i \tag{4}$$

Where, $Pi$ is an index of policy in country $i$, and where $A$ may be redefined to distinguish between different categories of Aid.

Subsequent work post-2000 has in general tended to find a stronger impact of Aid on growth than was found in earlier studies, although there is also evidence of diminishing returns.[7] Of the additions to the basic model the incorporation of the variable reflecting an interaction with policy created the greatest debate. At one level it simply reflects the intuition that whilst Aid may be ineffective in poor policy environments (where for example it adds to consumption, not saving, in equation 1), it is much more likely to be effective where it is used in support of sound policies. This commonsense approach became part of a debate on the reorientation of aid priorities towards countries that could use Aid most effectively. The original paper (Burnside and Dollar 2000) finding a positive growth effect in a good policy environment (so $\beta_5$ in 4 is positive and significant) used a contentious measure of policy based on macro balance and open trade, which was replaced in a revised version of the approach in which the policy environment was based on World Bank staff judgements on a wide range of criteria (Collier and Dollar 2004).

The link between Aid's growth impact and policy confirmed in both of these studies was cited widely by donors; however the debate was not conclusive in that reworking of the original Burnside-Dollar analysis with an extended data set and alternative specifications of policy failed to replicate their results (Easterley 2003). In turn, the response was that the relationship between growth and the Aid–policy interaction term works once one introduces external shocks in terms of trade changes (Collier and Dehn 2001). Several other studies found Aid to be effective even without a good policy environment (Dalgaard and Hansen 2001, Hansen and Tarp 2001, Clemens et al. 2004). However the debate still remains unresolved with other studies using different theoretical specifications and time periods continuing to report different results, with some still finding no impact on growth. Rajan and Subramaniam (2008), for example, find no link between Aid and growth whether or not a policy interaction term is included, and the same authors suggest this result may be due to the impact of Aid in appreciating the recipient country's real exchange rate with consequent negative effects on exports (Rajan and Subramaniam 2011). Alternatively in what the authors claim is the most rigorous assessment of the macro consequences of Aid to date Arndt et al. (2010) find a positive long-run (but not short-run) effect on growth.[8]

Given the relationship between growth and poverty, the expectation is that if aid genuinely raises an economy's growth rate it will reduce poverty. Part of the problem has been that the poorest countries have grown only slowly over several decades and even a positive growth effect from Aid has added relatively little to overall growth. Collier (2008) argues that in the thirty years or so from the early 1970s the income per capita of the world's poor (the "bottom billion") was virtually static but that Aid added one percentage point to this growth rate, so that without any Aid the global poor would have seen an annual cumulative decline in income. This decline in turn would have raised poverty rates above those actually experienced. Along these lines in their analysis of data pre-2000 Collier and Dollar (2003) argue that overall aid had succeeded in removing around 16 million people from poverty annually and that, were its allocation between countries to be changed to favor those with the highest poverty and an effective policy environment, this annual figure would rise to thirty million. The impact on poverty follows even if (as in sub-Saharan Africa) over periods of time many countries experienced negative per capita growth. Clemens et al. (2004), for example, suggest that without Aid Africa's growth record would have been even poorer and thus poverty rates would have been even higher. If Africa had received the developing country average Aid allocation, rather than the higher than average figure the continent actually received, they find that its per capita GDP growth rate would have been approximately 0.5 percent lower annually 1973–2001 (−0.8 percent as compared with the actual of −0.23 percent). If the poverty elasticity is as much as −2 (which is the figure used by Collier and Dollar 2003) this suggests that poverty in Africa would have been considerably higher each year.[9]

Macro estimates based on cross-country studies must be treated with caution given the inherent technical difficulties associated with the general approach. Aid flows are not the only category of inflow for which it is often difficult to establish a firm link with growth in cross-country analyses. The inability to link higher revenues from favourable prices for natural resource exports has been labelled the "Resource Curse" and even foreign direct investment, seen by some as a critical source of management skills, marketing links and new technology, is not inevitably found to have a significant growth effect in this type of study.[10] The lack of evidence for a positive Aid effect found in some studies, combined with the diverse types of Aid flows, donor practices and policy environments in which Aid is applied suggest that it is not sensible to think that aid always works in terms of raising growth (as opposed to merely increasing national income in the short-term).

There is also evidence of diminishing returns to Aid so that beyond a certain share of GDP it ceases to add to or in some cases reduces growth—potential explanations range from political economy arguments on the impact of very large Aid inflows on governance encouraging corruption and conflict over resources to narrowly economic mechanisms such exchange rate appreciation.[11] Furthermore some of the growth effects from the macro literature imply implausibly high internal rates of return from individual Aid-financed investments, raising doubts about the validity of the overall results.[12] Nonetheless the intuition that Aid works best where it is supportive of a government committed to effective policy change, where it is aimed at relieving key bottlenecks faced by recipient countries and where it does not leak out to non–priority or inappropriate uses is no more than commonsense.[13] In relation to poverty reduction this conclusion implies the need to identify the types of Aid-funded interventions that have had the greatest impact in addition to any positive growth effects that might or might not have occurred.


## Aid and Poverty-Focused Interventions

Moving away from macro discussions of Aid there are a number of key micro issues relating specifically to poverty impact. There are many anecdotes both positive and negative that have been recounted about the effect of aid projects and other interventions. Lack of donor coordination may slow implementation and raise costs, and Aid given largely to meet commercial interests of donor country firms may add little to longer-term development.[14] On the other hand, well-designed projects may supply inputs to or create markets for poor people who would otherwise be linked only peripherally with economic activity. Aid provides examples of both with different types of Aid body, whether multilateral development banks, bilateral donors or NGOs often behaving differently. Easterley and Pfutze (2008) rank different agencies in terms of best practice and in particular criticise fragmentation and lack of specialization, with agencies often attempting to spread modest funds across too many countries or sectors. They also highlight what they term "ineffective aid channels" of tied aid, food aid and technical assistance, with the first two used by a number of bilateral donors to promote their own exports rather than at poverty reduction in the recipient country.[15]

Only a minority of the expenditure headings listed in table 7.2 are likely to have a direct poverty targeting objective, although in recent years some donors have attempted to redesign projects so that poverty spillovers are maximized, for example by adding rural feeder roads to highway projects or by extending urban water supply schemes to slum areas. In terms of assessing effectiveness, simply telling a story of how particular groups are affected by Aid-funded interventions charts outcomes, not necessarily impact. A positive outcome means that after the intervention the living standard or welfare of poor people has improved—this is to be welcomed but in itself does not prove that the intervention caused the improvement. Impact studies which aim to isolate the impact of the intervention from other factors at work are required to show that aid has reduced poverty. There is now a large, often highly technical, literature on such impact studies.[16]

Ideally a comparison can be based on two sets of households chosen randomly, one which has benefited from the Aid intervention concerned and the other which has not. Once features of these households, like size, education level and location, have been allowed for, any remaining difference in income or a measure of welfare can be attributed to the intervention. This "experimental approach" which mirrors clinical health trials is seen by many as the new best practice. However randomization is not always possible and in its absence a common practical alternative is to compare a group of the poor who have been exposed to the intervention (the treatment group) with a comparable group who have not (the control group). Selection of the comparison group will be critical and can be on the basis of pre-determined criteria (like land ownership), observation of the actions of members (for example willingness to take out a loan) or observable characteristics (like income or education).[17] Since initial pre-intervention differences must be allowed for a different approach is required so impact is estimated as

$$PI \; = \; (T_{t1} \; - C_{t1}) - (T_{t0} - C_{t0}) \tag{5}$$

Where, $PI$ is poverty impact, $T$ is a monetary or welfare measure for the treatment group and $C$ is the comparable measure for the control group, $t_0$ is time period before the intervention and $t_1$ is time period after the intervention.

Conducting such impact studies is potentially complex because of the need to ensure that equivalent groups are being compared. If, for example, more enterprising individuals seek out Aid (for example in the form of micro credits) then self-selection bias will distort the comparison if these more enterprising individuals would have been able to obtain the resources necessary to improve their position even without the intervention concerned. For this reason it is now argued frequently that members of treatment and control groups should be selected randomly to allow an unbiased comparison.[18] Where a rigorous approach fails to identify a positive impact from the intervention concerned, but the living standard of the target group nonetheless has improved, the improvement is likely to be due to favorable factors that also affect the control group, for example a favorable relative price shift or macro-economic growth that affects most households in some way.

This type of impact study is best suited to specific projects or programmes where it is possible to identify beneficiaries and to trace how their welfare or income has changed over a period of time. However there is a diverse of set of mechanisms through which Aid may have a direct impact on poverty, not simply through projects. These can be categorized in various ways but a helpful framework separates out four broad mechanisms

- Measures to increase the access of the poor to productive assets (for example, land reforms and microfinance)

- Measures to raise the rate of return to the assets of the poor (for example, poor area infrastructure programmes)
- Government expenditures from which the poor benefit disproportionately (for example primary education and rural health clinics)
- Specifically targeted interventions to transfer resources to the poor (for example food subsidies, workfare and conditional cash transfers).

## Increased Access to Productive Assets

### Land

By definition the poor have few assets relative to the rest of society. In rural areas a key correlate of poverty is lack of land ownership and self-perceptions of poverty are often based on lack of land.[19] Major land reforms have in the past been associated with social upheavals such as war or revolution and Aid programmes have tended not to stray into controversial areas such as inequality in land ownership. This is despite the fact that as cross-country macro studies have revealed inequality significantly dampens the impact of macro economic growth on poverty reduction. A frequently heard explanation for the much lower poverty elasticity with respect to economic growth in India as compared with China, for example, is the much greater inequality in land ownership in India (Chaudhuri and Ravallion 2007). Land reform has figured relatively little on the Aid agenda due to its political sensitivity. Where it has occurred in recent years (for example in Brazil and South Africa in the 1990s) this is in response to a clear internal political constituency which governments wish to address.

Land reform can occur through a variety of mechanisms including forced expropriation with or without full compensation, privatisation of state land, auctions of land from bankrupt enterprises, compulsory land ceilings or heavy land taxes to force the break-up of large estates. In turn beneficiaries may be previous tenant farmers, estate workers or targeted or favoured groups. The consensus of the large volume of research on the topic is that the post-1945 land reforms in East Asia—in Japan, Korea and Taiwan—and more recently the reforms associated with the ending of collective agriculture in China and Vietnam were highly beneficial to the economies concerned and helped in the rapid poverty reduction experienced in all these countries, although the evidence does not meet the rigorous standards implied by equation 5, because of the complex social, political and economic issues that land reform gives rise to. Elsewhere however the record appears more mixed with some negative experiences in terms of both productive efficiency and the welfare of the poor.

Evaluations of land redistribution programmes have highlighted several key lessons from a poverty reduction perspective:[20]

- If the poor are to be helped by land reform it will be critical to provide not merely access to land, but technical support and credit to allow poor households to farm productively.
- Grant financing for land acquisition and working capital should be strictly targeted at the poor.
- A balance needs to be struck between reaching as many poor households as possible and the creation of viable plot sizes.
- Programme design should be clear and transparent and should encourage participation by stakeholder communities.
- A variety of paths to land reform can be tried but a key condition is to act speedily and decisively to prevent vested interest creating obstacles.
- Land redistribution can be combined with measures to increase the spread of land rental markets, as an alternative means of changing land use.
- Property rights of poor beneficiaries need to be protected, including the right to sell or rent the land they receive at a later date.
- Compensation for expropriated land should be fair but not excessive and land taxation as an alternative means of increasing land supply should be considered.

In short, land redistribution is a possible vehicle for poverty reduction but the political conditions necessary for it to work are not always present.

## Credit

Credit and land markets can be linked because land is the key collateral that the poor can offer to access the credit market. The recognition of this basic point has led to an emphasis on securing formal titles over land as a means of unlocking credit markets for the poor (de Soto 2000). As with land, a shortage of credit for either investment or consumption to meet one-off or unforeseen events is a key characteristic of the poor. Aid donors originally addressed this by channeling funds to financial institutions—typically state-run Development banks—who would on-lend, passing on the interest rate subsidy inherent in Aid to the borrower. This model of financial support came to be seen as a prime example of "financial repression" policies with subsidized interest rates encouraging inefficient investment with high rates of non-repayment and moral hazard on the part of both lending institutions (whose losses would be covered by central governments) and borrowers (whose debts would not need to be repaid).

Microfinance emerged with donor support as an innovative alternative financing model that offered a means of linking the poor with financial markets. Following the approach pioneered by the Grameen Bank of Bangladesh in the late 1970s, the original microfinance model involved channelling Aid through microfinance institutions (MFIs) that lent very small amounts of money to groups of the poor, where the group accepted collective responsibility for the loan taken out by each member, held frequent meetings with loan officers of the MFI and where small regular savings deposits were required of all members. The model was costly in administrative costs per dollar lent but appeared to address the twin limitations of financial markets—information asymmetry (through the direct contact between groups and loan officers) and moral hazard (through group responsibility). Because of its high cost the original MFIs required Aid funds to cover the difference between their operating cost and the revenues collected in interest and loan repayments from their borrowers.

Microfinance has come a very long way over the last thirty years and has become an established component of the financial systems of a number of countries. Its rapid expansion has seen a multiplicity of institutional forms making generalisations about the industry difficult.[21] Aid funding was critical in the early years of the industry. More recently private sources, like savers deposits, external bank loans, collateralised loan obligations, foreign equity and bond issues, have become significant. The institutional form and lending methodology of MFIs have changed significantly over the years. Early MFIs largely followed a variant of the Grameen model dominated by nongovernment organizations (NGOs) operating on a non-profit basis. Over time the institutional form of the industry shifted to a for-profit emphasis with an accompanying shift to individual lending tailored to the needs to clients. Some NGOs were "transformed" into regulated deposit-taking MFI banks or non-bank financial institutions, new specialist microfinance banks emerged and commercial banks "downscaled" their operations to address the micro end of the market. Grameen itself moved to individual lending in its Grameen II methodology with a special window for the very poor.

The initial donor projects fell under the "poverty reduction" model of microfinance with the explicit goal of providing credit to the poor to transform their economic opportunities and allow them to grow out of poverty. Following the success of Grameen and other NGOs group lending was the most widely used lending methodology at this stage. However in the late 1990s donor thinking shifted to a broader 'financial services' perspective, where microfinance was viewed as a distinct segment of the financial sector, as far as possible operating on a financially sustainable basis, and offering the poor a range of financial services including savings and insurance products, as well as micro-credits. The key phrase in the policy statements of donors became the need to "mainstream microfinance." This new approach required continued dialogue between donors and governments on ways of reforming financial sectors to allow MFIs to offer a range of financial products and in particular to remove controls on interest rates that were perceived as a key obstacle to the financial sustainability of MFIs.

The Aid community has strongly endorsed and encouraged the shift in objectives towards greater profitability and commercialization on the grounds that only in this model will microfinance ever reach the large numbers of as yet unserved potential poor borrowers. The implication is that the poor should borrow at interest rates which reflect MFI costs plus a profit margin. However a move in this direction runs the risk of an increased concentration on the less poor as these are more likely to take out larger and hence lower cost loans from the MFI point of view. This "mission drift" argument has pre-occupied much of the more recent debate on microfinance. Loan size is usually taken as a proxy for poverty impact (as only the very poor will find it worthwhile to borrow small amounts) and it is well documented that banks lend larger loans. However the mission drift argument implies that as they evolve and shift towards greater profitability there is a risk that all forms of MFI lose their poverty focus.

Relatively little work has tested this hypothesis rigorously but one recent analysis, based on another sample of MFIs from the *Microbanking Bulletin* has done so with ambiguous results. The main findings are that

- Other things being equal within the group of MFIs using an individual lending methodology, the more profitable lend more to the poor (as proxied by average loan size) and more to women, although there is no relationship between profitability and either loan size or loans to women for group lenders;
- However, as individual lending MFIs get older, they tend to lend less to the poor and less to women;
- As group lending MFIs get bigger, they also lend less to the poor and to women; this negative effect of size is also found more weakly for individual lenders.

These trends work in opposite directions but the authors conclude that for larger and older MFIs the results are consistent with the view that "as institutions mature and grow they focus increasingly on clients that can absorb larger loans." (Cull et al. 2007). This is not necessarily "mission drift" in a strict sense since more poor borrowers could still be served under a commercial model, but it is a warning that even NGO MFIs may not be focusing primarily on the very poor any more.

Another aspect of the mission drift argument is that the entry of commercial banks in the small loan market will change the behaviour of MFIs. One possibility is that MFIs start to behave like conventional banks, focussing on better-off borrowers and raising loan size. However it is possible that competition will affect different types of MFI differently, with the greatest impact on those closest in operation to commercial banks—that is specialist micro-banks and those focussed on individual lending—rather than NGOs. Evidence in support of this latter hypothesis is found by Cull et al. (2009), who in another analysis across a large sample of MFIs find that measures of commercial bank penetration are negatively associated with average loan size and positively associated with the share of loans going to female borrowers for micro-banks and those using individual lending. For NGOs there is no such relationship. The implication is that the entry of commercial banks into the small loan market can push the commercially oriented segment of the industry in a pro-poor direction, thus at least partially counteracting opposite tendencies associated with mission drift.

## What Is the Evidence on the Poverty Impact of Microfinance?

There were great expectations in some circles that well designed microcredit programs targeted at the poor could replace earlier and largely unsuccessful financial sector interventions based in development banks. Small and frequent loans could allow both small investment by the poor in productive activities and also the phasing or "smoothing" of consumption over time to address cyclical or seasonal trends or one-off unexpected events like illness or bereavement. The poor typically need access to small savings deposits that will allow sudden lumpy withdrawals to meet unexpected expenditures. This explains the existence of informal savings schemes (in some cases where the poor are charged for saving), rotating savings and credit associations and savings in physical assets like jewelry and livestock. If the introduction of microfinance services can offer safer and cheaper savings products to these informal mechanisms, it can meet an important need. However it was pointed out relatively early on that access to micro-credits might not be appropriate for all of the poor, particularly the very poor (Hulme and Mosley 1996). The latter might be excluded by groups on the grounds that the risk of their default was too high. Equally in individual lending schemes they might be excluded by an MFI loan officer for the same reason.

The very poor themselves might also be too risk averse to take out loans. The inability of MFIs to reach the core poor has been recorded in several research studies.[22] The characteristics that made the very poor difficult to reach with conventional financial institutions have also restricted their access to MFIs. Thinking in the industry has shifted to recognize the need for additional promotional and training measures for the very poor before they can be considered for membership of MFI credit programs.[23,24]

Ambitious claims have been made for microfinance as a means of transforming the lives of the poor, however there is surprisingly little detailed analysis substantiating this. There is a considerable body of evidence from surveys and interviews that shows fairly unambiguously that people who receive micro loans frequently see an improvement in their living standards after taking out the loans and that those who receive them often do better than similar households who do not borrow. In this sense microfinance clearly works, although there is also evidence that the very poor may benefit proportionately less than those close to or just above the poverty line. However for reasons noted earlier simple before and after or borrower and non–borrower comparisons cannot prove impact in terms of isolating the influence of microfinance from all other factors at work.

Microfinance has high repayment rates. If loans are taken out and can be repaid by clients this is prima facie evidence of a positive impact. The story is more complex however since it is possible that some poor borrowers may struggle to repay and then drop-out of the program after repaying leaving with no significant change in their earning potential. Further, others may be forced by group pressure to borrow from moneylenders at higher interest rates to repay the microfinance loan, so that repayment forces them further into debt and worsens their position.[25] The possibility of very high interest rates reflecting the high lending cost and risk levels per dollar, making it difficult for borrowers to graduate out of poverty (even if the loans themselves can be repaid), is now attracting attention in discussions of "irresponsible microfinance."

Early empirical work on microfinance in Bangladesh (including the Grameen Bank and BRAC) discovered a very large impact on poverty from female, but not male borrowing, with a headline result that every 100 taka borrowed and repaid raised the consumption of the poor by eighteen taka.

However this widely cited and influential result has been the subject of intense scrutiny. Highly technical econometric arguments have been brought to bear to question the robustness of the results with a reworking of the data failing to replicate the original strongly positive effect.[26] A few studies applying versions of equation 5 have found ambiguous results; for example Montgomery and Weiss (2011) in study of the Khushhali Bank in Pakistan find a positive impact on an expenditure measure of poverty for rural but not urban borrowers, and ADB (2007) on MFI program in the Philippines finds a weak overall impact on income and consumption, but surprisingly a worsening for the bottom two quintiles of borrowers.

These studies do not use the best-practice approach of randomly selecting control and treatment groups and as yet only a small number of published studies are available to test for impact in this way. For example, Banerjee et al. (2009) apply the approach to the expansion of MFIs in slum areas in Hyderabad, India, finding that, in the short-term, access to microcredit helps business start-up and to fund investment. However, it appears to have no impact on social indicators relating to female empowerment or family health or education. The qualification is that these are short-term effects and through higher investment future monetary and non-monetary benefits may arise. The authors conclude that microfinance can be a useful means of helping the entrepreneurial poor, but it is not a miracle in the sense of transforming social conditions. Interestingly some initial results from randomized work question key tenets of the microfinance literature, with for example a study on Sri Lanka finding far higher returns to male-headed as opposed to female-headed micro enterprises (de Mel et al. 2008).

The rapid spread of MFIs across a range of countries and the still relatively high repayment rates micro loans achieve are signs that they are helping the poor even if many gains cannot always be attributed to microfinance in rigorous impact studies. Nonetheless, the qualification must be made that microfinance is unlikely to be suitable for all of the poor and is no simple panacea. Further, for MFIs to work effectively they need to operate within an overall financial sector with sound financial regulation, supervision and appropriate technical assistance.[27]

## Raising the Return to the Assets of the Poor

A central feature of Aid programs in most countries is the provision of physical infrastructure which provides services that complement the land, labor or credit assets of the poor and thus potentially can raise the return to these assets. In some instances such infrastructure is provided as discrete projects —for example in the form of electricity grid expansion or road construction—and in others as part of community development packages which provide a range of services often as part of "poor area development programs." Generalizing across this range of activities is difficult since the critical infrastructure bottlenecks that disadvantage the poor may differ between countries.[28] In some instances there is evidence of a clear trade-off between the financial viability of infrastructure provision and its poverty impact. In the case of rural electrification programs, for example, a relatively recent World Bank evaluation (World Bank 2008) points out that many Bank projects have not benefitted the poor significantly because

- Villages for electrification have often been selected on financial grounds relating to closeness to the existing grid and the affordability of the tariff for villagers—so remote and poorer villages are connected last
- Once a village is connected the poor in the village have often been unable to afford the flat-rate connection charge and in many cases even a significant period of time after a village is electrified many of the very poor are still unconnected.[29]

The conclusion is that because the poor are less likely to be connected electricity subsidies implicit in many projects have been regressive rather than progressive and that subsidies to bring down the connection charge will be a useful means of ensuring benefits from electrification go the poor. The implication is that poverty objectives are still not yet firmly established in all bank electrification projects, although once connected the poor do benefit.

Access to water and sanitation is established as a separate MDG and there is a strong expectation that if clean water and sanitation services can be provided to the poor there will be benefits in terms of health outcomes, time savings and possibly wider social effects such as improved education attainment. Whilst these effects are highly plausible there is perhaps less evidence than might be expected. On the one hand contingent valuation surveys of poor communities have found relatively high willingness to pay for access to piped water and sanitation suggesting respondents perceive these benefits to be important.[30] On the other hand the link between improved health and access to clean water has not always been found in empirical studies, particularly since poor hygiene behaviour linked with low education level amongst the poor can weaken the water–health improvement link.[31]

Rural roads that link the poor with markets are widely seen as an important intervention since they offer the opportunity for goods to be marketed, for reductions in the cost of inputs used by the poor and make it easier for workers to move to fill any job vacancies. There are many anecdotes of their success, but rigorous proof of impact must control for the fact that the decision on where roads are located is never random. New roads are often located or existing roads often improved in areas with the greatest economic potential. Comparisons of improved and unimproved roads, for example, must find a way of addressing this "placement bias" to isolate the effect due solely to the road improvement. Due to advances in econometric techniques a small number of rigorous impact studies have demonstrated positive poverty reduction effects from road projects. One way of doing this is to use observable features of the area in which different roads are located to design the control group in the application of equation 5.[32] Using alternative approaches recent rigorous studies for Bangladesh and Ethiopia, for example, that use panel data and household fixed effects, whilst controlling for placement bias show that rural roads have had a positive effect on the welfare of rural households, reducing poverty incidence by between five and seven percentage points (Dercon et al. 2009; Khandker et al. 2009). Positive effects from roads are not inevitable since much will depend on their location and potential for traffic growth. As yet there is little evidence that the poor have benefited significantly from highway or secondary road projects, although they have done so indirectly through any growth effect such projects have created.

Some poverty is geographically concentrated in disadvantaged areas and poor area programs have been used to provide a package of infrastructure services combined with access to credit, often with local community involvement. The rationale is that the returns to different infrastructure projects will be greater where they are provided in combination as a means of overcoming a location disadvantage. China, with its large land area and distinct clusters of poverty, provides the clearest example of this approach. Initial analyses of poor area programmes in China found positive results in terms income change and poverty reduction (Park et al. 2002; Ravallion and Chen 2005). However a more recent analysis of the SouthWest China Poverty Reduction Project applying a version of equation 5 augmented with propensity score matching to identify comparator households in what the authors suggest is the first rigorous assessment of a poor-area programme finds ambiguous results (Chen et al. 2008:3).

In an analysis over ten years Chen et al. (2008) find a positive effect on income over the disbursement period of the project, the bulk of which is saved. However these gains are transitory and over the full period there is no significant impact of the programme on either income or consumption. Thus although the areas studied are getting wealthier there is no support for a positive impact from the programmes as the comparator groups experience the same gain. There are two significant qualifications to this headline result. First, there is some evidence of an improvement in income in kind likely to be due to the investment of initial savings generated by the program in animal husbandry. Second, amongst the participating households surveyed those who are both poor and relatively well educated (to junior high school level) do experience an improvement in income and consumption attributable to the programme. This suggests that the better educated households in these areas can identify investment opportunities that can be both funded and made more productive as a result of the programme. The selection of participating households was based on community decisions and the results imply that, had the allocation of credit under the programme all gone to the educated poor, the mean impact on income would have been nearly four times higher. Again, as with standalone infrastructure projects, much will depend on the specific design and context of a program. A package approach has an intuitive appeal but as this evaluation reveals it is no automatic guarantee of success in raising living standards and bringing down poverty.

## Pro-Poor Expenditure

Another means of addressing poverty is to employ "broad targeting" by increasing government expenditure on activities from which the poor will benefit disproportionately. The initial discussions of Structural Programs, for example, highlighted the need to reduce government expenditure to address macro-imbalances, whilst protecting basic health and primary education which were seen as critical for poverty reduction. There has been much discussion of the potential pro-poor nature of certain types of public expenditure. In one of the most detailed tests for the impact of such expenditure Mosley et al. (2004) construct an index of pro-poor expenditure across countries based on public expenditure data on agriculture (as a proxy for rural roads and agricultural extension), water and sanitation, housing, social security and education. Individual expenditure categories are weighted by the coefficients from a regression analysis of the relationship between poverty levels and the different expenditure categories. This composite index of pro-poor expenditure is used to explain poverty levels across countries and in a regression including income per capita, the Gini measure of inequality and this index, the index is found to be negatively and significantly related to poverty levels. Whilst higher public expenditure on activities that, a priori, are expected to help the poor does indeed appear to reduce poverty, this evidence is slightly tautological given the way the index is constructed. Further there is the issue of how far the pro-poor index is influenced by Aid. It appears that there is significant relationship—with higher Aid leading to more pro-poor expenditure—but only in low-income countries in the sample, which is plausible given the greater importance of Aid in low-as opposed to middle-income countries.[33]

Nonetheless the insight that certain types of public expenditure are likely to be more beneficial to the poor than others is one that should not be forgotten and increasing real levels for certain basic types of expenditure will be important from a poverty reduction perspective. Also there is no automatic guarantee that expenditures on health and education expenditure will target the poor effectively, since this will depend on the nature of the programmes and how they are funded. However there is the potential for such benefits from well designed primary education and healthcare interventions. For example, there are estimates for Indonesia showing proportionately higher gains for the bottom quintile from primary health and education expenditure (Morrison 2002).

## *Social Protection and Targeting*

In addition to the other interventions that in various ways try to increase the productivity and access to assets of the poor, Aid has also funded a variety of support measures to "protect" the poor by maintaining consumption levels. These include free or subsidised basic commodities, cash transfers, and employment guarantee or workfare schemes. Given the existence of tight budget constraints these have become increasingly targeted in recent years with various approaches used to identify beneficiaries. Given the difficulty of obtaining data to allow full "means-testing" based on household income targeting, approaches have included

- Indicators—such as quality of dwelling or land ownership,
- Location—in poor area or poor village programs,
- Self-targeting—in schemes like workfare designed to be attractive only to the poor.

Targeting risks two distinct errors—under-coverage where many of the poor are missed—and leakage—where much of the benefits go to those above the poverty line. Although a very narrow focus on the poor through well-designed targeting appears intuitively sound it is not inevitably the best way of reaching the poor. Narrow targeting should minimize the problem of leakage of benefits to the non-poor, but may have high administrative costs reducing the resources actually available to transfer to the poor. Further it has been argued more broadly based schemes might attract more popular support and thus ensure larger budgets.

In practice the record of targeting programs is very mixed. Many have been relatively modest in scale, so under-coverage was inevitable. Other programs have been much larger, but were relatively poorly targeted.[34] Subsidized foodstuffs often leaked to the market for resale at a commercial price. For example, for the subsidized rice program in Indonesia at the time of the Asian Financial Crisis roughly one-quarter of beneficiaries were not poor and only roughly half the target group were reached (Perdana and Maxwell 2005). Even the *zakat* system of Islamic charity that provides cash transfers to the poor and destitute has been found to have a significant leakage, although probably less so than for other cash transfer schemes; in urban areas roughly two-thirds of recipients in Pakistan are outside the bottom quintile (Arif 2006). Employment guarantee programmes now have a major role in poverty alleviation in India although their past record is again disappointing with high costs per job and in places evidence that the level of wages set have encouraged those just above or around the poverty line to take up the schemes (Srivastava 2005).

Location targeting giving priority to expenditure in backward areas has been an important policy tool, particularly in geographically large countries. However it has also proved a relatively imprecise measure missing the poor who live in non-backward areas and not excluding all of the non–poor in the target area. In China, for example, between 30 and 40 percent of the poor were estimated to live in non-poor counties in the 1990s and although there was evidence that the poor county program raised county level income, how much of this went to the poor was unclear (Wang 2005). Regional policy in Thailand has operated so that there was no clear negative association link between provincial income per capita and receipts per capita under poverty reduction programs, implying a failure of targeting at the provincial level (Warr and Sarntisart 2005).

Conditional cash transfer schemes pioneered in Latin America and now replicated elsewhere provide cash and in some cases food and nutrients to poor households (often to the female head) in return for a commitment to keeping children in school or attending health clinics for regular check-ups for mothers and children. They are conditional in that they provide support in the form of cash or food in return for actions that improve children's education and health.[35] These conditional schemes have proved successful in Latin America particularly, in raising school enrolment rates and improving attendance at health clinics. There is less evidence that they have improved education and health attainment levels, however, with the implication that they may need to be combined with further investment in improving service quality in these sectors. Many of these schemes have been evaluated rigorously and Fiszbein and Schady (2008) bring together these results in a comprehensive assessment of their impact. In relation to poverty they present estimates that indicate a significant drop due to these schemes in, for example, Colombia, Honduras, Mexico and Nicaragua. The main caveat appears to be the high monitoring and targeting cost associated with some of these schemes. For example, even three years after its inception the original Progresa scheme in Mexico used 35 percent of its budget in monitoring and targeting with only 41 percent delivered to participants. A similar figure of 37 percent for administrative expenses was recorded for the Food for Education Program in Bangladesh (Son 2008).

Given the diversity of targeting mechanisms and the range of country experiences it is helpful to have a comparative survey that looks at a large number of such schemes across a large country sample. Coady et al. (2004) survey 122 targeted anti-poverty interventions in forty-eight countries. Their primary focus is on the effectiveness of targeting in reducing leakage. They measure this by the ratio of share of the poor in benefits to their share in national income or population, what they term the actual outcome relative to a neutral untargeted outcome. A value greater than 1.0 for this ratio implies progressive targeting, whilst a value of 1.0 implies neutrality. With neutrality the poor obtain no greater share of benefits than they would under a random allocation system. A value of less than a 1.0 for the ratio implies regressive targeting with the poor receiving less than they would under a random allocation. Using this targeting indicator programs can be ranked by targeting efficiency. The median value is 1.25 suggesting a modest degree of progression in the typical scheme. However there is a very wide variation between types of scheme and countries, and as much as 25 percent of the schemes for which the indicator could be calculated are regressive, with an indicator of below unity. Here targeting fails utterly and a random selection of beneficiaries would have reached the poor more effectively.

How effectively different types of scheme are implemented appears to matter at least as much as the type of scheme. For example, although as noted earlier, employment guarantee schemes in India have been criticized for waste and leakage the single most effective scheme in the sample here is the Trabajar workfare programme in Argentina, with a targeting indicator of 4.0. Similarly, cash transfer schemes are some of the most effective and also some of the least effective in this sample. Self-selection targeting schemes appear to work well where workfare programmes are involved—so only the poor self-select to work for low wages. However self-selection through consumption—so subsidies are offered on low-quality products that only the poor are likely to purchase—is the least efficient of any form of targeting. There appears to be a link between targeting efficiency and the quality of governance, and regression analysis to explain the targeting indicators suggest that countries with better capacity for implementation—as proxied by income per capita—and where government's are more likely to be held accountable for their behavior—will have better targeting performance.[36]

In summary, detailed accurate targeting is difficult to get right and there are leakages from all schemes. However self-targeting in turn has problems, particularly in relation to consumption goods. Some leakage is a cost that needs to be borne to ensure that at least some of the poor are reached. Realistically there are budgetary constraints on how far protection measures can be spread in a very poor country, which is why growth and the employment and resources it generates is seen as the primary route out of poverty, although protection and targeting measures can be critical at times of stress for the poor.

## CONCLUSION

The impact of Aid on poverty reduction is clearly uneven across countries, policy environments and forms of intervention, with much that is still not well understood. We have some evidence that on average it raises economic growth with consequent expected impacts on poverty. However this evidence is disputed and has been proved to be sensitive to changes in theoretical specification, data and time period. Even taken at face value recent empirical results suggest that large amounts of Aid of around 10 percent of GDP are needed to generate an average increase of little more than one percentage point in growth. Compounded over a number of years this higher growth can create a worthwhile reduction in poverty, but the scale of the change is hardly likely to be transformative. Further there are doubts about the desirability of very high long-run Aid levels in relation to problems of governance and macro-economic management of the exchange rate. In short, if used wisely, modest amounts of Aid can in principle raise growth, but they are unlikely to be a substitute for internal efforts at resource mobilization and institution building.

Some Aid interventions are explicitly distributional in that they aim to ensure that the poor share in the fruits of growth and are protected against market fluctuations and personal shocks. However for many, if not all, recipient countries they still relate to only a minority of Aid-funded expenditure.

This type of intervention has been subject to increasingly sophisticated forms of impact evaluation, although there is still not a large set of rigorous studies on what works and what does not. Microfinance, rural roads, rural electrification and conditional cash transfers, for example, have each been singled out in recent years as important interventions that can reach the poor. Rigorous studies with appropriate control groups have found support for this judgment in some instances, but not in all cases and there is a need to learn from past mistakes. Rural roads built in the wrong areas may do little to integrate the poor with markets, electrification schemes can exclude the poor due to high connection charges, some microfinance institutions may lend to poor borrowers unable to use the funds productively and conditional cash transfers may be of little use if basic education and health services are poor. Broadly speaking project design will be at least as important as the type of intervention chosen for a poverty reduction programme.

Clearly some donor policies in practice have little to do with poverty reduction, where for example aid is used to support donor exports or to ensure raw material supplies. However international private capital flows—whether foreign investment, loans or remittances—are not alternatives to Aid for most countries. Aid may not always help the poor, but there is enough positive evidence from the literature to encourage the search for improved forms of intervention and to continue the dialogue between donors and governments to maximize its positive effects.


NOTES

1. The Development Assistance Committee of the OECD uses the simple definition of Aid as grants and loans with a 25 percent or more grant element net of repayments.

2. Chaudhuri and Ravallion (2007) discuss poverty trends in these two large economies.

3. This distinction has been described in the earlier literature as a "macro-micro paradox," reflecting the fact that whilst many aid projects appeared to work well, their effects often failed to show up in analyses of macro data (Mosley 1987).

4. Thus, $\Delta Yp = + \Delta Y.d0 + Yo.(d1 - d0) + \Delta Y.(d1 - d0)$, which reduces to $\Delta Yp = Yt0.(d1 - d0) + \Delta Y.(d1)$, where Yp is the income of the poor, d is the share of the poor in total income Yt, 1 and zero refer to periods 0 and 1, so Yt0 is total income at the start of the period and Yt1 is total income at the end, and $\Delta$ refers to change. The growth effect is $\Delta Y.d0$, the distribution effect is Yo.(d1 - d0) and their interaction is $\Delta Y.(d1 - d0)$.

5. World Bank (2001) cites an average poverty elasticity of -2 from a range of studies. Lower estimates are possible; for example Jalilian and Weiss (2002) estimated a poverty elasticity of close to -1. Using data for individual countries Warr (2000) found an elasticity of -0.7 for the Philippines and -2 for Thailand with the difference explained by greater inequality in the Philippines and thus a less equitable distribution of the gains from growth.

6. Where for example budget deficit reduction programs demanded by donors are excessively deflationary and create significantly lower capacity utilisation this will raise v, at least temporarily. Similar effects could be created by rapid trade liberalisation.

7. See the discussion in the symposium in the *Economic Journal* (2004) where the general tenor of the papers presented is that Aid had had positive effects which were not always detectable in earlier analyses. Diminishing returns are found in some of the paper, for example Dalgaard et al (2004).

8. The analysis involves a reworking of data from Rajan and Subramanian (2008) which originally found no significant growth effect. Using what they judge to be an improved econometric specification Arndt et al. (2010) find a mean point growth elasticity of 0.13 with respect to the Aid/GDP ratio, so an increase in the Aid share in GDP of ten percentage points raises growth by over one percentage point annually.

9. Given if the growth effect last only for ten years, not the full period twenty-eight year period of the original analysis, a boost to growth of 0.5 percentage points annually implies a one percentage point reduction in poverty annually at this elasticity and thus a cumulative lower poverty level of over ten percentage points at the end of the ten year period.

10. The influential study by Borenzstein et al. (1998) found foreign direct investment only has a positive effect on growth when it is interacted with a measure of education attainment.

11. Clemens et al. (2004) for example suggest short-term aid reaches its maximum impact at 8.8 percent of GDP and that diminishing, but not negative, returns set in beyond this point. However in their analysis short-term Aid is only slightly more than half of all Aid so they place the total figure for Aid beyond which diminishing returns set in at 17 percent of GDP.

12. Collier and Dollar (2003) report that for a country with mean a Aid level, in their analysis, the impact of Aid on growth at just below 0.5 percentage points annually implies a project return of 30–40 percent.

13. This conclusion is now widely accepted. For example in an evaluation of the Poverty Reduction Strategy Paper (PRSP) Initiative designed to give more country ownership to Aid programmes and to integrate poverty concerns into macro policy World Bank (2005: 5), in commenting on the uneven success of PRSP concludes that "The Initiative has added the most value in countries where government leadership and management of aid processes were already strong."

14. The response to the Asian Tsunami provides an interesting case-study of the complexities and difficulties involved in donor co-ordination in response to a major emergency; see Jayasuriya and McCawley (2010).

15. Multilaterals—the World Bank and the regional Development Banks—and DFID of the UK come out best in their scoring by quite a long way (Easterley and Pfutze 2008, table 5).

16. Estache (2010) surveys literature as applied to infrastructure projects.

17. The technique of propensity score matching provides a means of constructing a comparable control group based on observable characteristics and is now being used in some impact studies of Aid projects.

18. Another important potential problem is placement bias, where aid interventions are located in villages or towns with greater economic potential and the control group is located in areas of lesser potential.

19. For example, Srivastava (2006) reports the result of village surveys in India which show clearly that villagers themselves define poverty in terms of land access rather than income or expenditure.

20. This draws on the results of evaluations summarised in World Bank (2003:154–156). The Brazilian land reform programme is cited as a good practice· example of a land reform that fits well with a broader policy on reducing rural poverty through the development of smallholder agriculture.

21. Fully comprehensive data are unavailable but the Microcredit Summit Campaign cites over 3,500 specialist microfinance institutions reporting to it by the end of 2007 with 155 million clients, of whom roughly 80 percent are women and two-thirds of which it terms "among the poorest," although it is unclear how accurate this measure of "poorest" borrowers in fact is (Daley Harris 2009).

22. See the surveys of Meyer (2002), Montgomery and Weiss (2005) and Karlan and Murdoch (2009).

23. This idea of "graduation" into microfinance has been strongly influenced by the successful examples of a few institutions of which the best documented is the case of the Income Generation for Vulnerable Group Development (IGVDG) program of the Bangladesh Rural Advancement Committee (BRAC) (Matin and Hulme 2004).

24. This idea of "graduation" into microfinance has been strongly influenced by the successful examples of a few institutions of which the best documented is the case of the Income Generation for Vulnerable Group Development (IGVDG) program of the Bangladesh Rural Advancement Committee (BRAC) (Matin and Hulme 2004).

25. For example, a rigorous study of female borrowing from Village Banks in Thailand found a positive association between taking out a loan and repayments to moneylenders suggesting this phenomenon is not just a theoretical possibility (Coleman 1999).

26. The original results are in Pitt and Khandker (1998) and the analysis is extended in Khandker (2005). The most recent critique of this analysis and re-working of the data is Roodman and Morduch (2009).

27. ECG (2010) surveys lessons from evaluations and the wider research literature for donors and multilateral development banks operating in the microfinance sector.

28. There is evidence for example that critical infrastructure bottlenecks have ranged from rural roads in India, to rural electrification in the Philippines and irrigation in parts of China (Weiss 2008).

29. World Bank (2008:xv) points out that even in villages connected for fifteen to twenty years it is not uncommon for 20-25 percent of households to remain unconnected, citing Lao PDR as an example.

30. Gunatilake et al. (2007) explain how mean willingness to pay for water and sanitation can be estimated from household surveys.

31. Jalan and Ravallion (2003) is a widely cited study for India showing access to piped water reduces intestinal infections overall but not for the poorest households presumably due to lack of hygiene awareness.

32. This is the propensity score matching technique. Its use in relation to rural roads is explained in van der Walle and Mu (2007).

33. The results are not completely robust however since with a different specification of the pro-poor expenditure index the same authors find no impact of the index on non-monetary measures of welfare, the infant mortality rate and the Human Development Index; see Gomalee et al. (2005).

34. In India one estimate put all subsidy programmes with a nominal poverty focus at 11percent of central government expenditure in 2001. A comparable estimate for China was 5 percent; see the chapters in Weiss (2005) which survey experience with poverty targeting in Asia.

35. For example, the *Bolsa Escola* scheme in Brazil provides monthly payments to mothers of poor households with children in the age range of six to fifteen provided children maintain an 85 percent attendance rate at school.

36. The index of accountability comes from Kaufmann et al. (1999).

*The author is grateful for the comments of David Potts and Peter McCawley.

## BIBLIOGRAPHY

Asian Development Bank (ADB). "Evaluation Study Effect of Microfinance Operations on Poor Rural Households and the Status of Women." Manila: Asian Development Bank, Operations Evaluation Department, September 2007.

Ardnt, Channing, Samuel Jones and Finn Tarp. "Aid, Growth and Development: Have We Come Full Circle?" *Journal of Globalization and Development* 1, no. 2 (2010).

Arif, G. "Poverty Targeting in Pakistan: The Case of Zakat and Lady Health Workers" *in Poverty Strategies in Asia: a growth plus approach*, ed. J. Weiss, and H. Kahn. Cheltenham, UK; Northampton, MA: Edward Elgar publication, 2006.

Banerjee, Abhijit, Esther Duflo, Rachel Glennester and Cynthia Kinnan. "The Miracle of Microfinance: Evidence from Randomized Evaluations." MIT Department of Economics Working Paper May 2009.

Borenzstein, E, J. De Gregorio, and J-W. Lee. (1998) "How Does Foreign Direct Investment Affect Economic Growth?" *Journal of International Economics* 45, (1998): 115-35.

Burnside, C., and D. Dollar. "Aid, Policies and Growth." *American Economic Review* 90, 4 (2000): 847-68.

Chaudhuri, S., and M. Ravallion. "Partially Awakened Giants: Uneven Growth in China and India" in *Dancing with Giants: China, India, and the global economy*, ed. Winters, L.A and S. Yusuf. Washington DC: World Bank, 2007.

Chen, S., R. Mu, and M. Ravallion. "Are There Lasting Impacts of Aid to Poor Areas: Evidence from China?" World Bank Policy Research Working Paper no. 4084, 2008.

Clemens, M., S. Radelet, and R. Bhavnani. "Counting Chickens When They Hatch: The Short-term Effect of Aid on Growth." Center for Global Development working Paper no. 44, 2004.

Coady, D., M. Grosh, and J. Hoddinott. "Targeting Outcomes Redux." World Bank Research Observer 19, no.1. Washington DC: World Bank, 2004.

Coleman, B. "The Impact of Group Lending in Northeast Thailand." *Journal of Development Economics* 60, (1999): 105-142.

Collier, P. *The Bottom Billion.* Oxford, UK: Oxford University Press, 2008.

Collier, P., and D. Dollar. "Development Effectiveness: What Have We Learnt?" *Economic Journal* 114, (2004): F244-271.

———. "Aid Allocation and Poverty Reduction" *European Economic Review*, 2002: 1475-1500.

Collier, P., and J. Dehn. "Aid Policy and Shocks." Policy Research Working Paper no. 2688. Washington DC: World Bank, 2001.

Cull, R., A. Demirguc-Kunt, and J. Morduch. "Banks and Microbanks." Policy Research Working Paper no.5078. Washington DC: World Bank, 2009.

———. "Financial Performance and Outreach: A Global Analysis of Leading Microbanks." *Economic Journal* 117, February (2007): F107-F133.

Dalgaard, C., H. Hansen, and F. Tarp. 'On The Empirics of Foreign Aid and Growth.' *Economic Journal* 114, (2004): F191-F216.

Dalgaard, C., and H. Hansen. 'Aid, Growth and Good Polices.' *Journal of Development Studies* 37, (2001): 17-41.

Daley-Harris, S. State of the Microcredit Summit Campaign Report. Microcredit Summit Campaign: A Project of RESULTS Educational Fund. Washington DC: Published by Microcredit Summit Campaign, 2009.

De Mel, S., D. McKenzie, and C. Woodruff. "Returns to Capital in Microenterprise: Evidence from a Field Experiment." *Quarterly Journal of Economics* CXXIII, no.4, November 2008.

De Soto, H. *The Mystery of Capital. Why Capitalism Triumphs in the West and Fails Everywhere Else*? Reading, London: Black Swan, 2000.

Dercon, Stefan, Daniel, O Gilligan, John, Hoddinott and Tassew Woldehanna. "The Impact of Agricultural Extension and Roads on Poverty and Consumption Growth in Fifteen Ethiopian Villages." *American Journal of Agricultural Economics* 91, no. 4 (2009): 1007-21.

Easterley, W. "Can Foreign Aid Buy Growth?" *Journal of Economic Perspectives* 17, 3 (2003): 23-48.

Easterley, W., and T. Pfutze. "Where Does the Money Go? Best and Worst Practices in Foreign Aid." *Journal of Economic Perspectives* 22, 2 (2008): 29-52.

Eastache, A. "A Survey of Impact Evaluations of Infrastructure Projects, Programs and Policies." ECARES Working Paper, University of Brussels, 2010.

Evaluation Consultation Group (ECG). "Making Microfinance Work for the Poor: Evidence from Evaluations." Manila, Philippines: Independent Evaluation Department, Asian Development Bank. Evaluation Consultation Group paper no. 2 (2010): 1-51.

Fiszbein, A., and N. Schady. "Conditional Cash Transfers: Reducing Present and Future Poverty." Washington DC: World Bank Policy Research Report, 2008.

Gomanee, K., O. Morrissey, P. Mosley, and A. Verschoor. "Aid, Government Expenditure and Aggregate Welfare." *World Development* 33, no.3 (2005): 355-370.

Gunatilake, H., J-C Yang, S. Pattanayak and K. A. Choe. "Good Practices for Estimating Reliable Willingness-to-Pay Values in the Water Supply and Sanitation Sector." ERD Technical Note 23, Economics and Research Department. Manila: Asian Development Bank, 2007.

Hulme, D., and P. Mosley. *Finance Against Poverty*. London: Routledge, 1996.

Hansen, H, and F. Tarp. "Aid Growth Regressions." *Journal of Development Economics* 64, 2001: 547-70.

Jalan, J., and M. Ravallion. "Does Piped Water Reduce the Incidence of Diarrhoea for Children in Rural India?" *Journal of Econometrics* 112, no.1 (2003):153-173.

Jalilian, H., and J. Weiss. "Foreign Direct Investment and Poverty in the ASEAN Region." *ASEAN Economic Bulletin* 19, no.3 (2002): 231- 53.

Jayasuriya, Sisira, and Peter McCawley. *The Asian Tsunami: Aid and Reconstruction after a Disaster*. Cheltenham, UK; Northampton, MA: Edward Elgar publication, 2010.

Karlan, D., and J. Morduch. "Access to Finance." Chapter 2 *in Handbook of Development Economics*, ed. D. Rodrik and M. Rosenzweig. Amsterdam: North-Holland, 2010.

Kaufmann, D., A. Kraay, and P. Żoido-Lobaton. "Aggregating Governance Indicators." Policy Research Working Paper 2195, Washington DC: World Bank, 1999.

Khandker, S. "Microfinance and Poverty: Evidence Using Panel Data from Bangladesh." *World Bank Economic Review* 19, 2 (2005): 263-86.

Khandker, S., Z. Bakht, and K. Koowal. 'The Poverty Impact of Rural Roads: Evidence from Bangladesh.' *Economic Development and Cultural Change* 57, no. 7 (2009).

Matin, I., and D. Hulme. "Programs for the Poorest: Learning from the IGVD Program in Bangladesh." *World Development* 31, no. 3 (2004): 647-665.

Meyer, R. "Track Record of Financial Institution in Assisting the Poor in Asia." Tokyo: ADB Institute Research Paper 46, 2002.

Montgomery, H., and J. Weiss. Great Expectations: Microfinance in Asia and Latin America. *Oxford Development Studies* 33, 3-4 (2005).

———. "Can Commercially Orientated Microfinance Help Meet the Millennium Development Goals: Evidence from Pakistan." *World Development*, January 2011.

Morrison, C. *Education and Health Expenditure and Development: The Cases of Indonesia and Peru.* Paris: OECD Development Centre, 2002.

Mosley, P., J. Hudson, and A. Verschoor. "Aid, Poverty Reduction and the New Conditionality." *Economic Journal* 114, 2004: F217-243.

Mosley, P. *Overseas Aid: Its Defence and Reform.* Brighton: Wheatsheaf, 1987.

Mu, R., and D. van der Walle. "Rural Roads and Poor Area Development in Vietnam." Policy Research Working Paper 4340, Washington DC: World Bank, 2007.

Park, A., S. Wang, and G. Wu. "Regional poverty targeting in China." *Journal of Public Economics* 86, (2002): 123-153.

Pitt, M., and S. Khandker. "The Impact of Group-based Credit Programs on Poor Households in Bangladesh." *Journal of Political Economy* 106, no. 5 91996): 958-996.

Perdana, A., and J. Maxwell. "Poverty Targeting in Indonesia" in *Poverty Targeting in Asia,* ed. Weiss John. Cheltenham. UK; Northampton, MA: Edward Elgar, 2005.

Rajan, R., and A. Subramanian. "Aid and Growth: What Does The Cross-country Evidence Really Show?" *Review of Economics and Statistics* 90, no. 4 (2008).

———. "Aid, Dutch Disease, and Manufacturing Growth." *Journal of Development Economics,* Elsevier, vol. 94, no. 1, 2011:106-118.

Ravallion, M., and S. Chen. "Hidden Impact: Household Saving Response to a Poor-area Development Project." *Journal of Public Economics* 89, 2005: 2183-2204.

Roodman, D., and J. Morduch. "The Impact of Microcredit on the Poor in Bangladesh: Revisiting the Evidence." Centre for Global Development Working Paper, 2009.

Son, H. "Conditional Cash Transfer Programs: An Effective Tool for Poverty Alleviation?" ERD Policy Brief 51, Manila: Asian Development Bank (ADB), 2008.

Srivastava, Pradeep. "The Role of Community Preferences in Targeting the Rural Poor: Evidence from Uttar Pradesh" in *Poverty Strategies in Asia: a growth plus approach,* ed. Weiss, John and Haider, Kahn. Cheltenham, UK; Northampton, MA: Edward Elgar, 2006.

———. "Poverty in Targeting in India" in ed. John Weiss, *Poverty Targeting in Asia.* Cheltenham, UK; Northampton, MA: Edward Elgar, 2005.

Wang, S. "Poverty Targeting in the People's Republic of China" in *Poverty Targeting in Asia,* ed. J Weiss. Cheltenham, UK; Northampton, MA: Edward Elgar, 2005.

Warr, P. "Poverty Reduction and Economic Growth: Evidence from Asia." *Asian Development Review* 18, 2 (2000).

Warr, P., and I. Sarntisart. "Poverty Targeting in Thailand" in *Poverty Targeting in Asia*, ed. John Weiss. Cheltenham, UK; Northampton, MA: Edward Elgar, 2005.

Weiss, John. *Poverty Targeting in Asia*. Cheltenham, UK; Northampton, MA: Edward Elgar, 2005.

———. "The Aid Paradigm for Poverty Reduction - Does It Make Sense?" *Development Policy Review* 26, 4 (2008).

World Bank. "Land Policies for Growth and Poverty Reduction." World Bank Policy Research Report 26384. Washington DC: World Bank, 2003.

———.World Development Report. Washington DC: World Bank, 2001.

———."The Poverty Reduction Strategy Initiative." Operations Evaluation Department. Washington DC: World Bank, 2005.

———. "The Welfare Impact of Rural Electrification: A Reassessment of Costs and Benefits." Independent Evaluation Group, Washington DC: World Bank, 2008.

## 8

BOLSA FAMÍLIA: ITS DESIGN, ITS IMPACTS AND POSSIBILITIES FOR
THE FUTURE

*Sergei Soares*

Conditional Cash Transfer programs, including the *Bolsa Famíla* program, have been extensively studied over recent years. The books, working papers, and articles that have been written on the subject, if placed one upon the other, would pile up very high indeed. What excuse do I have for spending my time writing about it and asking you to spend yours reading it? My excuse is twofold.

The first excuse is that, in spite of the aforementioned pile of studies, much about the Program is still not common knowledge. In the different forums in which I have been I have seen that many elementary facts about *Bolsa Família* are still relatively unknown to audiences beyond (some) Brazilian policy-makers and government officials.

How did it come about? What exactly was the Lula government's role in its creation? What have been its impacts on poverty, inequality, education, health, and labor supply? Did it have significant political effects? What are its contradictions and possibilities for the future? My objective is to give answers that are brief and, if possible, conclusive to all these questions in a single text.[1]

My second excuse is timing. Brazil is at the end of an era. After eight years leading the country towards less poverty and more equality, President Lula has passed the reins of power to his handpicked successor, the tough Dilma Roussef. Like the end of any era, it is also a time of new beginnings, and President Dilma has already declared that one of her main objectives in power is not the mere reduction of extreme poverty, but its complete eradication. To this end, *Bolsa Família* will certainly be an important item in the policy toolkit and will consequentially face some kind of change.

To tell the story as quickly as possible, the chapter is divided into three parts. The first part is purely descriptive and chronicles the story of *Bolsa Família*'s origin and goes into the details of how it works. The second part attempts to evaluate the program's impacts. Finally, I argue that, in spite of its success, *Bolsa Família* must change, and discuss possibilities for the future.

# A BRIEF HISTORY OF BOLSA FAMÍLIA

Brazil, like most of Latin America, discovered social policy through its formal labor markets. This was in keeping with our longstanding traditions of benefits for only a few and exclusion of the rest. We adapted social protection models from societies that did not share our history of slavery and built a system that perpetuated inequality and exclusion. The 1930s *Estado Novo* model left us with reasonably good social protection for urban, formal, predominantly white industrial workers and nothing at all for the rest. Many of our Latin American neighbors did the same, copying European models into societies that had only recently left behind black slavery or indigenous servitude, had never seen agrarian reform[2] and were characterized by extreme inequalities and exclusion of the majority of the population from any and all public services.

For those excluded from this formal social protection, which were mostly black or indigenous, the only hope for a future outside of poverty was their gradual inclusion into the formal labor market through its expansion. For a period after WWII it seemed as if this might work out, but only Argentina and Uruguay ever came even close to incorporating a large majority into the formal labor market and its social protection. Even in these countries, social protection never really made it into agriculture or demographic groups that fell outside the contributory paradigm, such as single mothers.

In Brazil, the first step away from the contributory social protection paradigm was the 1971 Rural Pension scheme, which provided non-contributory pensions for all rural workers. In addition to the contributing to the welfare of rural elderly, this was the first time need, and not only contribution, entered into the social protection equation.

It was the 1988 Constitution, however, that really paved the road to a new social protection paradigm. The new Constitution established Brazil's first targeted benefit: the *Benefício de Prestação Continuada* (BPC), which is minimum wage benefit to elderly or disabled people living in poverty. The BPC recognizes poverty as a condition to be taken on by social protection, even if only when in tandem with other conditions such as old age or disability.

Brazil would have to wait for 1991 for the next step leading to *Bolsa Família*. In December of that year the Senate approved a minimum income program, authored by Senator Eduardo Suplicy. According to the project, all Brazilians twenty-five or older living on less than Cr$ 45.000,00 (about $100) a month would receive a bonus so as to bring their income to that value. Never mind that the project explicitly leaves out children, never mind that it talks about individual and not family income, never mind that the Chamber of Deputies never voted it. It was a turning point in the social protection debate in Brazil, recognizing that poverty is a serious problem to be solved by government action. The political and academic atmosphere was ready for (Conditional Cash Transfer) Programs.

## *Conditional Cash Transfer Programs*

Although the Suplicy Project proposed unconditional targeted cash transfers, what was implemented were Conditional Cash Transfer Programs. The Brazilian debate on this is divided between those who consider this difference to be a minor point and those who consider the Suplicy Project to have been vilely betrayed. I count myself among the former: what matters is that the poor are getting money. In any case, all subsequent history revolved around Conditional Cash Transfer Programs.

It all started in 1995. In that year, three local and independent conditional cash transfer experiences sprang up simultaneously in Brazil. Perhaps serendipity occurs not only in science but also in social policy. The three were in the cities of Campinas, in March; Riberão Preto; in December, and in the Federal District (Brasília) in May. Although the eligibility lines and transfer values varied, all three were limited to families with children not yet fifteen and required these families to send their children to school.

The next year, 1996, saw the birth of the first Federal Conditional Cash Transfer Program in Brazil: the Child Labor Eradication Program (*Programa de Erradicação do Trabalho Infantil*—PETI). It was highly targeted at children from 7–15 who worked or were in risk of working in a list of activities considered dangerous, unhealthy, or degrading. Examples were sugar cane harvesting or burning wood to make coal, which are really very dangerous and unhealthy for anyone and even more so for a child. The PETI benefit was R$25 for children in rural areas and R$40 in urban areas. These values amount today to about $35 and $55, respectively. Children under fifteen were to stop working and have school attendance of at least 75 percent. The PETI program was run by the Social Assistance Secretariat.

In the following years, Brazil saw an explosion of CCTs as they appealed to the imagination of politicians, the press, and policy analysts. Between 1997 and 1998, Belém, Belo Horizonte, Boa Vista, Catanduva, Ferraz de Vasconcellos, Franca, Guaratinguetá, Guariba, Goiânia, Jaboticabal, Jundiaí, Mundo Novo, Limeira, Osasco, Ourinhos, Paracatu, Piracicaba, Presidente Prudente, Santo André, São Francisco do Conde, São José do Conde, São José dos Campos, São Luiz, Tocantins, and Vitória (Lavinas 1998) all created some kind of Conditional Cash Transfer. In 1998, the federal government started supporting poor municipalities that were implementing minimum income programs with education conditionalities; this programme was named *Bolsa Criança Cidadã*. It would cover up to 50 percent of the transfer. In 2008, 1,373 municipalities had a co-shared CCT programme (Sposati 2010).

In 2001, the *Bolsa Criança Cidadã* became the *de facto* second Federal Cash Transfer Program. It was called *Bolsa Escola Federal* and was clearly inspired by the Brasília Program. The target group was composed of families with children between 6 and 15 and the conditionality was 85 percent attendance. The eligibility line was R$90 (about $105) and the benefit was R$15 per child, not to exceed R$45 per family. This program was run by the Ministry of Education.

Soon afterwards came the *Bolsa Alimentação* (Food Scholarship), run by the Health Ministry. The eligibility line and benefits were identical to those of *Bolsa Escola Federal*, but families were required to get prenatal exams and vaccines for their children between 0 and 6. As if that was not enough, in 2003· the newly sworn in Lula administration created a fourth program, the *Cartão Alimentação* (Food Card)[3] which was a flat R$50 transfer to families with no behavioral conditionality, but the money could be used only to buy food.

So by mid-2003, we had four federal Conditional Cash Transfer Programs, each with its own implementing agency, its own financing scheme, its own benefits and eligibility lines. The federal government was transferring different values to different families, under virtually the same arguments. There was no communication between the different agencies. Each program had its information system and they were not in any way interlinked. One family could receive all four programs and a neighboring one, living in identical circumstances, could receive nothing. And did I mention the thirty-something state and municipal programs? It does not take a genius to see that this was the definition of administrative chaos.

In spite of the chaos, keep in mind that the federal, state, and municipal governments all had recognized poverty as an issue and the poor as deserving social protection in the form of monetary transfers. The conceptual revolution was complete, but some good administrative housekeeping was in order.

In October of 2003, the Federal Government created the *Bolsa Familia* Program,[4] whose objective was to organize and unify the four existing federal CCTs. *Bolsa Familia* also incorporated the *Vale–Gas*, an unconditional targeted transfer run by the Mines and Energy Ministry (yes, you read it right, Mines and Energy). The keys to unifying the different programs were the Single Registry (*Cadastro Único*), which existed since 2001[5] but had never before been really effective, and the creation of the Social Development Ministry, which became the implementing agency of all targeted cash transfers.[6] I now turn to how *Bolsa Família* actually works.

## How Does It Work?

*Bolsa Família*'s basic layout has not changed significantly since 2003, and although it is not likely to remain untouched for long, a good description is worth the time it takes to read.

## Who Runs the Program?

The National Citizenship Income Secretariat (*Secretaria Nacional de Renda de Cidadania*, from now on referred to as SENARC) of the Social Development Ministry (*Ministério do Desenvolvimento Social e Combate à Fome*—also known as MDS) is responsible for the Program. It is responsible for establishing norms and regulations for program execution; dialogue with states and municipalities; defining how much each family is paid; defining conditionalities, how they are monitored, and the sanctions for noncompliance; establishing coverage targets and, therefore, the program's budget; establishing municipal targets and limits; dialogue with other parts of the Federal Government; and, finally monitoring program execution and regular evaluations.[7] In short SENARC decides rules for deciding who is to get paid and how much, what they have to do to keep getting paid, and what will happen to them if they do not keep their side of the bargain.

SENARC, however, is composed of only a couple hundred people in three floors of the Ministry. A heavyweight is needed to actually make thirteen million payments every month. The *Caixa Econômica Federal*, a Federal Bank, is in charge actually running the program. It is responsible for receiving information collected by each municipality on its population in poverty, processing this information, calculating per capita incomes and, thus, defining how much each particular family will receive, printing the ATM cards and sending them to each family and, last but not least, actually making the payments every month.

*Caixa*'s role is not trivial. It is not only in charge of making payments, but also receives and processes information. This limits anyone else's role in effectively choosing who gets paid and how much. While it is true that *Caixa* processes information following the rules set down by SENARC, it is relevant that *Caixa*, not SENARC, processes the information.

## The Single Registry

As has been mentioned, the information base for *Bolsa Família* is the Single Registry. It is difficult to overstate the Registry's importance. If one were to define *Bolsa Família* as composed of two elements, they would be the Single Registry and an ATM card. The Registry is a rolling census, albeit imperfect and always incomplete, of the poor in Brazil. This means that the Registry is how the government knows who the poor are, where they live, and how they make (or fail to make) a living. It also goes beyond the *Bolsa Família* and is used by various complementary programs, to be explained further later in this chapter, as well as, in principle, any new targeted programs, whatever their relationship to *Bolsa Família*.

The information in the Registry is collected by the municipal agents of *Bolsa Família*. In most municipalities, these are the social workers working for the local social assistance secretariat, but in smaller municipalities this task can fall to the education, health, other secretariats, or even the mayor's office. These municipal agents receive a standardized questionnaire, decided upon by SENARC and *Caixa*, and transfer the information back to *Caixa*, either in paper or (increasingly) online. There have been seven versions of the Registry, each with its own questionnaire, each an improvement over the previous one.

Municipal agents and families both know the criteria as they are public and there is little doubt that these two groups of people game the system, often in agreement with each other. For example, informal labor incomes are much smaller in the Registry than they are in household surveys. This is not, however, as big a problem as it might seem. For example, if a family happens to have high transitory incomes that take them beyond the eligibility line when they are interviewed for the Registry but the agent knows they are a highly vulnerable family, he (or more likely, she) might nudge the figure downward to guarantee they are covered.

## Coverage

*Bolsa Família* is not an entitlement. The law that creates the Program explicitly states that the number of beneficiaries must be adjusted to the available budget. Once the budget is exhausted, new families are included only if other families leave the program, unless a supplementary credit is voted by Congress. This is why the establishment of national and municipal quotas has been so important over the last six years.

What is odd is that *Bolsa Família* has eligibility criteria and not ordering criteria. These criteria are public and generate a strange category of family, eligible but not beneficiary. Some authors, such as Medeiros, Britto, and Soares (2008), describe *Bolsa Família* as a quasi-right. There have even been reports of families suing the government to receive the benefit and winning, although I never analyzed such a case.

In any case, the first coverage target was created with the program itself in 2003. The target was 11.2 million families and was based upon the number of poor counted in the 2001 National Household Survey. For operational reasons, the government decided to expand coverage gradually, and the eleven million mark was reached only three years later in 2006. In 2007 and 2008, the program's coverage remained at eleven million with new families entering only as old ones left. This made *Bolsa Família* one of the largest social policies in Brazil in coverage. It is surpassed in number of beneficiaries by only the Single Health System that covers the whole Brazilian population, by public education with its 52 million students, and by social security with its 21 million benefits.

Table 8.1 shows how big *Bolsa Família* is. Depending on the indicator you choose, it is either big or not so big. *Bolsa* reaches almost a quarter of the Brazilian population but does not amount to even half a percentage point of GDP. This, of course, is because the benefits are modest.

## Benefits

*Bolsa Família* is composed of two types of benefit and two eligibility lines. The lines always refer to per capita family income as measured in the Single Registry. A "family" is defined by the *Bolsa Família* law as a group of related people living under the same roof. This definition is virtually identical to the usual household survey definition of "household." Those families whose incomes fall beneath the extreme poverty eligibility line are entitled (or quasi-entitled) to a fixed benefit and variable benefits in accordance to the number of children they have. From 2003 to July 2008, each family received one benefit per child aged less than 15, with a maximum of three per family. Since July 2008, the variable benefit has been expanded to include up to two teenagers fifteen or sixteen. The benefit is usually paid to the mother and, if she is not present in the household, to the father or other adult.

**Table 8.1.–Bolsa Família: A Little Big Program**

| Criterion / Year | 2003 | 2005 | 2007 | 2009 |
|---|---|---|---|---|
| **Size (in people)** | | | | |
| Brazilian population (Household survey) | 171.6 | 180.1 | 182.4 | 185.1 |
| Individuals in beneficiary families (Household survey) | 27.4 | 31.0 | 33.1 | 41.2 |
| Percent of population beneficiary | 16.0% | 17.2% | 18.1% | 22.2% |
| | | | | |
| **Size (in households)** | | | | |
| Households in Brazil (Household survey) | 48.7 | 52.0 | 54.5 | 56.9 |
| Beneficiary households (Household survey) | 5.75 | 6.62 | 7.16 | 9.2 |
| Beneficiary / Total (Household survey) | 11.8% | 12.7% | 13.2% | 16.2% |
| Beneficiary households (Single Registry) | 7.50 | 10.59 | 11.07 | 12.38 |
| Beneficiary / Total (Single Registry) | 15.4% | 20.4% | 20.3% | 21.8% |
| | | | | |
| **Size (% of national income)** | | | | |
| Average Household Income (Household survey) | 482 | 523 | 587 | 632 |
| Participation in Household Income (Household survey) | 0.3% | 0.5% | 0.6% | 0.7% |
| Program Expenditure (R$ billion of 2009) (Budget) | | 8.077 | 10.112 | 12.462 |
| GDP (R$ billion of 2009) (Budget) | 2,490 | 2,716 | 2,995 | 3,143 |
| Bolsa / GDP | | 0.30% | 0.34% | 0.40% |

*Source*: Soares, Souza, Osório, and Silveira (2010).

**Table 8.2.–Eligibility Lines and Benefits**

| Data | January 2004 | July 2007 | June 2008 | July 2009 | March 2011 |
|---|---|---|---|---|---|
| Law | Lei 10.836 | Decreto 6157 | Lei 11.692 and Decreto 6491 | Decreto 6917 | Decreto 7447 |
| Extreme poverty line | R$ 50 | R$ 60 | R$ 60 | R$ 70 | R$ 70 |
| Poverty line | R$ 100 | R$ 120 | R$ 120 | R$ 140 | R$ 140 |
| Variable benefit | R$ 15 (0-14) | R$ 18 (0 -14) | R$ 20 (0 -15) R$ 30 (16-17) | R$ 22 (0 -15) R$ 33 (16-17) | R$ 32 (0 -15) R$ 38 (16-17) |
| Fixed benefit | R$ 50 | R$ 58 | R$ 62 | R$ 68 | R$ 70 |

*Source*: SENARC/MDS; Casa Civil/Presidencia.

Those families whose incomes falls above the extreme poverty line but below the non-extreme poverty line are (quasi) entitled only to the variable benefits and receive no fixed benefit. *Bolsa Família* benefits enjoy no formal indexation, but both they and eligibility lines have been periodically updated more or less to keep up with inflation.

Benefits are conceded for a two-year period, which means that municipal agents must visit the family, or otherwise update the Registry information, every two years. In practice, far from all municipalities can keep their registries updated and there are some families that have not seen a social worker in a long time.

In addition to periodic visits, other administrative records are used to regularly screen beneficiaries. Social security and formal labor market registries are matched with the Single Registry every year to check incomes. If someone in a beneficiary family receives an undeclared social security benefit or a formal labor market paycheck, both the family and their municipal administration are informed of the new situation and it is hoped the social worker will visit the family (or the family visit the social assistance office) immediately to check whether they are still eligible. One month after the second notice is given, the benefit is blocked until the situation is sorted out.[8] Once the Registry is updated, the benefit is unblocked if the family is still eligible and cancelled if not.[9]

Finally, there is a surprisingly high number of families that voluntarily ask to leave the program. According to the Benefit System, forty-four thousand families requested to leave the program since their living conditions improved. This, however, is a gross underestimate since most municipal agents record families asking to leave under the "improved income" category. This is true, their income did improve, but the variable that shows the family's honesty is lost in the process.

## Conditionality

This is a tough issue. One of the most divisive issues among the legion of *Bolsa Família* fans is conditionality. Law 10.836 states that families, in order to receive the benefits, must send their kids to school and get their health check-ups and vaccines on time. Many of the Program's defenders maintain that conditionality is as important as the benefit itself. They really emphasize the first C in CCTs. According to their vision, *Bolsa Família* is, first and foremost, an incentive to human capital accumulation for poor families. In the public policy arena, this group of people emphasizes more emphatic monitoring of conditionalities and even the establishment of new ones. Their paradigm is *Progresa/Oportunidades*, which is explicitly a human capital accumulation program. Another group, among whom I count myself, defends that *Bolsa Família* is, first and foremost, social protection. If the number of conditionalities is excessive or their monitoring too draconian, will be the most vulnerable families first to be sanctioned for noncompliance.

The Brazilian government initially kept a healthy distance from both extremes, keeping conditionality as a part of the program, but being quite lax in their monitoring. Since 2006, their attitude has become increasingly harsh. Although no new conditionalities have been created and, in relative terms, few families have been kicked out for noncompliance, monitoring has become increasingly draconian.

The most important conditionality is the one that the Constitution requires of all Brazilian parents or guardians—to send their children to primary school. The health conditionalities are not legal obligations for non-beneficiaries, although they are strongly recommended for any citizen, whatever his or her income.

An important fact to keep in mind is that complying with apparently simple conditionalities is not as easy for families living in highly vulnerable conditions as for those with stronger links to formality. They live far from schools and clinics and often beyond the reach even of the Post Office. They are often fragile families or households headed by single women, and certainly have low social capital.

**Table 8.3. – Conditionality Monitoring**

|  | Education | | Health | |
| --- | --- | --- | --- | --- |
|  | In million children | In % | In million families | In % |
| 2005 1 semester |  |  | 0.4 | 6.8 |
| 2005 2 semester |  |  | 1.9 | 36.0 |
| 2006 1 semester |  |  | 3.0 | 43.1 |
| 2006 2 semester | 9.6 | 62.8 | 3.4 | 40.3 |
| 2007 1 semester | 12.0 | 78.9 | 4.8 | 51.1 |
| 2007 2 semester | 13.2 | 84.7 | 5.2 | 54.6 |
| 2008 1 semester | 13.0 | 84.9 | 6.1 | 62.7 |
| 2008 2 semester | 12.7 | 84.8 | 5.7 | 63.6 |
| 2009 1 semester | 13.0 | 85.7 | 6.1 | 63.1 |
| 2009 2 semester | 14.0 | 89.5 | 6.3 | 64.5 |
| 2010 1 semester | 13.6 | 85.7 | 6.8 | 67.5 |

*Sources*: Sistema de Acompanhamento da Freqüência escolar do PBF and Sistema de Vigilância Alimentar e Nutricional (SISVAN). DEGES/SEANRC/MDS.

How has monitoring become harsher? Until September 2006 there was no effective monitoring of conditionalities. In October 2006, however, the Social Development, Education, and Health ministries set up a monitoring system that has become increasingly complete and effective.

Table 8.3 shows that the educational conditionality monitoring has gone from 62 percent of children in 2006 to 85 percent in 2008. Health monitoring has also risen: from 6 percent in 2006 to almost 60 percent in 2008. Table 8.3 shows the monitoring effort.

Conditionality monitoring is a complex and successful effort by three different areas in three different federative levels in Brazil. The educational conditionality is monitored by municipal and state educational secretariats and consolidated by the Education Ministry; health conditionalities are monitored by municipal and state health secretariats and consolidated by the Health Ministry. The Social Development Ministry receives the data and gives the beneficiaries feedback.

In more detail, the Single Registry is used to generate a list of children, indexed by their Social Information Number and school code. The Ministry of Education then distributes this list to municipal and state education secretariats, which then pass them on to school principals. Principals with online access receive a login to pass the information directly to the Ministry. Schools without internet access provide the information in paper forms, which are consolidated by their education secretariats and then sent by internet to the Education Ministry. Education then consolidates the information and every two months passes it to Social Development. The number of children whose school status changes because they either changed schools or grades is, evidently, quite large. It is up to the municipal secretariats to find these children and they do a very good job. About 85 percent of children are successfully found five times per year.

Table 8.3 shows that we are successfully monitoring the attendance of close to 14 million children from 6 to 15. This is about 40 percent of the 33.7 million children in Brazil. If sending your kids to school is a Constitutional requirement for all parents and guardians, not using the system to monitor attendance of all the children in Brazil defies logic. Health monitoring is analogous to education monitoring, except that only about 60 percent of the children are found and the conditionalities are monitored twice yearly.

So what does Social Development do with this information? It depends on why the conditionality is not being met. If, for example, a child misses school because she broke her leg or because the bridge from her house to the school was washed away by a flood, no action is taken because none is required. But if the reason is not acceptable or unknown, the municipal *Bolsa Familia* agent and the family are both informed of the situation. Let me rectify that. The Social Development Ministry attempts to inform the family through both a letter and a message displayed in the bank terminal when they withdraw their benefit, but many of the families are so marginally inserted into our society that they neither receive the letter nor are able to read the message.

If after the first warning, attendance continues below 85 percent and no acceptable justification is given, two months later a second letter is sent and the benefit is temporarily blocked. When the family attempts to withdraw the money, they will get a message telling them this time that their benefit is blocked. The benefit will be paid as soon as the attendance situation is solved, but while not solved it remains blocked. If the child remains with low attendance for another two months, the benefit gets suspended—meaning that it will not be paid even when the situation is rectified. Finally, after one year of noncompliance the benefit is permanently cancelled and passed on to another family.

**Table 8.4. – Families Suffering Sanctions for Noncompliance**

| Action | No of Families | Percentage (%) |
| --- | --- | --- |
| Warning | 2,092,394 | 100.0 |
| Blocked benefit | 765,011 | 36.6 |
| 1st Suspension | 339,205 | 16.2 |
| 1ndSuspension | 149,439 | 7.1 |
| Cancellation | 93,231 | 4.5 |

*Source*: DEGES/SEANRC/MDS.

Noncompliance on health conditionalties leads to the same sanctions and both sets are monitored simultaneously. This means that a family can make an effort to boost attendance only to lose the benefit due to vaccines, for example. Table 8.4 shows how many families passed through each of the steps above. The information refers to the 2006 to 2008 period.

In spite of the government's attitude toward monitoring conditionality becoming clearly harsher over time, only 4.5 percent of families that at one or another moment did not comply with their side of the bargain lost their benefit for this reason. This is little over 4 percent of all families that left the program to date. However, for those of us who have a social protection view of Bolsa *Família*, these are unjustifiably draconian, since it is likely that these 4 percent are precisely the 4 percent that most need more support from the State.

Finally, conditionality monitoring is part of Decentralized Management Index, explained next.

## Federative Relations

Municipalities are absolutely central in the design of *Bolsa Família*. Mayors and their employees, and no one else, are responsible for identifying who the poor are. If the Single Registry is the heart of *Bolsa Família*, mayors are the veins, and without them the Registry would be nothing but empty data banks. A large part of health and education service delivery is also done by municipalities and municipal agents are also in charge of conditionality verification. It is a crucial role and without competent municipalities, the whole program will suffer.

This decentralized design contrasts with that of other CCTs, such as *Progresa*, whose targeting and monitoring strategies are much more centralized. I will show below that, although the decentralized approach is much cheaper since it counts on a pre-existing municipal administrative infrastructure, the two strategies lead to more or less equivalent outcomes.

Although this pre-existing municipal administrative infrastructure is one of the keys of *Bolsa Família's* success, federative relations in Brazil have never been easy. With relation to CCTs this is even more complex since the federal government was a latecomer. When it waded into the CCT ring with heavyweight budgets, states and municipalities had already been there for several years. After the 2003 unification, there was some initial confusion about duties and responsibilities and the federal government began to sign agreements with municipalities laying out who was to do what.

In 2006, the federal government became a little more daring and decided to pay municipalities for their role in *Bolsa Família*. The payment formula is simple: a maximum of R$2.50 per beneficiary family and the first 200 families in each municipality are paid in double (Lindert et al. 2007). This maximum value is then multiplied by the Decentralized Management Index (*Índice de Gestão Descentralizada*, henceforth referred to as IGD). This index is the simple average of four percentages:

1. The proportion of families in the Single Registry with complete and coherent information.
2. The proportion of families in the Single Registry whose last visit was less than two years ago.
3. The proportion of beneficiary children with complete education conditionality monitoring.
4. The proportion of beneficiary families with complete health conditionality monitoring.

The median value of the IGD is 75 percent, which means that many municipalities lose resources due to poor quality upkeep of the Registry or poor quality conditionality monitoring.

Finally, the federal government encourages those states and municipalities who wish to put their own budgets into the program to do so. They can either top up the value of the admittedly modest benefits or they can try to find unassisted families. If they choose to increase the value of their citizen's benefits, they pass the money to the federal government who pays out the additional sum at the ATM and acknowledges the financial support. The Federal District, the State of Mato Grosso, and the Municipality of Belo Horizonte, for example, all chose this approach. If they choose to find unassisted families, the Ffederal government gives them the Single Registry and the state or municipalities try to find those who fell through the cracks. The Municipality and State of São Paulo both follow this approach.

## Exit Strategies

Another divisive issue is exit strategies. For some reason I cannot possibly fathom, many people in Brazil and elsewhere find the idea of giving money to poor people abhorrent. Many will only accept the idea if it is viewed as a temporary measure to people fallen upon hard times who will soon claw their way out of poverty by their own devices.

More academic and sophisticated critics of long-term transfers argue that there may be long-term negative consequences for the families themselves. If people get used to living on the dole, their human capital may depreciate, they may lose their social networks, and reduce long-term prospects for themselves and even their children. This, of course, presupposes that the main reason (or at least an important reason) they are in poverty is lack of ambition or low expectations.

An extreme solution to this perceived conundrum is to impose limits to how long a family may stay in the program, such as those that exist for unemployment insurance. Some CCTs, such as one of the benefits of *Chile Solidario*,[10] have maximum staying times, although these are usually not enforced with an iron hand.

A less extreme solution would be to charge *Bolsa Família* managers with finding the exit doors from poverty and from the program. In other words, it is the social assistance network and not the families that is responsible for seeing the poor out of poverty and out of income transfers. Once again *Chile Solidario* is an example. Considerable effort, time, and money are spent on support for the families so they can find a job or other source of income. They are provided job training, labor intermediation, and microcredit, among other services. Once again, this approach is coherent with the hypothesis that some of the main causes of poverty are individual characteristics that can be changed by public policy.

Note that time limits and exit strategies are not at all coherent with CCTs as a means for human capital accumulation. Building up human capital requires time and is best done by children, not adults. The CCT program that most clearly identifies with human capital accumulation, *Progresa/Oportunidades*, has nothing that remotely resembles time limits. The program actually makes it hard for families to leave.

Another contrary position is that poverty is caused either by the labor markets to which the poor have access or individual characteristics that public policy is powerless to change, at least in the short run. If poor families are families with virtually no human capital, no social capital, and no social networks, then the search for exit strategies is pointless.

The wise position of the Brazilian Government has been to reject time limits or exit strategies, but leave the door open to public policies that may help families to say goodbye to poverty through their own effort. This line of attack is called the complementary programs approach. Other ministries provide adult education, opportunities for youth, job training, labor intermediation, subsidized electricity, rural electric grid expansion, rural extension, or microcredit to those who either are or may soon be *Bolsa Família* beneficiaries. Once again the key is the Single Registry. A host of programs provide their services either exclusively or with priority to families or individuals in the Registry. The immense majority of these have been pre-existing programs that have been reoriented toward the Single Registry. To date, only the *Plano Setorial de Qualificação* (PLANSEQ), whose objective is to prepare poor workers for jobs in the construction sector, has been made from scratch for *Bolsa* beneficiaries.

## Targeting

Targeting is one of the inescapable issues for all CCTs or non–universal income transfers. If the country is targeting its efforts toward the poor, it is vital that they go to those who are, in fact, poor. Tandem with targeting, coverage is really important. Unless we are talking about pilot programs, it is important not only to reach some poor people, but as many as possible. Otherwise, impacts upon poverty or inequality will be quite limited.

*Bolsa Família* coverage did not begin at zero. When the program began in 2003, there were 6.7 million beneficiary families in the four programs that were merged to create *Bolsa Família*. In 2003, the eleven million family target was also set.

As I have mentioned above, in the three years that followed, *Bolsa Família* coverage increased continuously until eleven million were covered in 2006, but in 2007 and 2008 the total amount of benefits remained where it was. Throughout this period, there was queue of about two million families waiting to get in, which was pretty good evidence that more benefits were required. These were families that, in the Registry, fulfilled all the requirements for being a beneficiary of *Bolsa Família* but never received one cent from the program. One of the reasons that kept the federal government from increasing the number of benefits was that eleven million families was considerably more than the 8.6 million families in poverty according to Household Surveys.

What it took an unjustifiably long time for anyone to figure out was that the number of poor families (according to the *Bolsa Família* eligibility line) in the Household Surveys and the number of families were two different concepts and thus should never be the same. Why?

Because poor people's incomes are volatile. Using a four-month panel from the Metropolitan Employment Survey, Soares (2009)[11] showed that if poverty was defined as being poor in at least one of the four months rather than the current month, poverty rose from 15 percent to 25 percent. The study was limited to a four-month panel in the Metropolitan Employment Survey, but *Bolsa Família's* rules allow each family to stay two years in the program before their income is re-evaluated. Soares estimates that over two years twice as many families fall into poverty as are poor in a given month. This means that in order to reach all the poor, *Bolsa Família* should have about twice as many beneficiary families as those that show up as poor in a cross–section household survey.

In June of 2009, due in part to stubborn evidence of lack of coverage and in part due to the international crisis, the federal government decided to increase coverage to 12.9 million families.

Things become more complicated when we broaden our brush and let targeting interact with coverage. A standard tool for analyzing targeting is the Incidence Curve and the number associated with it, the Incidence Coefficient. The steps to drawing an Incidence Curve are:

1. Order the population by incomes without the benefit whose incidence you wish to analyze;
2. On the horizontal axis, accumulate population ordered by income;
3. On the vertical axis, accumulate the benefits (or other variable whose incidence you wish to analyze).
4. The area between the Incidence Curve and the line of perfect equality (a straight line linking the lower left and upper right hand corners), with areas above the line counting negatively, is called the Incidence Coefficient.

The closer to the upper left hand corner the Curve, the more pro–poor the transfer and the closer to −1 the Incidence Coefficient; the closer to the lower right hand corner, the more pro-rich and the closer to +1 the Coefficient. If the Incidence Curve is close to the line of perfect equality, the transfer is equally divided among the population and the Coefficient is close to zero.

The population must be ordered by income net of benefits (and not total income) because that the criterion for getting a benefit is family income without the benefit. In other words, we must know how poor a family is when it does not count on a *Bolsa Família* benefit.

**Figure 8.1. Incidence curves in 2004 and 2006**

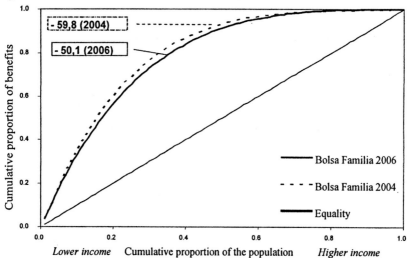

*Source*: Soares et al. (2008)

Figure 8.1 shows that there was a small deterioration in *Bolsa Família* targeting during the expansion from 2004 to 2006. If diminishing returns apply to targeted income transfers, this was to be expected and is not really a negative result per se. If the 2006 *Bolsa Família* Incidence Coefficient is compared to that of other CCTs such as *Progresa/Oportunidades* in Mexico (−0.56) and Chile's *Chile Solidário* (−0.57), we can see that the three values are quite close.

Another way to look at targeting is the hit miss ratio. Table 8.5 shows that in 2004 and 2006, respectively, 42.5 percent and 49.2 percent of the families receiving *Bolsa Família* counted on per capita incomes higher than the eligibility line.

**Table 8.5. Beneficiary and Eligible Families**

| 2004 | Not eligible (%) | Eligible (%) | Total (%) |
|---|---|---|---|
| Non-beneficiary | 77.9 | 9.6 | 87.5 |
| Beneficiary | 5.3 | 7.2 | 12.5 |
| Total | 83.2 | 16.8 | 100.0 |
| Among beneficiaries | 42.5 | 57.5 | 100.0 |
| 2006 | Not eligible (%) | Eligible (%) | Total (%) |
| Non-beneficiary | 76.6 | 6.6 | 83.2 |
| Beneficiary | 8.3 | 8.5 | 16.8 |
| Total | 84.9 | 15.1 | 100.0 |
| Among beneficiaries | 49.2 | 50.8 | 100.0 |

*Source*: Soares at ali (2008)

Figure 8.2 shows the same categories, but the unit of analysis is people and not families.

The horizontal axis shows income centile and the vertical axis shows two coverage ratios: an ideal one (bold line) and a moving average of what was observed (dotted dash line).

The bold numbers show the proportions of the people who are in the category they should be in: people in non-eligible families not receiving benefits and people in eligible families receiving them.

The italicized numbers show the opposite: either those in non-eligible families receiving benefits or those in eligible families but not getting a penny.

**Figure 8.2. Eligible and beneficiary population, by income centile**

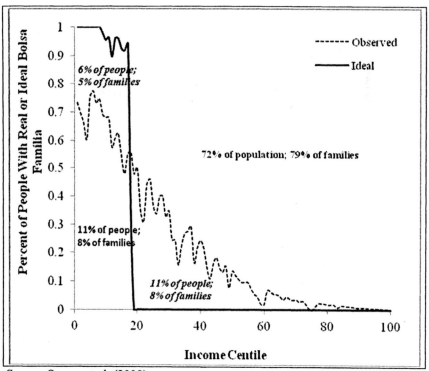

*Source*: Soares et al. (2009)

The dotted dashed line shows that the families in poorer centiles have a much higher chance of counting on a *Bolsa Família*. This is coherent with the good targeting shown by the Incidence Curves. Close to the eligibility line, however, the targeting error becomes quite large. Almost half of beneficiary families did not meet program criteria—a large number apparently in contradiction with the Incidence Curve results. What could possibly explain both the disparity and the apparently poor targeting?

Three candidates come to mind: (i) fraud, (ii) income measurement error, (iii) income volatility. I will comment on each in turn.

It is common knowledge that fraud exists. Putting it in other terms, it would be very strange if there was no fraud in a program distributing eleven million benefits in almost six thousand municipalities across Brazil. Relatives and friends of mayors and political appointees who clearly fall nowhere near eligibility criteria have been found on *Bolsa Família* payrolls. The problem with this explanation is magnitude. Quantitative analyses of fraud, such as those by our prestigious Accounting Office (*Tribunal de Contas da União*),[12] have found that they number in the hundreds or at most thousands of cases. Not very relevant if compared to eleven million benefits.

Measurement error and volatility appear to be more relevant. Families close to but above the eligibility line have a clear incentive to nudge their incomes downward. Their social workers may also identify a family as vulnerable even if their current income falls above the eligibility line and nudge their income to make sure they get in. Perhaps even more important is that many people, especially poor people who live from day to day, do not know precisely what their monthly income is. An illiterate informal street vendor may know exactly how much she sold on a particular day, but calculating her net profits for the whole month may be quite beyond her ken.

It is hard to estimate the prevalence of this type of targeting error because it gets mixed up with income volatility of poor families, which I have already advanced as an explanation of why the Program's targets cannot be calculated using cross-sectional poverty measurements. This happens because household surveys[13] without panels cannot distinguish between errors in measurement and volatility. This means that, in addition to not knowing exactly what her income was in a given month, our hypothetical street vendor also makes different amounts on different months. Someone who falls beneath the poverty line and requests a *Bolsa Família* benefit will receive it for two years before being re-interviewed. This means that: (a) the number of beneficiaries will be much larger than estimated using a cross-section household survey and (b) if the survey interviews them during the two-year period, their income may well be momentarily above the poverty line, thus leading to an over-estimate of targeting error.

## The Registry, Information Management, and the Role of Caixa

It is my belief that *Caixa Econômica Federal*, the Federal Bank that handles the *Bolsa Família* benefits, clearly has a role in managing the Program. This is because *Caixa* does much more than pay benefits. It also operates the Registry, which means that it controls the database on which all of *Bolsa Família* is based. Logically, SENARC, the National Citizenship Income Secretariat, should operate the Registry.

Why is this a problem? Because *Caixa* is a bank.

*Caixa's* computers are excellent and its IT people extremely competent, but they are used to dealing with information for other purposes, not social policy. Using Registry data for analysis or even run of the mill monitoring is harder than it should be. The Registry is therefore a payment system which happens to have social information that can be used for targeting, and not a social information system which, by the way, can also be used for payment.

In all justice, there has been much improvement. In recent years SENARC has been increasingly present in the design of the Single Registry system and Version 7, which was inaugurated in December of 2010, is much better than any of the previous versions. But the computers that collect the data and the people who run them are still in a bank and not in the Ministry of Social Development.

The Registry also suffers from several shortcomings common to many administrative record systems. There is a serious selection bias since some mayors go out of their way to find all their poor people while others apparently could not care less. This means that great care should be used when using Registry information for inter-municipal comparisons. Large municipalities seem to encounter great difficulty in their search for poor people. This is not a great surprise since anyone, not only the poor, is hard to find in a crowd of millions. What is surprising is that wealthy municipalities in the South and Southeast—Brazil's rich areas—are surprisingly timid in their search for their poor people. Mayors of small mostly rural municipalities in the Northeast— Brazil's really poor area—do the best job of finding their poor.

The Registry is thus obviously incomplete and some families do not receive benefits because they have never been registered. The IGD shows that about one-fifth of the families in the Registry have incomplete and incoherent information and another fifth have not seen a social worker in over two years.

In spite of these shortcomings, the Single Registry is a good database of poor people. It has a good questionnaire, closely modeled upon Brazilian Household Survey questionnaires, and a good IT system, particularly user-friendly to those updating their municipal Registry. By far more important, thousands of municipal social workers, teachers, health agents, and mayors believe in the Registry and do their best to guarantee that social protection is carried out with the best information possible.

The Single Registry is the first time any systematic attempt has been made to collect information on millions of families who had previously been completely absent from the radars of the State in Brazil. It is an admirable effort that has opened an important communication channel between the poor and the State that should care for them.

# IMPACTS

Perhaps because it was a new animal in the Brazilian social policy zoo, perhaps because people who strongly believe in evaluation were among its idealizers, perhaps because of its political visibility, *Bolsa Família* has been more extensively evaluated than almost any other social program.[14] While an experimental evaluation was never included in its design, all kind of techniques, both quantitative and qualitative, have been used to evaluate the Program. Since the universe of *Bolsa Família* evaluations in Brazil is vast, presented here are only some of the main results.

## *Inequality*

There is a considerable literature documenting the contribution of *Bolsa Família* to the surprising reduction of inequality in Brazil from 2001 onward. While the general conclusion—*Bolsa Família* contributed mightily to the reduction of inequality, but was far from its main source—is the same, the numbers vary quite a bit.

The reasons for the variation are two groups of methodological differences. Most of the studies decompose changes in the Gini coefficient by factor components. It is a parametric approach that is not only valid for the popular Gini coefficient, but also for the Mehran and Piesch indices, although only Hofmann (2010) calculates them.

Barros, Carvalho, and Franco (2007) follow another approach, using micro-simulations to create counterfactual distributions and then calculating different indices based on them. While very flexible, this micro-simulation approach has not been very popular due to difficulties in reproducing results.[15]

Another difference is how the benefit income is identified. The Pnad household survey will only count on a specific question for *Bolsa Família* income only from September 2011 onward. Previous to that date, *Bolsa Família* income was in the category "other income" that also includes dividends from stocks, interest payments, and income from other benefits such as some old-age non-contributory benefits. This is an absurdly heterogeneous category and further statistical treatment is necessary to identify *Bolsa Familia* benefits.

One solution is not to attempt any type of identification and just inform the reader that transfers income is reported with interest payments and stock and bond dividends. As Table 8.6 shows, these studies conclude that the "other income" variable was responsible for a reduction of about 0.6 to 0.9 Gini point (x100).

A second solution is to use the 2001, 2004, and 2006 CCT supplements to the survey. This is no doubt the most precise solution but it limits the number of years that can be analyzed to three. Those that follow this approach conclude that *Bolsa Família* lopped 0.6 Gini point (x100) off Brazilian inequality between 2001 and 2004 and another 0.2 point between 2004 and 2006.

What do to for the other years? A creative solution was developed by Barros, Carvalho, and Franco, which is to use the value of the "other income" variable to identify *Bolsa Família* income. Basically, while small values or values that are multiples of benefit values are likely to be *Bolsa Família* benefits, large values are likely to be interest or dividends. Foguel and Barros (2010) show that the typical value identification strategy slightly under-estimates targeting of *Bolsa Família*, relative to the use of supplements. Souza (2010) shows that typical values over-estimate, again slightly, coverage. This is what would be expected.

Table 8.6 summarizes the literature to date. *Bolsa Família* had an important effect on inequality reduction, but was clearly not the main reason behind it. What is surprising is that a program that amounts to less than 1 percent of household income could be responsible for up to a quarter of a non-trivial reduction in inequality. Recent periods also show a tendency towards more modest results. Once again, this is not surprising to those who believe in diminishing returns.

**Table 8.6. – Bolsa Família Impacts on Inequality**

| Study | Identification strategy | Methodology | Periods | Absolute impact in Gini Points (x100) | Change of Gini in percentage points (%) |
|---|---|---|---|---|---|
| Soares (2006) | No identification | Decomposition by factor components | 1995–2004 2001–2004 | - 0.64 - 0.86 | 27 30 |
| Hoffmann (2006) | No identification | Decomposition by factor components | 1997–2004 2002–2004 | - 0.79 - 0.41 | 25 31 |
| Soares F. et al. (2007) | Supplement | Decomposition by factor components | 1995–2004 | - 0.57 | 21 |
| Soares S. et al. (2007) | Supplement | Decomposition by factor components | 1995–2004 | - 0.57 | 21 |
| Barros, Carvalho, and Franco (2007) | Typical values | Micro-simulation | 2001–2005 | - 0.32 | 12 |
| Hoffmann (2010) | No identification | Decomposition by factor components | 2001–2007 | - 0.80 | 19 |
| Soares, Ribas, and Soares (2009) | Supplement | Decomposition by factor components | 2004–2006 | - 0.20 | 21 |
| Soares, Souza, Osório, and Silveira (2010) | Typical values | Decomposition by factor components | 1999–2009 (Odd years) | - 0.81 | 16 |

*Source*: See first column.

## *Poverty*

*Bolsa Família* has had a modest impact upon poverty and even in extreme poverty defined as the percentage of people living on less than the program's eligibility lines. This should not be too surprising since the values transferred are also quite modest. Equally modest are the number of studies addressing the issue, particularly when compared to the number of studies on inequality. The best of the three studies I know of is Soares et al. (2010), and this section will closely follow their discussion.

Figure 8.3 shows the evolution of poverty as a percentage of the population, according to the Program's two eligibility lines in 2003, updated using the inflation index. The lower line will be called extreme poverty. The fall in poverty in Brazil over the last decade is quite impressive, but uneven over time. Poverty fell from 26 percent to 14 percent of the population, and extreme poverty from 10 percent to 5 percent, but almost all of this fall was from 2003 to 2009.

**Figure 8.3. Poverty and extreme poverty in Brazil, 1999 to 2009, odd years**

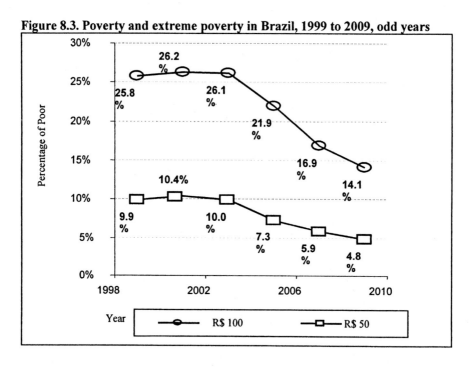

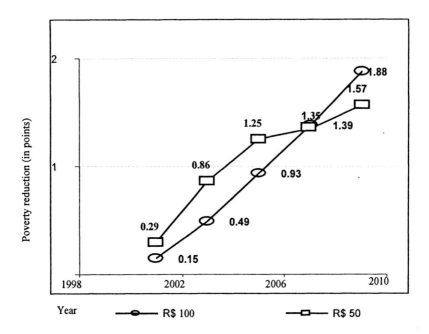

**Panel 1—Poverty and extreme poverty     Panel 2—Reductions due to *Bolsa Família***
*Source*: Soares et al. (2010).

To calculate *Bolsa Família's* effects on poverty, Soares et al. (2010) calculate for each year a counterfactual poverty figure that is what poverty would have been if transfers did not exist. In 1999, when the programs were almost non-existent, there was almost no effect. In 2001, the programs that preceded *Bolsa Família* made extreme poverty 0.1 percentage point less than it would otherwise have been. Not terribly impressive.

From 2005, the effect doubles and in 2009, the program's benefits brought poverty and extreme poverty down 1.9 and 1.6 percentage points, respectively. These numbers represent 13 percent of poverty and 32 percent of extreme poverty in that year, which is much more impressive.

*Bolsa Família's* impacts upon other poverty measures are stronger. The poverty gap and the severity of poverty, also known as FGT (1) and FGT (2), have been more responsive to the Program's benefits than just the number of poor people. This is to be expected since these benefits are able to push over the poverty line only those families who were previously close to it. These same benefits, however, represent a large percentage of poor people's income, strongly improving their lives even if they remain in poverty. This explains why the poverty gap and the severity of poverty fell more than the percentage of poor.

**Table 8.7. —Impacts Upon FGT Poverty Measures**

| Effect / Measure | P0: Percent Poor | P1: Poverty Gap | P2: Severity |
|---|---|---|---|
| With *Bolsa Família* | 21.7% | 9.4% | 5.9% |
| Without *Bolsa Família* | 20.0% | 7.8% | 4.6% |
| Absolute Reduction | 1.64 | 1.68 | 1.30 |
| Percent Reduction | 8% | 18% | 22% |

*Source*: Soares and Satyro.

Table 8.7 shows that while *Bolsa Família* reduces the percentage of poor, P(0), in only 1.6 percentage point, it reduces the poverty gap by 18 percent and the severity of poverty by almost a quarter.

In spite of the improvement of welfare among those who remain poor, bigger impacts on the percent in poverty require higher benefits.

## *Schooling*

There are two basic sources of information on schooling: household surveys and educational administrative records. Silveira Neto (2010) uses the former and Glewwe and Kassouf (2008) the latter to estimate the impact of *Bolsa Família* on schooling.

Silveira Neto uses Propensity Score Matching to estimate *Bolsa Família*'s impact upon school attendance/enrollment.[16] He finds that the program increases attendance/enrollment by about two to three percentage points. This is because attendance/enrollment in Brazil was already close to universal—about 97 percent —before Conditional Cash Transfer Programs really took on scale.

Glewwe and Kassouf use a richer educational dataset—the yearly School Census—to estimate the effects of the program on enrollment, grade promotion, and drop-out. The limitation of their study is that the unit of analysis is the school and they know only if a school has at least one *Bolsa Família* beneficiary enrolled, but not how many. Their results on enrollment are unimpressive and hard to interpret: having a beneficiary increases enrollment by about two percentage points, but this could mean that *Bolsa Família* schools are bigger. The results on grade promotion and drop-out, on the other hand, are so impressive as to be fishy: having a beneficiary in your school reduces your drop-out probabilities by 31 percent and increases your promotion probabilities by 53 percent. A possible interpretation of these immense effects is that movement of students between schools is often recorded as drop-out and *Bolsa Família* students, due to conditionality monitoring, are followed from school to school. Glewwe and Kassouf's is an excellent effort, but is limited by the limitation of available data.

There are no evaluations of the program's effect on learning or cognitive skills. Clearly, this is an area in need of more work.

## *Nutrition*

One of the main reasons for fighting poverty is nutrition. Most scientific poverty lines define poverty as someone who does not have enough to eat. Furthermore, *Bolsa Família* itself was created within the Zero Hunger initiative. Last but not least, there is ample evidence that poor nutritional conditions of young children can hobble their productive capacity and well-being for the rest of their lives. Segall-Corrêa et al. (2008) evaluate a program's impact upon perception of food insecurity and Andrade, Chein, and Ribas (2007) evaluate its impact upon *de facto* measured child nutrition.

Segall-Corrêa et al (2008) estimate a probit to show that every R$10 transferred by *Bolsa Família* reduces a perception index of food insecurity by approximately 8 percent. Searching for more objective measures of impacts, Andrade, Chein, and Ribas use Propensity Score Matching, comparing nutrition of children in *Bolsa Família* families with those in comparable families with no benefit. The authors analyzed the usual indicators for children from 6–60 months: (i) height for age; (ii) weight for height; (iii) weight for age; and (iv) Body Mass Index for age. Separate analyses were undertaken in the Northeast, North/Center–West, and South/Southeast and for poor and very poor families separately. The results show no impacts at all. Children in families with a *Bolsa Família* benefit had the same nutritional profile as those with no benefit.

This is not a novel result. The nutrition effect for *Progresa/Oportunidades* beneficiaries, for example, is unclear. Significant positive impacts were observed on the height of children who were 12–36 months old (Behrman and Hoddinott 2005). However, there is no way of knowing whether this positive impact was due to the nutritional supplements given by the program or to the cash transfer itself.

The lack of nutritional impact is particularly serious when we consider that increasing the next generation's human capital is the main or one of the main objectives of these types of programs. Due to these poor results, the nutritional component was redesigned with nutritional supplements being distributed to the beneficiaries.

This result suggests that a nutritional complementary program may be necessary for *Bolsa Família*.

## *Work*

One of the criticisms repeated *ad nauseam* by the press is that *Bolsa Família* provides adverse labor market participation incentives to its beneficiaries. This is actually an excessively nice way of putting it because what we see in the press is more along the lines of saying that the poor are poor because they are lazy slobs and free money will make them even lazier. Returning to polite conversation, the criticism is that a means-tested transfer will create a labor disincentive, particularly in those extremely poor families that receive the fixed benefit, for which the only condition is being poor.

Microeconomic analysis of the labor participation effects of a transfer conditioned both upon school attendance and low income is not trivial. An unconditional transfer can increase or reduce labor supply since it involves only a wealth effect and does not change relative prices. When the benefit is means-tested, or conditioned to family income, relative prices change in a perverse way and there is an unambiguous incentive toward reducing labor supply.

However—and this is important—job search is a costly activity and it may be that an exogenous income will increase the chances of finding a job. This means even a means-tested transfer may actually increase employment. Finally, once child labor is considered, the inclusion of the education conditionality will lead to an unambiguous reduction of their labor supply, which may or may not affect the labor supply of the adult members but could conceivably increase it. In conclusion, economic theory does not shed much light on the existence of labor market disincentives, transforming it into a purely empirical issue.

By luck or by providence, there are at least eight good evaluations of the impact of *Bolsa Família* on labor supply. Given the importance of the issue, we will provide a brief overview of each.

Ferro, Kassouf, and Levinson (2009) and Ferro and Nicollela (2007) apply probits and other regression methods to 2003 household survey data and find that child labor falls and adult labor rises. The effects, while significant, are small.

Cardoso and Souza (2008) use the 2000 Census and Propensity Score Matching (PSM) to estimate the impact of those programs in existence in 2000 upon school attendance and child labor. Their results are a significant 1.0 percentage point reduction in female child labor and a 0.5 point reduction in male child labor.

Tavares (2008) also uses PSM to estimate the changes in the labor participation behavior of mothers. She uses the 2004 supplement that identifies beneficiary families. She concludes that the benefit reduces the workweek in about 5 percent to 10 percent, which amounts to something between 0.8 to 1.7 hours. Statistically significant but not exactly impressive.

Teixeira (2010) uses the intensity of treatment in a very creative paper. In other words she estimates if the size of the transfer has any impact upon the supply of hours in informal and formal activities. The effects are zero for both men and women in the formal labor market, but negative and significant for those in the informal sector. Despite their statistical significance the impacts are quite small.

Foguel and Barros (2010) make a panel of municipalities covered by the Pnad Household Survey from 2001 to 2005 to investigate the labor market impact of the Program. The authors divide the panel by gender and income groups, so that results are obtained for men and women separately. They find that a 10 percent increase in the proportion of beneficiaries in a municipality increases the female participation rate by a negligible 0.1 percent. For men, the effect is even smaller, such that a 10 percent increase in the beneficiary population increased their supply by only 0.05 percent. For the poorer men, the elasticity is slightly higher (0.01), since a 10 percent increase in the beneficiary population increases the supply by 0.1 percent. For hours worked the elasticity for women in general is about -0.01. Among men, there is no impact at all.

Ribas and Soares (2010) use the same database, at a more disaggregated level (census tract), with a slightly different methodology, and look separately at rural areas, urban areas, and metropolitan areas. They find that *Bolsa Família* increases informality—people do not become lazy but they do try to hide their income from the government—and that rural women reduce nine hours the time they dedicate to paid work. However, in the metropolitan areas, their results point to a reduction in labor supply, not only labor market participation, but also a reduction in labor force participation. This is slightly at odds with previous findings, although the effects are quite small and only borderline significant.

In conclusion, although the exact size and statistical significance of effects varies from study to study, they are always small. The possible exception is the number of hours worked by women with children. Given the importance of early childhood development, this is hardly a negative effect.

## *Fertility*

Another specter that cash transfers to poor kids always brings from the dark is that their parents will breed like rabbits in order to get more money. This fear led to the limits on the number of benefits of child benefits per family: three for kids up to fourteen and another two for teenagers fifteen and sixteen. This can hardly be a regarded as a serious fear when Brazil's birthrate is far below replacement level and we are looking at potentially serious demographic contraction effects in a few decades. A more justifiable ghost is that maybe young girls failing in school will be encouraged to get pregnant when they are young and this will severely handicap their human capital accumulation, frustrating the program's very objectives.

Rocha (2009) estimates *Bolsa Família's* impacts on fertility using a very clever approach in which he uses as an instrumental variable: the Program's three child limit. He builds a treatment group composed of women with two children and a comparison group of those with three or more. Those with three or more have no monetary incentive from *Bolsa Família* to have more kids. The impact variable is whether the woman bore a child in the last twelve months. The results are that *Bolsa Família* has absolutely no effect on fertility decisions.

I know of no studies on teenage pregnancy and *Bolsa Família*. This is clearly another area that needs more work.

## *Citizenship, Gender Roles and Social Isolation*

Quantitative measurement of the effect of an income transfer upon citizenship, gender roles, and social isolation is notoriously hard. There are, however, qualitative studies, such as Suarez and Libardoni (2007), who use semi-structured questionnaires and focus groups composed of both beneficiaries and municipal agents in ten different municipalities. Suarez and Libardoni conclude that the program has strong impacts upon feelings of empowerment and citizenship, strong impacts upon gender relations through empowerment of women in the household negotiations, and also reduces the social isolation of women.

Beneficiaries and municipal agents state strongly that *Bolsa Família* had strong impacts on both possession of official documents such as birth certificates and identity cards and on notions of what citizenship means.

In gender relations the transformation is clear. Although there are undoubtedly cases of violence against women to get the benefits they control, none of the interviewees had ever seen or been exposed to any. Almost all of them stated that the benefit increased their autonomy with relation to their husbands. Now that they have their own money, they no longer have to beg their husbands for money to buy food or clothes.

Finally, one of the conditions that poor people suffer and that both reduces their welfare and keeps them in poverty is their social isolation or small social networks. The poor live in isolated areas and have little contact with neighbors or relatives. Both the money they receive, which provides them with the means to leave home, and also *Bolsa Família* beneficiary groups reduce this isolation.

## *Media Impacts*

Considering the exposure *Bolsa Família* has had in the media, it is surprising that there are not many studies of the relations between the two. The only quantitative study I could find is Lindert and Vanina (2008), who analyze this relationship from a variety of angles.

Their most important conclusions are that both *Bolsa Família* and its most relevant predecessor, *Bolsa Escola,* have been and still are highly visible in the Brazilian press, and media coverage increases with the scale of the program. The authors do not calculate the correlation coefficient, but Figure 8.4 shows that it must be elevated.

A second message is that elections, both national and municipal, bring increased scrutiny of highly visible cash transfer programs. Articles focused specifically on *Bolsa Família* or *Bolsa Escola* become more critical as elections approach (Lindert and Vanina, 2008). This is expected, because, after all, the proximity of elections will raise worries of any program being used for political patronage and a discretionary cash transfer to millions of families is really a prime suspect. The obvious question that follows is whether *Bolsa Família* has, in fact, any impact upon electoral behavior.

**Figure 8.4. Press and program coverage of *Bolsa Escola* and *Bolsa Família***

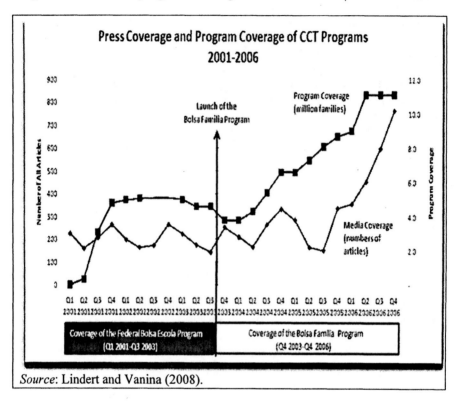

*Source*: Lindert and Vanina (2008).

## *Electoral Impacts*

This is a not as touchy a subject as one would expect. Perhaps because *Bolsa Família* has been subject to strong academic and other scrutiny, perhaps because all municipalities—government or opposition aligned—receive benefits, or perhaps for some other reason, the accusation that *Bolsa Família* is an electoral stunt has been mostly absent from mainstream media (in extremist blogs, on the contrary, it has been quite present).

A strong electoral impact of *Bolsa Família* does not, *per se*, mean the program is an electoral stunt. It can be seen simply as the electorate rewarding good government, much the same way that a government that fosters price stability and employment should be rewarded by the electorate for doing so. This is fortunate, because all studies show a statistically significant impact on voting behavior. Zucco (2011), Marques et al. (2009), Nicolau and Peixoto (2007), Shikida et al. (2009), and Soares and Terron (2008), among others, all analyze the impact of the *Bolsa Família* benefit on President Lula's vote in the 2006 re-election using municipalities as units of analysis.

Although all four come to more or less the same conclusions, probably the best analysis are those of Shikida et al. and Soares and Terron. The two studies consider spatial correlation in their model and use more covariates than either Zucco or Nicolau and Peixoto, so we will reproduce their results here. The main difference between the two is the number of covariates and how the *Bolsa Família* variable is built: Soares and Terron use the ratio between *Bolsa Família* income and total income, and Shikida et al. use the ratio between *Bolsa Família* beneficiaries and total population.

**Table 8.8. —Bolsa Família's Electoral Impact**

| Variable | Soares e Terron | | Shikida et al. | |
|---|---|---|---|---|
| | OLS | Spatial | OLS | Spatial |
| *Bolsa Familia* Variable | 1.63 | 0.81 | *0.07* | 0.04 |
| Constant | 30.08 | 33.27 | -0.04 | -0.45 |
| Per capita income, 2000 | -0.04 | -0.02 | -0.15 | -0.11 |
| % Lula vote in 2002 | 0.47 | 0.37 | *0.02* | 0.02 |
| Urbanization rate | 0.06 | 0.04 | | |
| North (dummy) | 9.82 | 11.33 | | |
| Northeast (dummy) | 12.69 | 17.13 | | |
| Southeast (dummy) | 6.07 | 6.01 | | |
| Center-West (dummy) | *0.10* | *-0.79* | | |
| Distance | | | 0.01 | *0.00* |
| Inequality | | | 0.12 | *0.04* |
| Demographic density | | | 0.05 | 0.04 |
| Child mortality | | | 0.04 | 0.04 |
| Illiteracy | | | 0.02 | 0.05 |

*Source*: Soares and Terron (2008), p. 296; Shikida et al. (2009), p. 12, 13.
*Note*: Numbers in *italics* are not significant at 10%.

The first result, patently visible in all articles, is a strong correlation between any indicator of *Bolsa Família's* presence and the electorate's preference for Lula. This, of course, may be due to the fact the benefit was targeted to poor people who, in 2006, tended to side with Lula.

When regression analysis is performed, however, the result remains strong. Both Soares and Terron and Shikida et al. estimate two models: an Ordinary Least Squares regression with no spatial correlation and a model in which the residuals are spatially correlated. Spatially correlated residuals amount to supposing that there are spatially correlated unobservable factors influencing upon the Lula vote.

Table 8.8 shows that the inclusion of spatially correlated residuals reduces the impact in about half, but the impact remains significant (in Shikida et al.'s analysis it becomes significant with the inclusion of spatial correlation). But how strong is it?

Using Soares and Terron's coefficient (0.81), if we consider that in 2006 *Bolsa Família* accounted for around 0.7 percent of household income, this means that the somewhat less than 0.5 percent of GDP Lula spent on program' benefits won him about 0.5 percent of the vote. Using Shikida et al.'s coefficient (0.04), if we consider that *Bolsa Família* benefits about 22 percent of people in Brazil, then the program won Lula about 0.9 percent of the vote. In both cases, the electoral effect of the program is not irrelevant but neither is it responsible for an electoral tidal wave.

Shikida et al.'s and Soares and Terron's analysis considers only the direct effect the value of the benefit has upon voting. It does not consider an indirect story such as *Bolsa Família* establishing Lula as a candidate who feels the pain of the poor, who in turn vote for him, whether or not they receive a benefit. In any case, once the proper model is run, *Bolsa Família's* electoral impact is undoubtedly real but hardly overwhelming.

The conclusion that I draw from the studies reviewed in this section is that *Bolsa Família* is unquestionably a success story. It is relatively cheap and very well targeted; it has important impacts upon income distribution, reducing inequality in disproportion to its modest contribution to average household income; its impacts upon the poverty rate are not as impressive as some might expect due to the low benefits, but they are greater upon extreme poverty and upon poverty measures that distinguish higher and lower incomes among the poor, such as the poverty gap and the severity of poverty; it does all the above with no noteworthy negative impacts upon labor markets; and finally, it has become an important issue in all of Brazil and slightly changed voting behavior. But what now? A new administration has recently been sworn in which has made eradication of extreme poverty one of its main objectives. Will this change *Bolsa Família*?

## THE FUTURE IS NOW

Brazil saw considerable changes in how cash transfers managed from the mid-nineties until about 2004. We went from decentralized CCTs to chaos to a very large program with both centralized and decentralized aspects. The last six years, however, have seen little more than marginal increases in coverage and benefits getting updated to keep pace with price inflation. There have been no major design or conceptual changes since 2003, but from the first of January of this year (2011), Dilma Roussef is the new president of Brazil and she has made it clear that ending extreme poverty is a top item in her agenda. This cannot be done without a significant change in *Bolsa Família* or the creation of a similar program.

There are two non-exclusive paths to the eradication of extreme poverty: social protection and generation of opportunities. While it leans clearly more toward social protection than opportunity generation, *Bolsa Família* remains a hybrid program whose clear definition Brazilian society preferred to leave to the future. *Bolsa Família* is really not an opportunity generation program such as *Chile Soldiario*: it does not kick out its beneficiaries and its articulation to things such as job training or microcredit is relatively weak. It is not a human capital accumulation program such as *Oportunidades/Progresa*: it does not necessarily keep beneficiaries in for the time needed for a generation of kids to complete school and the conditionalities are still less than draconian. It is not a full blooded social protection program such as Urugay's reformed *Asignación Familiar* (the inheritor of *Plan de Asistencia Nacional a la Emergencia Social*: PANES) or the original Suplicy Basic Income Program: it is not an entitlement and has only weak links to the remainder of the social protection system in Brazil.

Being none of these things, *Bolsa Família* is a hybrid creature being pushed and pulled in different directions by different constituencies, both inside and out of government. If *Bolsa Família*'s hybrid nature were a mere problem of classification, it would give social policy analysts yet another issue to wrangle about. Unfortunately, the problem is more profound: while this chimera remains all things and none, its internal contradictions become more acute and the pursuit of one objective frustrates the others.

Each of the three directions *Bolsa Família* can go in leads to a different program layout and leads to different answers to questions that have haunted the program since its conception. Should conditionalities be draconian or a minor detail? Should they be extended to include chasing opportunity? If it is social protection, should its information system not be integrated to those of social security and unemployment insurance? Should families be kicked out after a certain period? Should exit strategies be a worry at all? In short, *quo vadis, Bolsa Família*?

The Mexican model has been largely absent in the media debate and defended only by a minority of people in the academic debate. Draconian conditionalities and intentionally long benefit periods are not positions much defended in Brazil. In our country, the two paradigms that vie for supremacy are the opportunity generation model and the pure social protection model. Since the extreme poverty eradication debate caught fire, there is little space for a program that promises its real benefits only many years down the road. The choice is between either social protection or generation of opportunities.

What are the consequences of each? There are important design issues that depend upon the road chosen: the Single Registry, conditionality, and the entitlement issue.

*The Single Registry*. Right now the Single Registry, for all its shortcomings, is a surprisingly good registry of poor people. This is not a problem for an information system if the objective is generation of opportunities as no one expects the rich to sign up for microcredit and the like. If the rich are well off, it is precisely because they had opportunities and took advantage of them, so by definition they do not need any more. Furthermore, if the objective of the benefit is only to pave the way to opportunity, temporary illicit accumulation with other cash transfers becomes less serious.

The rich, however, do participate in social protection and, in fact, are some of its main beneficiaries. This means that if *Bolsa Família* is to become 100 percent social protection, integrated into social security and unemployment insurance, the Single Registry should also be integrated into the social security and unemployment insurance information systems. In other words, it should become a registry of citizens, of all the people, and not only of the poor.

Furthermore, if *Bolsa Família* is to be only social protection, illicit accumulation becomes a much more serious issue.

*Conditionality.* Here the choice is stark: while going to school and getting check-ups are very weak conditionalities for an opportunity generation program, they are adequate or even excessive for social protection. There are arguments for conditionality both for social protection and generation of opportunities, but its intensity and scope varies.

If the objective of *Bolsa Família* is social protection, then it should protect the poorest and most vulnerable families and, within these, it should protect the most vulnerable individuals. Vulnerable families are exactly those that will have the most difficulty meeting even mild requirements. In this case, program conditionality should be limited to ensuring that the social protection is passed on to the most vulnerable family members—children—and does not get captured by rotten parents.

On the other hand, if Bolsa Família becomes generation of opportunities, its conditionalities should be closely linked to the services offered. Families should be required not only to send their kids to school, but also to sign up for labor intermediation, go to job training, get microcredit, or otherwise get off the couch and look for work.

*Should it be an entitlement?* Once again the choice is clear–cut. If the objective is to get people to make their own money, of course it should not. Making the benefit an entitlement will only weaken the incentives to get out of poverty by your own effort. If the objective is to protect those most in need, then there is no doubt that the benefit should be an entitlement.

Where will Brazil go now that eradication of extreme poverty is a presidential promise? My own opinion is that *Bolsa Família* should follow in the steps of the *Asignación Familiar* in Uruguay and become integrated into the social protection system. This does not mean that we should give up on helping people leave poverty through their own effort, but that this should be undertaken by another program. To date, Brazil has developed an efficient system of institutions, laws, and norms for transferring money to poor people that works surprisingly well. We can identify the poor with reasonable precision and get the money to them with almost no loss to attrition. Taking advantage of this to rescue all the extreme poor from extreme poverty and incorporating it into social security is little more than common sense.

Brazil has a long and glorious history of failure in government-led generation of opportunities. Precious little has been evaluated and when there are rigorous evaluations, the results show meager results. If we are to do "opportunities" seriously, everything we try should be done with control groups to ascertain whether it works. This requires a program that is substantially different from *Bolsa Família*.

Do I think this is where Brazil will go? To a certain extent and through tortuous paths, yes I do. I believe that there will be a wave of integrated programs, in addition to various and uncoordinated efforts to allow the poor to exit poverty on their own. They will fail in the same way that previous attempts have failed and, apart from some improvement in the incomes of the poorest 5 percent due to economic growth, labor incomes will not change much. Once this becomes apparent, *Bolsa Família* or its successor will step in and fill the poverty gap with money from transfers. And Brazil will no longer have extremely poor people.

$^v$ I would like to thank Natália Sátyro, Sônia Terron, Cleyton Domingues de Moura, Rafael Guerreiro, Pedro Souza, Rafael Ribas, and especially Fábio Veras Soares and Russell Bither-Terry for their help or comments on this chapter. Any errors of fact or interpretation are, of course, mine alone.

## NOTES

1. Soares, Ribas, and Osório (2010) also review the literature on *Bolsa Família's* design and impacts. While it is a smaller piece than this text, it is certainly complementary since their focus is on comparisons with *Chile Solidario* and *Oportunidades*.

2. With a few exceptions, such as Mexico, which saw quite comprehensive land reform a little less than a century ago.

3. Actually the *Cartão Alimentação* was inspired more by the American Food Stamp program than our CCTs.

4. Medida Provisória N° 132, de 20 de outubro 2003.

5. Decreto N° 6.135, de 26 de junho de 2007. (Revogados: Decreto no 3.877, de 24 de julho de 2001, and Decreto de 24 de outubro de 2001).

6. Lei 10.836, de 09 de janeiro de 2004 and Decreto n° 5.209, de 17 de setembro de 2004.

7. The Evaluation and Information Secretariat (Secretaria de Avaliação e Gestão da Informação – SAGE), of the same ministry, is in charge of more analytical evaluations.

8. The only administrative record that entails immediate benefit cancellation is when a beneficiary shows up in the *Sistema de Controle de Óbitos* (SISOBI), which is a Registry of dead people.

9. These screenings could be much improved. Social security and labor market registries are available about one year after the events they register occur. If the social security, labor market, and the Single Registry were to be merged into a really "Single" Registry, this could be done in real time.

10. *Chile Solidário* is not to be confused with income transfer policy in Chile, which counts on on other programs, such as the *Subsidio Único Familiar* (SUF) and the *Pensión Básica Solidaria*, that are paid unconditionally to all poor families in Chile.

11. This Soares is me, but there are four different Soares who have published works on *Bolsa Família* cited here.

12. The TCU is an accounting office linked to the Congress almost completely outside the influence of the executive branch.

13. The Household Survey used for almost anything relating to *Bolsa Família* is the *Pesquisa Nacional por Amostragem de Domicílios* – PNAD. The PNAD was, from 1974 to 2009, a cross-section survey. It will become a panel survey in 2011, but this had not yet happened when this article was written.

14. Much remains to be evaluated, and the institutional evaluations by the Social Development Ministry are far from complete or satisfactory, but remember that Brazil is a country that is only now beginning to have a culture of evaluation.

15. Micro-simulation of an entire household income distribution is very complex and requires many "administrative" decisions. Without access to the computer code itself, it is very hard to reproduce results.

16. The Pnad question on enrollment/attendance is not very precise. It asks attendance but the consensus is that most people answer enrollment.

## BIBLIOGRAPHY

Andrade, Mônica Viegas, Flávia, Chein and Rafael Perez Ribas. "Políticas de Transferência de Renda e Condição Nutricional de Crianças: Uma Avaliação do Bolsa Família." (CEDEPLAR Discussion text no 312) Belo Horizonte: Cedeplar, 2007.

Barros, Ricardo Paes de, Mirela, Carvalho, Samuel, Franco and Mendonça Rosane. "A Queda Recente da Desigualdade de Renda no Brasil." IPEA Discussion text no. 1; 258. Rio de Janeiro: Ipea, 2007. http://www.ipea.gov.br/sites/000/2/publicacoes/tds/td_1258.pdf

_____. "A importância das cotas para a focalização do Programa Bolsa Família." UFF Discussion Text no. 238. Niterói: Faculdade de Economia, Universidade Federal Fluminense, 2008.

Cardoso, Eliana, and André Portela Souza. "The Impact of Cash Transfers on Child Labor and School Attendance in Brazil". Working Paper no. 04-W07, Department of Economics. Nashville, TN: Vanderbilt University, April 2004.

Castro, Jorge Abrahão de, and Lúcia Modesto (Organizers) "Bolsa Família 2003-2010: Avanços e Desafios," v.2. Brasília, 2010.

Ferro, Andrea R., and Nicollela C. Alexandre. "The Impact of Conditional Cash Transfers on Household Work Decisions in Brazil." Presented in the IZA/World Bank Conference on Employment and Development, 2007. http://www.iza.org/conference_files/worldb2007/ferro_a3468.pdf

Ferro, Andrea, Ana Lúcia, Kassouf and Deborah, Levison. "The Impact of Conditional Cash Transfer Programs on Household Work Decisions in Brazil," in *Anais do XXXVII Encontro Nacional de Economia,* Proceedings of the 37th Brazilian Economics Meeting, 2009. http://www.anpec.org.br/encontro2009/inscricao.on/arquivos/000-dde4869521f17def1b2e6c61111aa203.pdf

Foguel, Miguel N., and Ricardo, Paes de Barros. "The Effects of Conditional Cash Transfer Programmes on Adult Labour Supply: An Empirical Analysis Using a Time-Series-Cross-Section Sample of Brazilian Municipalities," pp. 259-293 in *Estudos Economicos* 40, no. 2 (2010). http://www.scielo.br/

Glewwe, Paul, and Ana Lucia, Kassouf. "The Impact of the Bolsa Escola/Familia Conditional Cash Transfer Program on Enrollment, Grade Promotion and Drop out Rates in Brazil." In: Anais do XXXVIII Encontro Nacional de Economia. Salvador: Anpec, 2008.

Hoffmann, Rodolfo. "Transferências de Renda e Redução da Desigualdade no Brasil e em Cinco Regiões, entre 1997 e 2005," pp. 17-40 in *Desigualdade de Renda no Brasil: Uma Análise da Queda Recente,* vol. 2. ed. Barros, Ricardo Paes de; Foguel, Miguel N.; and Ulyssea, Gabriel. Brasília: Ipea, 2007.                    Available                    online                    at http://www.ipea.gov.br/sites/000/2/livros/desigualdaderendanobrasilv2/Cap15.pdf

Hofmann, Rodolfo. "The Evolution of Income Distribution in Brazil: What Promotes and What Restricts the Decline in Inequality." Presented at the conference; *A comparative analysis of growth and development: Argentina and Brazil,* April 22-23, 2010.Chicago: University of Illinois, 2010.

Lavinas, Maria Helena. "Programas de Garantia de Renda Mínima." IPEA Discussion Text no. 596. Rio de Janeiro: Ipea, 1998. http://www.ipea.gov.br/pub/td/1998/td_0596.pdf

Lindert, Kathy, and Vincensini Vanina. "Social Policy, Perceptions and the Press: An Analysis of the Media's Treatment of Conditional Cash Transfers in Brazil." World Bank Social Policy Discussion Paper no. 1008. Washington, DC: The World Bank, 2010.

Lindert, Kathy, Anja, Linder, Jason, Hobbs and Benedicte de la, Briere. "The Nuts and Bolts of Brazil's Bolsa Família Program: Implementing Conditional Cash Transfers in a Decentralized Context." World Bank SP Discussion Paper no. 0709. Washington, DC: World Bank, 2007.

Marques, Rosa Maria, Marcel Guedes, Leite, Áquilas, Mendes, Mariana Ribeiro Jansen, Ferreira. "Discutindo o Papel do Programa Bolsa Família na Decisão das Eleições Presidenciais Brasileiras de 2006," pp. in Revista de Economia Política 29, no. 1 (2009).

Medeiros, Marcelo, Tatiana, Britto, and Fabio Veras, Soares. "Targeted Cash Transfer Programmes in Brazil: BPC and the Bolsa Familia," IPC Working Paper no. 46. Brasília: IPC, 2008. http://www.ipc-undp.org/pub/IPCWorkingPaper35.pdf

Nicolau, Jairo, and Vitor, Peixoto. "As Bases Municipais da Votação de Lula em 2006" presented at Fórum Nacional. Cadernos do Fórum Nacional, no. 6, 2007.

Ribas, Rafael Perez, and Fabio Vera, Soares. "Is the Effect of Conditional Transfers on Labor Supply Negligible Everywhere?" Presented at 6th IZA/World Bank Conference: Employment and Development, May 30-May 31. ITAM, Mexico City, Mexico: ITAM, 2011.

Rocha, Romero. "Programas Condicionais de Transferência de Renda e Fecundidade: Evidências do Bolsa Família." Presented at Sociedade Brasileira de Econometria, 2009.

Segall-Corrêa, Ana Maria, Leticia, Marin-Leon, Hugo, Helito, Rafael Perez, Ribas, Leonor Maria Pacheco, Santos, Rômulo, Paes-Souza. "Transferência de Renda e Segurança Alimentar no Brasil: Análise dos Dados Nacionais." Pp. 39-51 in Revista de Nutrição v. 21 (supplement) 2008.

Shikida, Claudio Djissey, Leonardo Monteiro, Monastério, Ari Francisco de Araujo Jr, Andre, Carraro, Otávio Menezes, Damé. "It is the Economy, Companheiro!": An Empirical Analysis of Lula's Re-election Based on Municipal Data." Economics Bulletin 29, 2 (2009). http://ideas.repec.org/a/ebl/ecbull/eb-08c30073.html.

Sposati, Aldaíza. "Bolsa Família: Um Programa com Futuro(s)," pp. 273-305 in Bolsa Família 2003-2010: Avanços e Desafios - Volume 2. ed. Castro, Jorge Abrahão de and Modesto, Lúcia. Brasília: IPEA, 2010.

Silva, Maria Ozanira, Maria Carmelita, Yazbeck, Geraldo di, Giovanni. A política social brasileira no século XXI: a prevalência dos programas de transferência de renda." São Paulo: Cortez, 2004.

Silveira Neto, Raul da Mota. "Impacto do Programa Bolsa Família Sobre a Freqüência à Escola: Estimativas a Partir de Informações da Pesquisa Nacional por Amostra de Domicílios (PNAD)," pp. 53-71, in *Bolsa Família 2003-2010: Avanços e Desafios,* Volume 2. ed. Castro, Jorge Abrahão de and Modesto, Lúcia. Brasília: IPEA, 2010.

Soares, Fabio Veras, P. Rafael, Ribas, Rafael Guerreiro, Osório. "Evaluating the Impact of Brazil's Bolsa Família: Conditional Cash Transfers in Perspective." *Latin American Research Review* 45, no. 2 (2010): 173-190.

Soares, Fabio Veras, Sergei, Soares, Marcelo, Medeiros, Rafael Guerreiro, Osório. "Programas de transferência de renda no Brasil: impactos sobre a desigualdade." IPEA Discussion Text no.1; 228. Brasília: Ipea, 2006. http://www.ipea.gov.br/sites/000/2/publicacoes/tds/td_1228.pdf

Soares, Gláucio Ary Dillon, and Sonia Luiza, Terron. "Dois Lulas: A Geografia Eleitoral da Reeleição (Explorando Conceitos, Métodos e Técnicas de Análise Geoespacial)," pp. 269-301 in *Opinião Pública* 14, no. 2 (2008).

Soares, Sergei, and Natália, Sátyro. "O Programa Bolsa Família: Desenho Institucional, Impactos e Possibilidades Futuras," IPEA Discussion Text no.1, 424. Brasília: Ipea, 2010.

Soares, Sergei. "Volatilidade de Renda e a Cobertura do Programa Bolsa Família," Texto para Discussão, no. 1459. Brasília: Ipea, 2009.

Soares, Sergei. Análise de bem-estar e decomposição por fatores na queda da desigualdade entre 1995 e 2004. "Econômica: revista do programa de pós-graduação em Economia da UFF." Rio de Janeiro: UFF, v. 8, no. 1 (2006): 83-115.

Soares, Sergei, Rafael Guerreiro, Osório, Fabio Veras, Soares, Marcelo, Medeiros, Eduardo, Zepeda. "Conditional Cash Transfers in Brazil, Chile and Mexico: Impacts upon Inequality," pp. 207-224 in *Estudios Económicos,* Número Extraordinario. Mexico, 2009. http://ideas.repec.org/

Soares, Sergei, Ribas, Rafael Perez, Soares, Fabio Veras. "Targeting and Coverage of the Bolsa Família Programme: Why Knowing What You Measure Is Important In Choosing the Numbers." IPC Working Paper no, 71. Brasília: IPC, 2010. http://www.ipc-undp.org/pub/IPCWorkingPaper71.pdf

Soares, Sergei, Pedro H. G. F, Souza, Rafael Geurreiro, Osório, Fernando Gaiger, Silveira. "Os Impactos do Benefício do Programa Bolsa Família Sobre a Desigualdade e Pobreza," pp. 27-52 in *Bolsa Família 2003-2010: Avanços e Desafios* - Volume 2. ed. Castro, Jorge Abrahão de and Modesto, Lúcia. Brasília: IPEA, 2010.

Souza, Arnaldo Machado, and Ana Maria Medeiros, Fonseca. "O Debate Sobre Renda Mínima: a experiência de Campinas," pp. 22-32 in *São Paulo em*

*Perspectiva*, 11(4): 1997. http://www.seade.gov.br/produtos/spp/v11n04/v11n04_03.pdf

Souza, Pedro H. G. Ferreira de. "Uma Metodologia para Decompor Diferenças entre dados Administrativos e Pesquisas Amostrais, com Aplicação para o Programa Bolsa Família e o Benefício de Prestação Continuada na PNAD." IPEA Discussion Text no. 1, 517. Brasília, 2010.

Suarez, Mireya, and Marlene, Libardoni. "O Impacto do Programa Bolsa Família: Mudanças e Continuidades na Condição Social das Mulheres," pp. xx-xx in *Avaliação de Políticas e Programas do MDS – Resultados. Volume II: Bolsa Família e Assistência Social.* ed. Vaitsman, Jeni and Paes-Souza, Rômulo. Brasília: MDS, 2007.

Tavares, Priscilla Albuquerque. "Efeito do Programa Bolsa Família sobre a Oferta de Trabalho das Mães," in *Anais do XXXVI Encontro Nacional de Economia.* Proceedings of the 36th Brazilian Economics Meeting, 2008. http://www.anpec.org.br/encontro2008/artigos/200807211028050-.pdf

Teixeira, Clarissa Gondim. "A Heterogeneity Analysis of the Bolsa Família Programme Effect on Men and Women's Work Supply." IPC Working Paper no. 61. Brasília: IPC, 2010. http://www.ipc-undp.org/pub/IPCWorkingPaper61.pdf

Yaschine, Iliana, and Laura, Dávila. "Why, When and How Should Beneficiaries Leave a CCT Programme?" *Poverty in Focus* no. 15, 2008. http://www.ipc-undp.org/pub/esp/IPCPovertyInFocus15.pdf

Zucco, Cesar. "The President's 'New' Constituency: Lula and the Pragmatic Vote in Brazil's 2006 Presidential Elections." *Journal of Latin American Studies* 40, 2008: 29-49.

## 9

# THE BIODIVERSITY CONSERVATION: AN EFFECTIVE MECHANISM FOR POVERTY ALLEVIATION?

*Dilys Roe and Terry Sunderland*

There is an explicit assumption that conserving biodiversity (or reducing the rate of biodiversity loss) can help in efforts to tackle global poverty.[1] Evidence of this assumption lies in the target that the Convention on Biological Diversity (CBD) set itself in 2002: "to achieve by 2010 a significant reduction of the current rate of biodiversity loss at the global, regional and national level *as a contribution to poverty alleviation* [emphasis added] and to the benefit of all life on earth."[2] The development community also bought into this assumption: when the Millennium Development Goals (MDGs) were formulated in 2000, for example, Goal 7 included a target to "reverse the loss of environmental resources," one indicator of which was the area of land under protection for biodiversity. Subsequently, the CBD "2010 Target" was included as a new target within MDG7 following the 2006 UN General Assembly[3] with additional biodiversity indicators.[4]

The reduction in the rate of biodiversity loss anticipated in the 2010 target was not achieved.[5] This continued loss of biodiversity is lamented not just for its own sake but for its implications for continued human well-being and poverty reduction. The latest progress report on the MDGs, for example, notes "The irreparable loss of biodiversity will also hamper efforts to meet other MDGs, especially those related to poverty, hunger and health, by increasing the vulnerability of the poor and reducing their options for development.[6] A high level meeting at the September 2010 UN General Assembly further stressed the linkage, claiming: "preserving biodiversity is inseparable from the fight against poverty."[6] The CBD's new Strategic Plan (2011–2020), agreed at the 10th Conference of Parties in Nagoya, Japan, in October 2010 continues to emphasise the link between achieving conservation goals and reducing poverty: its mission being to "take effective and urgent action to halt the loss of biodiversity in order to ensure that by 2020 ecosystems are resilient and continue to provide essential services, thereby securing the planet's variety of life, and contributing to human well-being, and poverty eradication."[7]

In this chapter we explore the nature of this assumed positive relationship between conserving biodiversity and alleviating poverty. We highlight where and how biodiversity contributes to poverty alleviation and explore how effective this contribution is compared to "mainstream" approaches. We draw specifically on the example of great ape conservation to review the experience of biodiversity conservation organisations in tackling poverty.

# The Nature of the Beast: Understanding the Biodiversity–Poverty Relationship

Much has been written about the contribution that biodiversity can make to poverty alleviation goals. This ranges from broad-based overviews to narrow case studies. One body of work in this regard focuses on the MDG framework as a proxy for poverty (see for example Koziell and McNeill[8]; Pisupati and Warner[9]; Roe and Elliott[10]; Hazlewood et al.[11] Sachs et al. 2009[12]). These studies note that biodiversity underpins the delivery of the MDGs beyond MDG 7 as illustrated in Box 9.1.

---

**Box 9.1. Biodiversity and the MDGs**

**MDG 1: Eradicate extreme poverty and hunger**
Biodiversity and ecosystem services are essential to the productivity of agriculture, forests, and fisheries. The soil fertility, erosion control, and nutrient cycling provided by ecosystems enables people to derive food, water, fibres, fuel, and income and livelihoods from natural and managed landscapes. Degraded ecosystems make the poor more vulnerable to increased frequency and impact of droughts, floods, landslides, and other natural disasters.

**MDGs 2 and 3: Achieve universal primary education; Promote gender equality and empower women**
When biodiversity and ecosystem services are degraded or destroyed, the burden falls disproportionately on women and girls, who are forced to travel farther and spend more time in the search for drinking water, fuel wood, and other forest products. This increased burden limits their opportunities for education, literacy, and income-generating activities.

**MDGs 4, 5, 6: Reduce child mortality; Improve maternal health; Combat major diseases**
Genetic resources are the basis for modern and traditional healthcare treatments. Some 80 percent of the world's people rely on traditional healthcare systems that use traditional medicines, mostly derived from plants found in the local environment. The global pharmaceuticals industry also depends on genetic diversity: of the 150 most frequently prescribed drugs, more than half are derived from or patterned after the natural world.

---

Also affecting maternal and child health is the increased spread of malaria, dengue fever, and other insect- and water-borne diseases linked to degraded ecosystems. Loss of biodiversity and ecosystem function can lead to economic disruption, population dislocation, and urban crowding, which encourages the spread of communicable diseases such as tuberculosis, hepatitis, and HIV/AIDS.

**MDG 8: Develop a global partnership for development**
Maintaining biodiversity and the integrity of critical ecosystem functioning will require global partnerships—encompassing government, the private . sector, and civil society in developing and industrial countries. MDG 8 embodies, among other things, the commitment of the developed countries to increase development assistance and open their markets to developing-country products—efforts that should be undertaken in ways that support rather than degrade the biological resource base on which achievement of the MDGs ultimately depends.

*Source*: Hazlewood et al.[11]

The Millennium Ecosystem Assessment (MA) changed the narrative from biodiversity to ecosystem services and from poverty to well-being. The conceptual framework of the MA views biodiversity as underpinning the delivery of a range of ecosystem goods and services on which human well-being depends (poverty being "the pronounced deprivation of well-being."[13] The MA highlights the different winners and losers from biodiversity use noting that while the use (and loss) of biodiversity has benefitted many social groups (largely in allowing for current levels of food production) "... *people with low resilience to ecosystem changes—mainly the disadvantaged—have been the biggest losers and witnessed the biggest increase in not only monetary poverty but also relative, temporary poverty and the depth of poverty.*"[14] It suggests that priority should be given to protecting those elements of biodiversity, and the services it provides, that are of particular importance to the well-being of poor and vulnerable people.   .

The World Resources report published in the same year[15] picks up on this theme. Its central thesis is that well managed ecosystems are productive ecosystems and can generate sustainable income streams. Under the right governance conditions, poor people can tap into these income streams as a route out of poverty. Unfortunately the report also notes that "an array of governance failures typically intervene."[15]

Most recently, the study on The Economics of Ecosystems and Biodiversity (TEEB), which was concluded in 2010, highlighted the contribution of forests and other ecosystems to the livelihoods of poor rural households, and therefore the significant potential for conservation efforts to contribute to poverty reduction. It estimates that that ecosystem services and other non-marketed natural goods account for 47 to 89 percent of the so-called "GDP of the Poor" in some large developing countries. It recommends, *inter alia,* that the role of ecosystem services as a lifeline for the poor should be more fully integrated into policy.[16] It seems clear from these international analyses—not to mention hundreds of documented case studies—that biodiversity is an important asset for the rural poor and, hence, biodiversity conservation has great potential as a mechanism for tackling poverty. There are, however, a number of problems with the body of evidence.

Firstly, the causal relationships are not so simple that one can confidently state that poverty is a direct cause of biodiversity loss, or that the conservation of biodiversity reduces poverty, either directly or indirectly. In exploring poverty–environment linkages Nadkarni[17] describes six different relationships: from a vicious cycle of poverty leading to environmental degradation and thence to more poverty; to a win–win scenario where environmental conservation contributes to poverty alleviation. The same applies to biodiversity. There is clearly a relationship between rural poverty and loss of forest cover and associated biodiversity in developing countries, yet there is much we do not yet understand about the cause-effect links between these two phenomena. For example, we do not yet know to what extent poverty alleviation and forest conservation are converging or diverging policy goals. Other commentators have noted the dynamic and context-specific nature of the biodiversity-poverty relationship.[18, 19, 20] In particular, cross-cutting determinants such as governance, policies on poverty and biodiversity protection, and population growth and density which are associated with the socio-economic context, are critical in determining whether or not biodiversity conservation leads to actual poverty reduction.[21] This suggests a need to be more specific in defining what types of poverty and biodiversity issues are being assessed[22,23] and what levels of integration at both policy and implementation scales are required for optimized outcomes.[24]

And herein a second issue is raised: there is a lack of clarity in terminology and a tendency to conflate terms or use them interchangeably—for example poverty/livelihoods; poverty reduction/poverty alleviation; biodiversity/natural resources; biodiversity conservation/biodiversity use. So claims may be made that poverty has been *reduced* (i.e. people have been transformed from poor to non-poor) when in truth it has been *alleviated* (i.e. some of the symptoms of poverty are addressed but people are not actually transformed from poor to non-poor) or *prevented* (i.e. people are prevented from falling (further) into poverty). Similarly claims may be made that *biodiversity* has delivered poverty benefits when in fact it was a single species (e.g. trophy hunting revenues from elephants) or it was the collective *biomass* of a number of species (e.g. fisheries, non-timber forest products). Furthermore, many of the beneficiaries of biodiversity conservation/use are rural communities who happen to be located close by but who may not necessarily be defined as poor.[25]

It is therefore necessary to make a number of clear distinctions in any discussion of biodiversity–poverty links:

- Poverty reduction *vs.* poverty prevention: It is important to differentiate between: a) the safety net role that biodiversity plays in supporting rural people's livelihoods in developing countries (e.g. see Vira and Kontoleon[26] for a meta-analysis of studies) and b) the role that biodiversity can play in actually lifting people out of poverty (e.g. WRI[15]; Leisher et al.[27]). Sunderlin et al.[28] make this same point with respect to the link between forests and poverty alleviation.
- Biodiversity *vs.* biomass *vs.* selected species: Biodiversity is defined by the CBD as "the variability among living organisms from all sources ..." Species abundance or biomass, or the economic value of a single species or habitat can often make a far more significant contribution to poverty reduction—at least in the short term—than variability. Abundance, biomass, individual ecosystems, species, and genes are all components or attributes of biodiversity but often the term biodiversity is used to refer to just one of these components. Vira and Kontoleon[26] reviewing the evidence for the dependence of the poor on biodiversity note: "the term 'nature's resources' better captures the generic categories of resources that have been studied in this literature. These include forests, both in terms of wood-based and non-timber forest products (NTFPs); mangroves; fish; wild animals (bushmeat) and wild plants (including herbs); and common pool resources (CPRs) more generally." Nevertheless, while only a limited number of species may be in direct use, they themselves are dependent on biodiversity to maintain the ecological systems of which they are part. Diversity is also important for food security, risk avoidance, livelihood insurance, and the delivery of critical ecosystem services.

- Biodiversity use *vs.* biodiversity conservation: Some use of biodiversity may deliver significant poverty benefits in the short term but may be unsustainable in the long term (e.g. felling of Brazil nut trees for timber). Biodiversity conservation implies maintenance, enhancement or restoration of biological resources—not depletion.

- Conservation as an outcome *vs.* conservation as an intervention. Because the rural poor depend on biodiversity for their day to day livelihoods, it is logical that if it is conserved (maintained, enhanced, restored) it can continue to provide livelihood support functions. However the conservation *intervention* that is employed to reach that outcome may itself have a negative poverty impact. For example strict enforcement of protected area and other exclusionary land uses may actually increase local incidence of poverty through the loss of resource access.[29] The way conservation interventions are designed and implemented is a key determinant of their poverty impacts—whether they are protected areas (e.g. West and Brockington[30]); payments for environmental services schemes (e.g. Wunder[31]) or any other mechanisms.[27]

Nevertheless, despite all the problems identified above, and despite caveats about the state of the evidence base,[26, 27, 32] it is clear that that billions of poor people living in rural areas of developing countries are directly dependent on "*biodi*versity"—whether it be individual components, or whether it really is the variability among living organisms.

At the very least, the spatial coincidence between areas of high biodiversity and chronic poverty suggests that the poverty alleviation potential of biodiversity be seriously investigated.[33]

## How Does Biodiversity Contribute to Poverty Alleviation and When Does It Actually Reduce Poverty?

Before answering the question of how biodiversity contributes to poverty alleviation, it is first important to understand what poverty alleviation means. What needs to change in an individual's or household's circumstances for them to be less poor?

"Poverty" is defined differently in different countries and contexts but it usually relates to some level of material wealth. In the MDGs the poor are defined as those living on less than a dollar a day. But it is also widely recognised that poverty is about more than a lack of money. In the MA it is defined as a "profound deprivation of well-being"—where well-being includes security, health, freedom of choice and action, as well as the basic materials for good life (food, shelter, livelihoods, access to goods).[13]

Bass et al.[34] identify a number of factors that contribute to poverty (and, hence, that need to be addressed in alleviating poverty):

1.  Inadequate and often unstable income—resulting in inadequate consumption of necessities that need to be purchased—including food, water, medicine, school fees)

2.  Inadequate, unstable or risky asset base—including both material assets (e.g. productive land, savings) and non-material assets (e.g. health, education)

3.  Poor quality, insecure, hazardous or overcrowded housing

4.  Inadequate provision of public infrastructure (e.g. sanitation, piped water, roads)

5.  Inadequate provision of basic services (e.g. schools, healthcare, communications)

6.  Limited or no safety net to ensure consumption needs are met and access to other necessities (e.g. health, schools) is available without the means to pay

7.  Inadequate protection of poorer groups' rights through operation of the law, particularly protection of land and resource rights

8.  Lack of political voice and power, resulting in little or no entitlements to goods and services; little or no ability to hold government, NGOs, etc. to account

Bass et al.'s analysis focuses on the links between poverty and environment and they note that the relevance of environmental management to poverty alleviation is not immediately obvious when poverty is narrowly conceptualized around income or food consumption. It does, however, become much more obvious with these broader understandings of poverty that focus on assets, security and political power. The same is true of biodiversity (Box 9.2).

**Box 9.2. Biodiversity as Social Safety Net**

**1. Natural Health Service**
Many medicines, both traditional and modern are based on wild plants or compounds extracted from them. It has been estimated that up to 70 000 plant species are used in traditional and modern medical systems throughout the world.[35] For poor people with no physical access to clinics or doctors, and with no cash to pay for commercial drugs, traditional medicine can be the only form of healthcare available. The World Health Organisation estimates that up to 80 percent of the rural populations in Asia and Africa use traditional medicine for primary health care.[36]
Biodiversity not only provides medicines but also reduces the risk of infectious diseases. Fully functioning diverse ecosystems are able to balance the abundance of disease vectors, parasites and parasitic hosts such that the spread of infectious disease is controlled. Changes in biodiversity can provide opportunities for vectors or hosts to flourish if, for example, their natural predators have declined. Many of the diseases whose spread has been associated with changes in ecosystem biodiversity are those that are common amongst poor people. These include malaria, dengue fever, encephalitis, cholera and rabies.[37]

**2. Diversity as Security**
There is a strong body of evidence that high genetic diversity amongst agricultural crops results in both higher, and more consistent, yields.[38] It also protects against crop failure—for example, while drought may affect soil fertility and quality and hence have serious implications for agricultural productivity, maintaining agricultural biodiversity can help to reduce the negative impacts. Traditional varieties of agricultural crops have a higher degree of genetic diversity than modern varieties (as well as requiring fewer inputs of labour and chemicals). In Ethiopia, farmers growing sorghum experienced more crop failures due to drought with the use of modern varieties compared to traditional landraces.[39] This biodiversity benefit is amplified in less fertile, more degraded lands—those that tend to be occupied by poorer people.
The Dutch development agency Hivos administers a biodiversity fund that supports agricultural biodiversity projects. Their experience from ten years of funding found that higher biodiversity in and around farms improves the resilience of food-production systems: soil fertility increases, water supply services improve and pests are better controlled.[40] Food is the primary and most immediate concern of poor people, and this contribution to food security—and in particular the buffering against climate impacts and other environmental and market "shocks"—is thus invaluable.

The contribution that biodiversity conservation makes to alleviating poverty at the individual or household level varies hugely from context to context. In some cases it is biodiversity itself that makes a direct contribution:

- Many components of biodiversity—fish, wildlife, crops etc.—are freely available and can be harvested and used with little processing and with low-cost technologies. Research by IUCN for example, found that villagers in north east Lao relied on over fifty-six types of medicinal plants, forty species of trees, thirty-four different kinds of wild vegetables, fifteen different bamboos as well as a variety of mushrooms, wild fruits, grasses, palms and vines to meet their everyday needs.[41] The same patterns of wild plant use are also recorded in Thailand.[42] Biodiversity thus acts as a form of natural capital—particularly important for individuals and households with little financial or physical capital. The World Bank estimates that forest products provide roughly 20 percent of poor rural families' "income"—of which half is cash and half is in the form of subsistence goods. [43]

- Biodiversity provides the poor with a form of cost effective and readily accessible insurance against risk, particularly food security risks, risks from environmental hazards, and health risks.[26] The evidence suggests that, as the poor have few alternative sources for protecting themselves, they have a higher dependency on biodiversity for dealing with risk. As such, harvesting wild biodiversity provides a safety net whereby the benefits provided by forest resources stop rural dwellers from becoming poorer and provide cash income at critical times of the year, particularly during times of low agricultural production.[44, 45]

- Biodiversity underpins the delivery of a range of critical ecosystem services—clean water, soil fertility and stabilization, pollination services, pollution management—on which we are all dependent, but the poor more so because of their inability to purchase technological substitutes (water filters, chemical fertilizers etc).[14] For example, as natural habitat for bees, bats and other critical taxonomic groups, forests provide pollination services to adjacent agricultural areas. Studies suggest that forest-based ental Services in Agricultural Landscapes, eds. Leslie Lipper, Takumi Sakuyama, Randy Stringer and David Zilberman. Springer Science + Business Media, 1forests were 20 percent and 27 percent higher, respectively, than on sites far from forests. This difference in productivity translated into an additional farm income of approximately US$60 per hectare.[46] Maintaining forests in landscape mosaics can thus increase agricultural productivity and rural incomes.

In other cases, organizations and institutions that act to conserve biodiversity can contribute—through interventions that are intended to generate income, improve governance, increase accountability (Box 9.3).

In a review of over 400 case studies (table 9.1), Leisher et al.[27] found empirical evidence for poverty reduction benefits from six different kinds of conservation interventions including community timber enterprises, nature-based tourism, fish spillover from marine protected areas, protected area jobs, agroforestry and agrobiodiversity conservation.

There is also evidence that four conservation mechanisms may not have been a route out of poverty but at least *contributed to reducing poverty* or *provided a safety net* in times of need: commercialisation of non-timber forest products (NTFPs), payments for environmental services, mangrove restoration, and grasslands management.

**Table 9.1. Summary of Poverty Reduction Evidence for Conservation Mechanisms**

| Mechanism | Number of studies?* | Poverty reduction benefits? | Which groups benefit? | Other benefits? | Is it biodiversity or biomass that is important for poverty reduction? |
|---|---|---|---|---|---|
| Non-Timber Forest Products | Many | Low | Very poor and better off | Nutritional benefits and medicinal properties | Biomass |
| Community Timber Enterprises | Many | Medium | Very poor, moderately poor, and better off | Stronger community organization | Biomass |
| PES | Moderate | Low | Landowners | Stronger property rights, capacity building, social organization | Biomass |
| Nature-based Tourism | Moderate | High | Moderately poor and | Infrastructure and social | Biodiversity |

| Mechanism | Number of studies?* | Poverty reduction benefits? | Which groups benefit? | Other benefits? | Is it biodiversity or biomass that is important for poverty reduction? |
|---|---|---|---|---|---|
| | | | better off | services | |
| **Fish Spillover** | Moderate | High | Very poor, moderately poor and better off | Stronger social cohesion | Biomass |
| **Mangroves** | Moderate | Medium | Very poor, moderately poor | Reduced coastal erosion, storm protection, and greater fish stocks | Biomass |
| **Protected Area Jobs** | Few | Low | Moderately poor and better off | Multiplier effect of local jobs | Biodiversity |
| **Agro-forestry** | Moderate | Medium | Moderately poor and better-off landowners | Helps even-out income fluctuations | Biomass |
| **Grasslands** | Few | Low | Not enough evidence | Stronger social cohesion | Both |
| **Agro-biodiversity** | Few | Medium | Moderately poor and better-off farmers | Global benefits to agriculture | Biodiversity |

*Source*: Leisher et al.[27]  * Many = more than 50 studies. Moderate = 10 to 50 studies. Few = less than 10 studies

---

**Box 9.3. National Policy Framework Enabling Income from Wildlife Management**

In Namibia, rural communities that organize themselves into so-called "conservancies"—geographically defined land units varying in size from 30 km$^2$ to over 9000 km$^2$- are legally entitled to the benefits from wildlife use on their land. In 2008 over fifty conservancies earned a combined cash income of more than N$26 million (about US$3.25 million)—predominantly from tourism and hunting—plus game meat to the value of N$3.06 million (about US$382, 500). Despite these impressive figures, few people were actually lifted out of poverty apart from the fairly limited numbers who gain full-time jobs. Nevertheless, individual conservancy members have received cash dividends of N$100 to N$300 which were still considered significant compared to other sources of income. In addition there are other benefits that are valued as much as—or more than—cash. The remote Marienfluss Conservancy on the Angolan border in the north west of Namibia uses its income to provide transport to clinics and other services nearly 200 km away. Several conservancies provide support to local schools and other social projects, such as soup kitchens for pensioners and support to HIV-AIDS affected orphans.

*Source*: Roe et al.[25]

---

There are however some major challenges to using biodiversity as a mechanism for poverty alleviation. Vira and Kontoleon[26] for example found that the poor tend to depend disproportionately on relatively low value or "inferior" goods and services from biodiversity, while the more affluent groups tend to capture the benefits of resources with higher commercial values, often crowding out the poor in the process. Leisher et al.[27] also found that better-off households with higher social capital were more likely to participate in a conservation programme and were often the main recipients of conservation-related economic benefits. As a result, conservation programmes can sometimes have the effect of widening income disparities.

Vira and Kontoleon[26] also found that risk dependence of the poor on biodiversity often takes the form of a last resort, in the absence of alternatives. This dependence of the poor on low-value activities (and on biodiversity as a last resort against various forms of risk) may confirm the suggestion in some recent literature of a resource-based "poverty trap." This may have important policy implications, as it suggests that the poor may need to break their dependence on biodiversity in order to improve their livelihood outcomes.

## CASE STUDY: *Linking Great Ape Conservation and Poverty Alleviation*

Great ape conservation is not synonymous with *biodiversity* conservation—nevertheless it provides a useful case study since species-focussed conservation is often a major strategy of conservation organizations and institutions. Kaimowitz and Sheil[47] suggest this is because charismatic, endangered species are attractive propositions for fundraising campaigns. However, large—top of the food chain—species can also act as an umbrella for a wide range of less attractive conservation targets, which are the *de facto* beneficiaries of interventions to conserve the wider habitat of the target. The great apes are a case in point, attracting a great deal of conservation interest and funding due to the close genetic relationship with humans but also acting as a charismatic global flagship species for conservation that also promotes public awareness of biodiversity loss.

The four taxa of great ape (gorilla, chimpanzee, bonobo and orangutan) are all members of the same zoological family as humans and we share between 96-98 percent of DNA. Aside from biological similarities, we are also increasingly sharing habitats often resulting in negative human–wildlife interactions. Great apes live in forested regions of sub-Saharan Africa and South East Asia where they are often protected through strictly controlled and enforced conservation areas that can—intentionally or otherwise—have negative impacts on the livelihoods of the already poor local communities, through restrictions on resource access and so on. At the same time, the economic benefits derived from great ape conservation—for example from tourism—are not often shared with local people at a level that generates real incentives for landscape-scale conservation. As a result a potentially valuable resource does not only fail to realize its full poverty reduction potential, but the actual, or perceived, negative impacts of conservation may result in local antipathy—or even outright hostility—to conservation efforts.

Great ape habitats in Africa can be divided broadly into those that are in relatively intact forests, with very low human population density, and those that are in forest fragments, with high human population densities between the fragments. The relationship between biodiversity and poverty is different in these different contexts, and this (alongside other contextual factors such as infrastructure, market development, governance) affects the choice of intervention employed. In the low forest–high population areas (such as Uganda, Rwanda, parts of Nigeria) the main threats to apes and their habitat are primarily poverty driven, represented by forest clearance by small farmers, subsistence hunting, redress for ape incursions onto farmsteads for crop raiding and so on. Here the conservation interventions employed tend to focus on compensation schemes such as social services (support for schools and hospitals, public health and family planning); income generation (enterprise development, tourism); reducing direct pressures on resources (fuel efficient stoves, agricultural improvements) and problem animal management. In the high forest–low population areas (such as the Democratic Republic of Congo, Gabon, parts of Cameroon) local pressure on land and resources is limited but major threats come from commercial forestry and commercial hunting rather than direct localized poverty. Here interventions focus around corporate responsibility, such as certification schemes, community conservation, and the provision of bushmeat alternatives.

Organisations concerned with biodiversity conservation are increasingly aware of the need to address poverty in combination with conservation efforts. Often this is for purely pragmatic reasons (to reduce the threat to target species or habitats). However for a number of organizations including development agencies and those working with or for indigenous/local community rights, poverty alleviation is a key objective and biodiversity conservation is a mechanism to deliver on that objective. Specifically related to great apes, the 2005 Kinshasa Declaration on Great Apes reinforced the connection between poverty alleviation and great ape conservation.[48] A review of efforts to link ape conservation with poverty alleviation in African countries[49] identified a wide range of different approaches taken by different organisations at different sites: from changing the behaviour/attitudes of communities towards conservation to changing the practice of conservation *vis a vis* communities; from finding alternatives to resources of conservation concern to generating benefits from resources of concern; from enforcing conservation priorities to paying for them. Specific examples include:

- Income generation:
o   As a means to incentivize investment in/tolerance of conservation: e.g. employment and/or revenue shares in tourism enterprises; revenue shares from park entrance fees; payments for ecosystem services
o   As a means to reduce pressure on natural resources though alternative livelihood strategies: e.g. beekeeping, improved agriculture, piggeries; facilitating market access for community products
- Providing for subsistence needs: e.g. alternative sources of protein to bushmeat; energy alternatives to firewood, fuel efficient stoves; multiple use zones within protected areas
- Providing social services: e.g. human health and family planning initiatives; support to schools, clinics and other community projects
- Sustaining the natural resource base: e.g. community involvement in protected area management; Risk management/insurance: strategies to avoid or mitigate damage from wildlife (e.g. crop raiding, livestock predation)
- Capacity building: e.g. enterprise training, book-keeping, agricultural extension
- Governance and empowerment: e.g. policy advocacy, community involvement in protected area management

The Dian Fossey Gorilla Fund International has used a number of these different interventions and its experience to date has highlighted a number of lessons[50]:

a)   Social infrastructure projects (such as health and education) have far greater outreach and numbers of beneficiaries than income-generating projects
b)   But income-generating projects have a higher level of impact; the value of money in household/individual pockets cannot be under-estimated
c)   The poorest households have benefitted least from interventions, yet these are the ones that are most likely to access the park and exploit the natural resources.

Overall, however, few initiatives have systematically documented the impact of their interventions although there is some reasonable data on income generation, particularly from tourism (Box 9.4).

**Box 9.4. Great Ape Tourism: A source of income and jobs for poor people**

In 2008 the Sabyinyo Silverback Lodge opened on the borders of Volcanoes National Park in Rwanda. It is a joint venture between Musiara Ltd, International Gorilla Conservation Programme (IGCP), African Wildlife Foundation (AWF), Rwanda Development Board (RDB) and Sabyinyo Community Livelihoods Association (SACOLA). In its first year of operation it generated US$300,000 for SACOLA. The Rwandan Development Board also has a policy (since 2005) of investing 5 percent of the revenue from park entry fees into community projects (such as schools, clinics) worth over US$100,000/year. Similarly in Uganda, parishes adjacent to Bwindi Impenetrable National Park received between $50–$75,000 in total per year between 2005 and 2007, and spent the money on a range of projects including roads and health facilities.

Although the figures sound impressive, the actual impact on poverty levels may however be limited given the extremely high population density around some of these sites. The 5 percent revenue share from park fees in Rwanda, for example, works out at less than US$0.5/year/person to the 300,000 park-adjacent people. Nevertheless, tourism can have one of the few opportunities available in remote rural areas and can more direct impacts on poverty, through the creation of jobs and opportunities to sell goods and services. These were worth around $360,000 in retained revenue for a single parish at the Bwindi tourism hub in 2004—about four times the value of all other sources of revenue to the area combined. Similarly the Sabyinyo Silverback Lodge employs forty-five local people, purchases local produce and supports local tourist service enterprises such as handicrafts, dancing, guiding etc. while the Volcanoes National Park employs nearly 200 people as guides, guards and trackers.

Other sites with ape tourism have fared less well in their ability to generate poverty benefits. A good example is the Dzanga Sangha site in CAR, which has been running since 1997 but has barely been able to cover running costs and has not generated poverty benefits beyond a small number of jobs. Similarly in Asia orangutan-based tourism has not taken off in the same way, despite the economic value of mainstream tourism in the region. The solitary nature of orangutans and their large home ranges make them extremely hard to habituate and thus guarantee sightings in the wild. For this reason, tourists are more likely to visit one of the many rehabilitation centres focused on the care of orphan individuals with a view to them being someday being re-released into the wild.

*Source*: Nielsen and Spencely[51]; Sandbrook and Roe[49]; Sandbrook[52]; Nantha and Tisdell.[53]

The experience of conservation organizations revealed a number of factors influencing poverty outcomes:

1. The scale of poverty: it is very difficult to have a meaningful impact in areas with huge populations of poor people (e.g. in Rwanda population density can reach over 800 people/km$^2$).[51] Moreover, benefits often end up in the hands of a small local elite
2. The availability of economic opportunities: where tourism is possible it can generate meaningful benefits but in remote/undeveloped areas this is very difficult (Box 9.4). Emerging financial mechanisms such as REDD+ or ABS payments may have some future potential.
3. Local capacity: community organisations often lack the capacity to address power imbalances and claim rights that enable them to benefit from land and biodiversity resources.
4. Conservation organisation capacity: conservation projects tend to be very focused on sites, and have limited engagement with national policy and governance processes that are critical for scaling up. Furthermore conservation professionals often lack the development skills that are needed to build capacity at the local level and to engage in political processes.

## How Does Biodiversity-Based Poverty Reduction Compare to the "Mainstream"?

It is hard to compare the effectiveness of biodiversity conservation interventions with "mainstream" development interventions since few organisations routinely monitor the impact of their interventions. We note above the lack of data from the great ape conservation sector. This is likely to be symptomatic of the biodiversity conservation sector as a whole. Equally, Baker[54] notes that "despite the billions of dollars spent on development assistance each year there is still very little known about the actual impacts of projects on the poor." Nevertheless it is possible to compare approaches and experience.

It is clear that reducing poverty is not just a challenge for conservation organisations but also poverty professionals. Economic growth is clearly fundamental to poverty reduction.[55] However, Steele et al.[22] remind us that national-level economic growth is not enough. Poverty in countries such as China and India may have been significantly reduced by rapid economic growth but it has left many behind. Steele et al.[22] point out the need for *targeted* interventions to reach those pockets of poverty bypassed by growth followed by the need to scale up to have a significant poverty impact.

Key lessons from the development sector[22, 34, 56, 57] are that reducing poverty is not all about increasing household income. *Assets* matter (including social assets) as does the *regularity* and *security* of income, more than its actual scale. Interventions that "work" include:

- Building assets and income: including employment; selling local goods and services; increasing access to land and resources; increasing productivity of existing resources.
- Providing or improving infrastructure and services in order to reduce environmental health risks (including clean water, sanitation, safe housing) or mitigate impact of risks (clinics, health services etc.).
- Securing safety nets through social protection and social assistance (including cash transfers) in order to protect people from shocks and reduce vulnerability, help conserve and accumulate assets, help transform economic and social relations.
- Increasing voice and visibility within national political structures and within own locality

Some of these interventions are short term, practical actions to meet immediate needs while others entail long-term support to organise and develop political power and voice.

It is clear from the preceding discussion and from our case study of great ape conservation that biodiversity conservation can contribute to all of these interventions (table 9. 2).

Table 9.2. Biodiversity Conservation as a Development Intervention

| Development interventions that work for poverty alleviation | Biodiversity conservation examples |
|---|---|
| Building assets and income | Maintenance/restoration/enhancement of natural asset base as a result of biodiversity conservation Employment in biodiversity-based enterprises e.g. jobs in wildlife lodges; tour guides, game guards etc.) Revenue sharing from biodiversity-based enterprise (park entry fees, tourism ventures Small enterprise development e.g. tourism, sales of NTFPs; handicrafts; wildlife trade). Increasing agricultural productivity (as a strategy to reduce pressure on biodiversity resources) |
| Infrastructure and services | Maintenance/restoration/enhancement of pro-poor ecosystem services (e.g. medicinal plants; soil fertility; agricultural biodiversity; water purification) Extension to local communities of infrastructure/services provided for conservation personnel and or tourists (e.g. roads, communications, piped water) Provision of infrastructure/services from conservation income (e.g. support to schools, clinics, market links) Conservation-linked human health and family planning initiatives |
| Securing safety nets | Maintenance/restoration/enhancement of biodiversity-based healthcare, wild foods, etc. Insurance/risk management value conferred by diverse resource base Regular cash from revenue shares; Compensation for wildlife damage Fuel-efficient stoves |
| Increasing voice | Community involvement in biodiversity management; Clarification/strengthening of land and resource rights Strengthening local institutions for sustainable resource management |

Many examples of development interventions that work are things that conservation organisations already do (or could do) but it is important to recognise that there are potential trade-offs between poverty reduction and biodiversity conservation. For example, where poverty reduction/economic growth is a priority, biodiversity may be overlooked (e.g. palm oil development in South East Asia is fuelling deforestation and habitat loss for already endangered species such as orangutans). Similarly, placing a high priority on biodiversity conservation–e.g. protecting endangered species or habitats in well guarded game reserves or national parks—can mean that poverty—for those who were reliant on the natural resources within that reserve—is increased or exacerbated. Win-wins are hard to come by, particularly in poor countries whose economies are heavily reliant on natural resource exploitation (particularly forests).

It is critical to realise that both biodiversity conservation and poverty reduction are inherently political processes. Decisions about how resources are used and by whom are often made independently from any concern with either poverty reduction or biodiversity conservation but rather reflect the political interests of the decision-maker. Promoting pro-poor biodiversity conservation thus implies addressing perverse political structures and processes: promoting good governance (or "good enough" governance) at all levels—international to local.

## CONCLUSION

### Should Development Pay More Attention to Biodiversity?

It is clear from this review that biodiversity conservation is not a panacea for poverty alleviation. Many of the claims and assumptions that are made about the linkage between these two societal objectives are based on weak evidence, conflated or confused terminology and inconsistent indicators of success. Nevertheless, mainstream development is equally not a panacea for poverty alleviation and biodiversity conservation can certainly contribute significantly to desirable development outcomes. Although there may be far more (and far more effective) pathways out of poverty, biodiversity does seem to have a particular role to play in *poverty avoidance*, which should be a key component of anti-poverty strategies. In many cases, in remote rural areas, it may be the only option for poverty avoidance, e.g. for people living beyond the reach of national social protection programmes. Furthermore, much of "what works" for poverty reduction—e.g. employment, regular income, access to health, education etc.—is—or can be—generated by conservation interventions (notwithstanding the poverty avoidance benefits of conservation in the first place).

At the very least, the overlapping location of chronic rural poverty and biodiversity provides a strategic opportunity. Given that biodiversity represents a natural asset to the rural poor—and one of the few they have—its potential role in poverty reduction, alleviation and/or preventions should be given serious consideration in strategic planning and should be seriously considered in strategic planning for poverty alleviation. Equally, biodiversity strategies and plans should take into account the needs—and rights—of the rural poor including in the types of conservation interventions that are employed, the components of biodiversity that are targeted, and the types of benefits that are delivered. A critical first step in this regard is more and better communication and collaboration between those that "do" poverty reduction and those that "do" conservation in order to: a) learn from each other as to what works and what doesn't, b) identify where seemingly different interventions and objectives can be mutually supportive and where they can't; and c) increase the effectiveness of individual efforts to tackle the governance issues that affect both poverty and biodiversity outcomes.

## NOTES

1.    Roe, Dilys, and Joanna Elliot. "Biodiversity Conservation and Poverty Reduction: An Introduction to the Debate." In Roe, Dilys, and Joanna Elliot (eds.) *The Earthscan Reader in Poverty and Biodiversity Conservation.* London: Earthscan, 2010.

2.    SCBD. "Decision VI/26: Strategic Plan for the Convention on Biological Diversity." Montreal: Secretariat of the Convention on Biological Diversity, 2002.

3.    United Nations. "Report of the Secretary General on the Work of the Organization." New York: United Nations, 2006.

4.    United Nations. "Official List of MDG Indicators, 2008." Accessed March                    10,                    2011. http://mdgs.un.org/unsd/mdg/Host.aspx?Content=Indicators/OfficialList.htm

5.    Mace, G. M., W, Cramer, S, Diaz, D.P, Faith, A, Larigauderie, P, Le Prestre, et al. "Biodiversity Targets After 2010." *Current Opinion in Environmental Sustainability* 2, 2010: 1-6.

6.    United Nations. Secretary-General, at High-Level Meeting, Stresses Urgent Need to Reverse Alarming Rate of Biodiversity Loss, Rescue. *Natural Economy* Press Release Sept 22, 2010. Accessed March 10, 2011. http://www.un.org/News/Press/docs/2010/ga10992.doc.htm

7. SCBD. "Global Biodiversity Outlook 3." Montreal: Secretariat of the Convention on Biological Diversity, 2010.

8. Koziell, Izabella, and Charles I. McNeill. "Building on Hidden Opportunities to Achieve the Millennium Development Goals: Poverty Reduction through Conservation and Sustainable Use of Biodiversity." London: IIED, WSSD Opinion Paper, 2002.

9. Pisupati, Balakrishna, and Emilie Warner. "Biodiversity and Millennium Development Goals." IUCN Regional Biodiversity Programme, Asia, Colombo, Sri Lanka, 2003.

10. Roe, Dilys, and Joanna Elliott. "Meeting the MDGs - is conservation relevant?" In Roe, Dilys (ed.), *The MDGs and Conservation: Investing in Natures Wealth for Society's Health.* London: IIED, 2004.

11. Hazlewood, Peter, Geeta, Kulshrestha, and Charles McNeill. "Linking Biodiversity Conservation and Poverty Reduction to Achieve the Millennium Development Goals." In Roe, Dilys (ed.), *The MDGs and Conservation: Investing in Natures Wealth for Society's Health.* London: IIED, 2004.

12. Sachs, Jeffrey D., Jonathan E. M, Baillie, William J, Sutherland, Paul R, Armsworth, Neville, Ash, John, Beddington, et al. "Biodiversity Conservation and the Millennium Development Goals." *Science* 325, 2009: 1502-1503.

13. Millennium Ecosystem Assessment. "Ecosystems and Human Well-being: Current State and Trends." Washington, D.C: World Resources Institute, 2005a.

14. Millennium Ecosystem Assessment. "Ecosystems and Human Well-being: Biodiversity Synthesis." Washington, D.C: World Resources Institute, 2005b.

15. WRI. "World Resources 2005 – The Wealth of the Poor: Managing Ecosystems to Fight Poverty." Washington D.C: UNEP, UNDP, World Bank and World Resources Institute, 2005.

16. TEEB. The Economics of Ecosystems and Biodiversity. "Mainstreaming the Economics of Natures: A Synthesis of the Approach, Conclusions and Recommendations of TEEB." Nairobi: UNEP, 2010.

17. Nadkarni, M.V. "Poverty, Environment, Development: A Many-patterned Nexus." *Economic and Political Weekly* 35, no.14 (2000): 1184-1190.

18. Kepe, Thembela, Munyaradzi, Saruchera, and Webster J. Whande. "Poverty Alleviation and Biodiversity Conservation: A South African Perspective." Oryx 38, no. 2 (2004): 143-145.

19. Redford, Kent H., and Steven E. Sanderson. "No Roads, Only Directions." *Conservation & Society* 4, no. 3 (2006): 379-382.

20. Bird Life International. *Livelihoods and the Environment at Important Bird Areas: Listening to Local Voices.* Cambridge: BirdLife International, 2007.

21. Tekelenburg, A., B.J.E, ten Brink, and M.C.H. Witmer. *How do Biodiversity and Poverty Relate? An Explorative Study.* Bilthoven, Netherlands: Netherlands Environmental Assessment Agency (PBL), 2009.

22. Steele, Paul, Neil, Fernando, and Maneka Weddikkara. *Poverty Reduction That Works: Experience of Scaling Up Development Success.* London: Earthscan, 2008.

23. Walpole, Matt, and Lizzie Wilder. "Disentangling the Links between Conservation and Poverty Reduction in Practice." *Oryx* 42, no. 4(2008): 539-47.

24. Adams, William M., Ros, Aveling, Dan, Brockington, Barney, Dickson, Joanna, Elliot, Jon, Hutton, Dilys, Roe, Bhaskar, Vira, and William Wolmer. "Biodiversity Conservation and the Eradication of Poverty." *Science* 306 (2004): 1146-1149.

25. Roe, Dilys, Matt, Walpole, and Joanna Elliott. "Linking Biodiversity Conservation and Poverty Reduction: What why and how?" *Summary Report of the ZSL Symposium: Linking Biodiversity Conservation and Poverty Reduction,* April 28-29th, London, 2010. Accessed 10 March 10, 2011, http://povertyandconservation.info/docs/Masindi_Workshop_Report-Final.pdf

26. Vira, Bhaskar, and Andreas Kontoleon. "Dependence of the Poor on Biodiversity: Which Poor, What Biodiversity?" In Roe, Dilys (ed.), *Linking Biodiversity Conservation and Poverty Reduction: A State of Knowledge Review.* Montreal: Convention on Biological Diversity, Technical Series no 55, 2010.

27. Leisher, Craig, M, Sanjayan, Jill, Blockhus, Andreas, Kontoleon and S. Neil, Larsen. "Does Conserving Biodiversity Work to Reduce Poverty?". In Roe, Dilys (ed.), *Linking Biodiversity Conservation and Poverty Reduction: A State of Knowledge Review.* Montreal: Technical Series no 55, Convention on Biological Diversity, 2010.

28. Sunderlin, William, Arils, Angelsen, Brian, Belcher, Paul, Burgess, Robert, Nasi, Levania, Santoso, and Sven Wunder. "Livelihoods, Forests, and Conservation in Developing Countries: An Overview." *World Development* 33, no. 9 (2005): 1383-1402.

29. Cernea, Michael M. "Restriction of Access' is Displacement: A Broader Concept and Policy," *Forced Migration Review* 23, 2005: 48-49.

30. West, Paige, James, Igoe, and Dan Brockington, "Parks and Peoples: The Social Impact of Protected Areas." *Annual Review of Anthropology* 35, 2006: 251-277.

31. Wunder, Sven. "Payments for Environmental Services and the Poor: Concepts and Preliminary Evidence." *Environment and Development Economics* 13, no. 3 (2008): 279-297.

32. Agrawal, Arun, and Kent Redford. "Poverty, Development and Biodiversity Conservation: Shooting in the Dark." New York: Wildlife Conservation Society, Working Paper No. 26, 2006.

33. Sunderlin, William, Sonya, Dewi, and Atie Puntodewo. "Poverty and Forests: Multi-country Analysis of Spatial Associate and Proposed Policy Options." Bogor: Centre for International Forestry Research, Occasional Paper No. 47, 2007.

34. Bass, Steve, Hannah, Reid, David, Satterthwaite, and Paul Steele. *Reducing Poverty and Sustaining the Environment: The Politics of Local Engagement.* London: Earthscan, 2005.

35. Herndon, Christopher N. and Rhett A. Butler. "Significance of Biodiversity to Health." *Biotropica* 42, no. 5 (2010): 558-560.

36. WHO. "Traditional Medicine." World Health Organization: Fact sheet No.134, December 2008.

37. CIFOR. "Forests and Human Health." Bogor: Centre for International Forestry Research, CIFOR Info brief 11, 2006.

38. Chappell, Michael J. and Liliana A. LaValle. "Food Security and Biodiversity: Can We Have Both." *Agriculture and Human Values* 28, no.1 (2009): 3-26.

39. Arnold, J. and E. Michael. *Managing Ecosystems to Enhance the Food Security of the Rural Poor. A Situation Analysis Prepared for the IUCN.* Gland: IUCN, 2008.

40. Hivos. *Biodiversity and Poverty Reduction. 10 Years Experience with the Biodiversity Fund.* Netherlands: Hivos, 2009.

41. Foppes, Joost, and Sounthone Ketphanh. "NTFP Use and Household Food Security in Lao PDR." Paper presented for the NAFRI/FAO Symposium on "Biodiversity for food security," Vientiane, 14th September, 2004.

42. Delang, Claudio O. "The Role of Wild Food Plants in Poverty Alleviation and Biodiversity Conservation in Tropical Countries." *Progress in Development Studies* 6, 2006: 275-286.

43. Vedeld, Paul, Arild, Angelsen, Espen, Sjaasrad, and Gertrude K. Berg. "Counting on the Environment: Forest Income and the Rural Poor." Environmental Economics Series No. 98. Washington, D.C: World Bank, 2004.

44. Angelsen, Arild, and Sven Wunder. "Exploring the Forest-poverty Link: Key Concepts, Issues and Research Implications." Bogor: Centre for International Forestry Research, CIFOR Occasional Paper No. 40, 2003.

45. Ros-Tonen, M.A.F, and K.F. Wiersum. "The Scope for Improving Rural Livelihoods through Non-timber Forest Products: An Evolving Research Agenda." *Forests Trees Livelihoods* 15, 2005: 129-48.

46. Ricketts, Taylor H., Daily, Gretchen C., Paul R., Ehrlich, and Charles D. Michener. "Economic Value of Tropical Forest to Coffee Production." *Proceedings of the National Academy of Sciences USA* 101, no. 34(2004): 12579–12582.

47. Kaimowitz, David, and Douglas Sheil. "Conserving What and for Whom? Why Conservation Should Help Meet Basic Needs in the Tropics." Biotropica 39, no.5 (2007): 567-574.

48. Specifically the Kinshasa Declaration states: "Encourage the provision of long-term ecologically sustainable direct and indirect *economic benefits to local communities*, for example, through the introduction or extension of carefully regulated sustainable ecotourism enterprises in areas of great ape habitat, and the creation of long-term research projects operating in or near these areas" (Target 7:) "Developing ecologically sustainable local poverty-reduction strategies which recognize and integrate the needs of local communities sharing great ape habitats, while securing the lasting health of the environmental resources upon which they depend" (Target 10d).

49. Sandbrook, Chris, and Dilys Roe. "Linking Biodiversity Conservation and Poverty Alleviation: The Case of Great Apes." PCLG Discussion Paper, 2010. Accessed March 10, 2011, http://povertyandconservation.info/docs/20100808Linking_Ape_Conservation_a nd_Poverty_Alleviation.pdf. PCLG. *Linking Great Ape Conservation and Poverty Alleviation: Learning from Experiences and Identifying New Opportunities.*

50. Nielsen, Hannah, and Anna Spenceley. "The Success of Tourism in Rwanda – Gorillas and More." Background paper for the African Success Stories Study. Mimeo. World Bank and SNV (The Netherlands Developmemt Organisation), 2010.

51. Sandbrook, Chris. "Putting Leakage in Its Place: The Significance of Retained Tourism Revenue in the Local Context in Rural Uganda." *Journal of International Development* 22, no. 1 (2009): 124-136.

52. Nantha, H. Swarna, and Clem Tisdell. "The Orangutan-oil Palm Conflict: Economic Constraints and Opportunities for Conservation." *Biodiversity and Conservation* 18, 2009: 487-02.

53. Baker, Judy L. *Evaluating the Impact of Development Projects on Poverty: A Handbook for Practitioners.* Washington DC: World Bank, 2000.

54. Chandy, Laurence, and Geoffrey Gertz. "Poverty In Numbers: The Changing State of Global Poverty from 2005 to 2015." Washington DC: The Brookings Institution, Policy Brief 2011-01, 2011.

55. Lawson, David, Hulme, David, Matin, Imran, and Karen Moore. "What Works for The Poorest: Poverty Reduction Programmes for the World's Ultra Poor." London: Practical Action, 2010.

56. Hanlon, Joseph, Armando, Barrientos, and David Hulme. "Just Give Money to the Poor. The Development Revolution from the Global South." London: Kumarian Press, 2010.

## 10

TRACING THE REDD BULLET: IMPLICATIONS OF MARKET-BASED
FOREST CONSERVATION MECHANISMS FOR POVERTY
ALLEVIATION IN DEVELOPING COUNTRIES

*Bryan R. Bushley and Rishikesh R. Bhandary*

## INTRODUCTION

Halting the destruction of forests and reducing rural poverty have both been high
on the development agenda during the past few decades, prompting intensive
responses from international organizations, donor agencies, governments and
non-governmental organizations. These include policy and program innovations
at both the national and international levels aimed at promoting both forest
conservation and poverty alleviation imperatives simultaneously. Recently,
forests have been recognized for their contribution to the global environmental
service of sequestering and storing carbon dioxide from the atmosphere, thereby
helping mitigate climate change. This has spawned various new international
mechanisms and markets for the trading of carbon offsets. This chapter
examines the implications of past market-based forest conservation mechanisms
and evolving forest-carbon offsetting schemes for poverty alleviation.

Since the birth of the Bretton Woods system in the mid–20th Century,
poverty alleviation has become increasingly central to international development
assistance. There have been many concerted efforts aimed at reducing poverty,
following the Millennium Development Goals and other global development
objectives. These include general assistance initiatives, such as loans for
infrastructure and economic development in remote areas, national poverty
alleviation funds, national debt-relief programs, and microcredit. They also
incorporate sector-specific programs targeting the poor, such as skills
development for local manufacturing and marketing, agricultural extension
programs, community health and sanitation initiatives, drinking water projects,
and community development and resettlement programs.

Some scholars and development practitioners argue that poverty is one of
the leading causes of environmental degradation and deforestation, stating that
there is a "vicious cycle" of mutual reinforcement between the two and that

broad-based economic development is the solution to both maladies.[1] Others claim that both poverty and environmental degradation are the result of unchecked development, resulting in the unsustainable exploitation of human and natural resources.[2] Whichever side of the poverty–environment debate one stands on, it has become increasingly difficult to deny that there is a connection between them.

Acknowledging this connection, recent conservation policies and programs have focused on providing economic incentives for conservation that help raise the income and livelihood standards of forest-dependent communities in developing countries, especially those communities living in close proximity to areas identified as having particular ecological significance. These programs include *international fiscal transfers* (multilateral or bilateral) in the form of targeted projects under the auspices of international conservation and development organizations and global funding mechanisms, such as debt-for-nature swaps and the Global Environmental Fund (GEF); *transnational market-based mechanisms* like sustainable forest management certification and carbon trading; and *domestic policies and programs*, including various incentive-based programs, conservation trusts, and economic assistance initiatives in ecologically threatened areas like buffer zones surrounding protected areas.

The convergence of conservation and development efforts has also led to the design and implementation of mechanisms to compensate local communities for the environmental services they provide by protecting forests and watersheds. These efforts have resulted in a rapidly evolving new conservation paradigm called payments for environmental services (PES), based on the establishment of mechanisms for compensating land managers for the continued provision of valuable environmental services such as water supply for drinking, irrigation and hydropower generation; scenic beauty and recreational opportunities; and the conservation of unique wildlife and plant species, ecosystems and biodiversity. By creating an economic value and markets for these services and amenities, it is hoped that they will be preserved, while also promoting the socioeconomic development of local communities.

Over the past decade, the sequestration and storage of carbon by forests has become recognized as a vital environmental service that can help address the global threat of climate change. As a result, new voluntary carbon-offsetting markets that compensate land managers for the provision of this service through concerted conservation efforts have emerged. Now, a new global forest–carbon trading mechanism known as reducing emissions from deforestation and forest degradation, and enhancing forest carbon stocks in developing countries (REDD+) is taking shape through the ongoing international negotiations of the

United Nations Framework Convention on Climate Change (UNFCCC). A preliminary agreement on REDD+ was one of the key outcomes of the COP–16 climate talks in Cancun in December, 2010. The agreement outlines a phased approach that focuses on building technical and institutional capacity toward the goal of implementing REDD+ programs. Subsequent meetings of the UNFCCC will decide how the necessary funds will be generated. Donors, governments and NGOs are currently implementing various pilot projects in developing countries to demonstrate the social, economic and ecological feasibility of REDD+ at the national and sub-national levels. In addition, a group of developed countries have instigated the "REDD+ Partnership," which has garnered pledges of over US$4 billion to support readiness activities in various developing countries worldwide. Although REDD+ has galvanized significant support internationally, among both developed and developing countries, its implications for poverty alleviation at the local level remain unclear.

This chapter traces the history of market-based forest conservation mechanisms from their origins to REDD+, and discusses the outcomes, synergies and implications of these mechanisms in relation to poverty alleviation goals. It concludes that such mechanisms provide no inherent benefits for the poor or other socially marginalized groups. Rather, the impact of such mechanisms on these groups depends to a large extent on institutional arrangements and incentives across multiple levels (international to local), local socioeconomic and political relations, and the degree to which new markets are integrated into existing forest management practices, livelihood activities and needs. Furthermore, emerging global mechanisms for carbon trading such as REDD+ entail a host of other political, financial, institutional and technical challenges that must be overcome if they are to function at all, let alone help reduce poverty. If these new schemes truly aim to support the development and aspirations of the poor and marginalized, they should heed the lessons from both the successes and failures of past market-based forest conservation mechanisms.

## SUCCESS AND IMPACT OF FINANCIAL AND MARKET-BASED MECHANISMS

Since recognition of the failings of the "fortress" or command-and-control approach to biodiversity conservation, governments and international donor organizations and NGOs have been searching for innovative ways to promote economic incentives for the conservation of tropical forests and to enhance the benefits to those who rely on them for their livelihoods. Some of the prominent mechanisms that have been tried include debt-for-nature swaps, payments for environmental services, forest products certification and marketing schemes for both timber and non-timber forest products. All of these involve the commodification of nature or sustainable forest management in some way, either directly or indirectly.

Market-based solutions have been promoted on the grounds that deforestation and environmental degradation result from market failure, stemming from the inability to account for the full value of nature and the services, both material and non-material, that it provides to humans. It is largely believed that this "market failure" or undervaluing of nature, coupled with poverty and high demand for forest products, leads to the excessive exploitation and degradation of natural ecosystems and the resources and services they provide.[3] McCauley (2006) argues that the value of all services provided by nature—including its aesthetic, cultural and evolutionary significance—is priceless and that, therefore, these services will always be undervalued by any attempt to attach a price tag to them.[4] While he does not deny that an ecosystem services approach can work in some circumstances, he argues, "Nature conservation must be framed as a moral issue and argued as such to policy-makers, who are just as accustomed to making decisions based on morality as on finances."[5] McCauley (2006) cites four reasons why the valuation of nature is undesirable: (a) Many aspects of nature provide no tangible services to humans and thus cannot be properly valued under an ecosystems services approach; (b) the value of ecosystem services is not constant and can shift with changes in land use and/or international markets, effectively devaluing nature; (c) human ingenuity can find substitutes for nature's services, rendering them less valuable and more vulnerable to exploitation; and (d) conservation of nature is not always compatible with enhancing biodiversity.[6]

In their edited volume entitled "Neoliberal Environments," Heynen et al. (2007) examine the impacts of market–based mechanisms and neoliberal policies on environmental governance and change in a broad range of scenarios and contexts, including wildlife management, water resources, common property resources, mining, wetlands, agrarian landscapes, forest management, and fisheries.[7] They highlight a number of problems associated with these various "experiments" representing the neoliberalization of nature, foremost of which are environmental degradation and destruction. Heynen et al. (2007) argue for a radical transformation of the current embrace of neoliberal environmentalism, away from reliance on market orthodoxies and toward alternative "environmental futures":

The failed logic of neoliberalism and its ravenous craving for markets, commodities, and sites of accumulation across the planet, propels a loss of species that it has promised to defend, a destruction of ecosystems it has claimed to value, and a reduction in the quality of life that it professed to maintain. It is in need of replacement! We require utopian forms of environmental praxis to help us imagine alternative possibilities, emancipatory projects and an end to social and environmental destruction at all scales.[8]

Despite these strong critiques of market environmentalism, there is considerable evidence from a growing number of programs, pilot projects, case studies and policy experiments that financial incentives, if carefully designed, can play an important role in regenerating and protecting valuable natural ecosystems such as forests; and that the conservation of biodiversity is indeed compatible with development and poverty alleviation goals. In a statistical comparison of ninety-seven World Bank projects with both environmental and development goals, and sixty-one projects with explicit biodiversity goals, with projects focusing strictly on development, Kareiva et al. (2008) found that the inclusion of environmental or biodiversity goals did not in any way compromise the accomplishment of development objectives.[9] Furthermore, they found that the most significant predictor of success for biodiversity projects was the inclusion of market mechanisms and sustainable financing approaches.[10] However, they cautioned that, regardless of their focus, less than 20 percent of the projects examined were deemed "highly satisfactory," so it is not easy to achieve win-win outcomes for biodiversity conservation and development.[11]

Di Leva (2002) takes the middle road, stating that, while market-based mechanisms alone probably cannot reduce the rapid rate of biodiversity loss around the globe, they may help to alleviate poverty and thereby reduce some of the pressure on natural resources and ecosystems.[12] He adds that market-based approaches are particularly relevant where private sector resources dwarf public sector funds, and argues that, under such circumstances, public-private partnerships should be formed to leverage these private resources.[13] Di Leva (2002) further stresses that conservation outcomes can also be enhanced by reducing existing disincentives for conservation, such as national agricultural subsidies, and bolstering regulatory efforts, especially in developed and rapidly industrializing countries that consume the bulk of the world's energy and resources.[14]

Kremen et al. (2000) conducted a comparative economic analysis of conservation values in and around a national park in Madagascar relative to financial returns from other, extractive land uses at multiple geographical scales: *local* (sustainable community forestry, ecotourism, NTFP production, hill rice farming, industrial logging); *national* (donor investments, ecotourism/park employment, sustainable community forestry/biodiversity products, sustained use of NTFPs, watershed protection value, internal benefit from ICDP project, industrial logging, hill rice farming); and *global* (carbon sequestration value, donor investments in ICDP).[15] They note that while local incentives matter, national and global incentives are key to the success of conservation efforts, because national governments often make large-scale natural resource decisions affecting conservation, and the international community supports conservation through foreign aid and technical assistance.[16] They found that conservation values for tropical forests exceeded returns from agriculture and logging at both the local and global levels, but that industrial logging was more financially attractive than conservation when viewed from a national perspective. They further claim that this inconsistency and corresponding differences in economic incentives across scales may exacerbate tropical deforestation; but that this problem could be overcome by the introduction of carefully crafted market mechanisms for protecting forests as a form of climate change mitigation, under the Kyoto Protocol or a similar global policy framework.[17]

Spiteri and Nepal (2006) stress that while some suggest abandoning the incentive-based approach to biodiversity conservation, perhaps the best solution is to work on improving the shortcomings of existing incentive-based programs to ensure the simultaneous fulfillment of multiple objectives: "the ultimate goal of [incentive-based programs] is to reduce conflicts between the social and economic needs of rural communities and the need to protect the environment."[18] In this regard, they note that such mechanisms can play a constructive role in making markets work for conservation.[19] However, they also cite some equity issues with incentive-based programs, including the leakage of benefits to elites or migrants and the corresponding marginalization of socioeconomically disadvantaged and indigenous groups, which are aggravated by the lack of residency requirements for participants.[20] Spiteri and Nepal (2006) conclude that benefits from incentive-based programs are of different value to different people within a community—depending on their socioeconomic status, their specific occupation and livelihood strategy, their status as indigenous or migrants, local tenure systems, and geographic location relative to protected areas and other land use designations—and that the failure to recognize this local diversity is a common cause of failure and a barrier to the equitable distribution of benefits:

> Designing benefits to fulfill the needs of all stakeholders of a conservation initiative is challenging, but ensuring that the benefits actually reach the intended beneficiaries has proven to be an even greater challenge in the global effort to conserve biodiversity. Conservation initiatives based on inaccurate assumptions and incomplete considerations of community are not likely to

succeed in creating sufficient incentives for conservation among residents. Benefit programs that acknowledge the heterogeneous needs of communities and account for inequities in the distribution of benefits on the local, regional, national and international scales are best able to generate local commitment to conserving natural resources among those most affected by limitations on its use.[21]

Agrawal et al. (2008) note three major trends (and challenges) for the future of forest governance: decentralization of management, commercial logging concessions, and growth in market-oriented certification efforts.[22] They argue that these trends are influenced from above by donor initiatives, government invest ments and policies and NGOs, and from below by the impacts of climate change, growing international concerns about deforestation, social pressure for more local governance systems, and changes in demographics, consumption patterns and living standards.[23] In addition, they point out that forest certification efforts have primarily been implemented in temperate regions, but are rapidly expanding into tropical regions; and that future progress in tropical forest conservation will depend on an increased role for financial incentives, civil society and market actors in forest management and governance.[24]

## ORIGINS AND OUTCOMES OF FINANCIAL AND MARKET-BASED MECHANISMS

We now turn to a review of the various types of incentive-based and market-based forest conservation mechanisms. These mechanisms include debt-for-nature swaps, sustainable forest product certification and marketing (of timber and NTFPs), payments for environmental services, and forest carbon trading. Di Leva (2002) cites a wide range of legal and market-based incentives for conservation that have been employed in both developing and developed countries, such as ecotourism, taxes and surcharges, user fees, concessions, sales and royalties from bio-prospecting, land donations, conservation easements, mitigation banking, zoning, eco-labeling and product certification, and financial incentives for carbon sequestration.[25] This chapter does not consider purely legal incentives, but instead focuses on financial and market-based ones. In general, there are two broad categories of such incentives, those that promote market integration of specific forest products, and those that involve the development of new markets and payment mechanisms that substitute for the extraction of such products. Examples of market integration include sustainable forestry certification schemes and trade in non-timber forest products. Substitute markets and mechanisms include debt-for-nature swaps and payments for environmental services initiatives. There is some scope for combining market integration and substitution mechanisms within the same forest area, through integrated schemes that focus on both the provision of valuable environmental services and the extraction of low-impact, high–value NTFPs. In addition, both market integration and substitution mechanisms can occur at multiple scales. However, many of these mechanisms require oversight and regulations at the national and/or international levels.

Building on the abovementioned conceptualizations, the following discussion provides an overview of three basic types of financial and market-based mechanisms that have been employed to promote forest conservation in tropical countries: *international transfer payments, transnational market-based mechanisms*, and *payments for environmental services*. These are presented more or less chronologically in terms of their development, as well as in order of their scale from (a) country-to-country fiscal transfers, to (b) global consumer-to-local producer transfers, to (c) payments from state/regional/local users to local providers. The findings of research on the implications and effectiveness of each for biodiversity conservation and poverty alleviation are also discussed.

## International Transfer Payments

One of the earliest examples of international transfer payments—and of incentive-based conservation mechanisms in general—are debt-for-nature swaps. They were first introduced in Latin America during the late 1980s in response to two related concerns: that tropical deforestation was accelerating at a rapid pace; and that this deforestation was driven in large part by the efforts of developing countries to pay off growing foreign debts through the extensive exploitation of natural resources via cattle grazing, timber harvesting, mineral extraction and agricultural expansion.[26] A debt-for-nature swap is not a purely market-based mechanism, but an international transfer payment, or a "non-market transfer of financial resources from consumer nations in recognition of the global public good values of forests."[27] More specifically, it is an agreement whereby a sponsoring country government and/or international conservation organization purchases a portion of a debtor country's external debt in exchange for the dedication of refinancing revenues to nature conservation initiatives,[28] and often the simultaneous adoption of macroeconomic (neoliberal) restructuring policies.[29]

Debt-for-nature swaps were initially greeted with great enthusiasm by both conservation organizations and the national governments of debtor nations—

pointing to the resulting funds generated for valuable and pressing conservation activities—but this enthusiasm was later tempered by higher debt prices in secondary markets and lower appropriations from sponsoring governments and conservation organizations.[30] As a result, funding for debt-for nature swaps trailed off considerably during the latter half of the 1990s. However, debt-for-nature swaps experienced somewhat of a revival in recent years, under provisions of the United States' Tropical Forest Conservation Act (1998).[31] Such bilateral transactions could become an important part of future climate change mitigation strategies as tropical forests comprise the world's largest carbon sinks.[32]

Those who advocate debt-for-nature swaps claim that reducing debt in developing countries will promote free-market systems, boost economic growth and trade liberalization, attract foreign investment, and enhance environmental protection.[33] They note that, although conservation funds generated by these transactions are small in global perspective, they represent significant sources of funding for domestic environmental protection efforts in many countries. Furthermore, they argue that these transactions enhance local environmental conditions, promote sustainable resource use, and preserve valuable biodiversity and ecosystem services.[34] Critics of this mechanism argue that debt reduction does not lead to less extraction of minerals or timber in developing countries with large foreign debts; that they generate insufficient funding to effectively address environmental problems, and that they may compromise national sovereignty through the adoption of external conservation priorities and stringent macroeconomic policies.[35]

Using a simple regression model and data from fifty-five tropical countries, (Didia 2001) established a statistically significant, positive relationship between debt and deforestation.[36] However, Didia (2001) acknowledge that there are also other important factors driving deforestation, such as population growth and demand for farmland, and that deforestation would probably not cease with the eradication of all international debts.[37] Therefore, they conclude that debt-for-nature swaps cannot mitigate tropical deforestation, though they may help to alleviate some pressure on forests associate with poverty; and that a lack of democratic principles, poorly defined property rights and ineffective markets and governing institutions are the real culprits behind widespread and ongoing deforestation trends.[38] Thus, they imply that market mechanisms and supporting institutions are more important than mere financial mechanisms like debt-for-nature swaps in securing positive conservation outcomes.[39] However, they also acknowledge that, "for markets to promote sustainable production of natural resources, there must be institutions that ensure a balance between the social costs and social benefits arising from the "selfish" actions of market participants."[40]

Other types of international transfer payments include the United Nations Global Environment Fund (GEF), which was set up in 1991 as a means of achieving the International Conventions on Climate Change (ICCC) and Biological Diversity (ICBD).[41] It has also been used to leverage private sector funds through venture capital investments, but critics claim that biodiversity projects have adopted an over-scientific and non-participatory approach, and have failed to influence donor programs and practices.[42] Following the lead of the GEF, some countries have set up "national environmental funds" to leverage resources from the GEF and debt-for-nature swaps, finance domestic biodiversity priorities, and meet their obligations as signatories to the ICBD. In addition, various forms of international taxation have been proposed to create revenue streams for sustainable forest management—such as the Tobin tax (on foreign exchange), carbon taxes, and air travel taxes—but these remain unenforceable without corresponding international environmental regulations.[43] While there has been considerable innovation and interest in transfer-payment approaches promoting international cooperation, Richards and Moura Costa (1999) note:

> Donor-driven finance-raising approaches such as the GEF, debt swaps and the associated national environmental funds have fewer political and technical constraints, but are not tied to specific values and have little or no impact on user incentives. They are not market-based, and there can be political and technical problems in ensuring the money is effectively spent.[44]

Integrated Conservation and Development Projects (ICDPs) are based on the principle that most of the conservation related benefits accrue at the national or international level while the costs of such benefits need to be borne by local people. ICDPs would explicitly address the needs of the local people while simultaneously trying to achieve conservation goals. They were initiated primarily after realizing that the command-and-control approach was seriously limiting access to livelihood resources and, as a result, was inherently unsustainable. ICDPs became popular among aid agencies during the 1990s. Programmatically, ICDPs have relied on three ways to achieve conservation and poverty reduction goals: (a) social *compensation*, by supporting schools, health facilities and other social infrastructure projects; (b) economic *alternatives*, offering trainings and opportunities to pursue alternative livelihoods that would reduce pressure on natural resources; and (c) financial *enhancements*, by increasing the economic value of the area through eco-tourism.

However, research by Christensen (2004) reveals that ICDPs have not achieved success due to a number of factors.[45] First, many ICDPs were based on misconceptions and "naïve assumptions."[46] Improving the livelihoods of those people living around protected areas did not necessarily help to improve biodiversity; in many cases the resulting higher incomes probably exacerbated the problem. Second, ICDPs did not recognize the heterogeneity of needs and interests within local communities and the threat of elite capture of benefits. Third, Christensen (2004) argues that ICDPs focused on small-scale drivers like subsistence farming and hunting while ignoring major drivers like road building or commercial resource extraction.[47] Finally, ICDPs were initially conceived of as programs that would become financially independent, however, most of the funds were provided up front and the programs lacked any sense of "conditionality."

## *Transnational Market-Based Mechanisms*

International and national schemes for the certification and commercialization of sustainable forest products are market-based mechanisms that encourage conservation by endeavoring to connect consumer's decisions directly to local management practices. These initiatives began to take shape in the early 1990s and include the production and marketing of both timber and non-timber forest products (NTFPs) at both national and international levels. They include extensive supply chain systems and, occasionally, direct producer-to-consumer arrangements, though there is typically a third party involved in verification and/or marketing and trade. Forest certification efforts have been around for over fifteen years, and have spread rapidly in their coverage. Compatible NTFP marketing initiatives have also flourished, thanks in large part to the support of international donor organizations and conservation organizations, and to their partner institutions in many countries.

Ramesteiner and Simula (2003) conducted a comprehensive study of forest certification efforts around the world since 1993.[48] They concluded that, although forest certification has the potential to enhance biodiversity conservation, there is no guarantee that it will achieve this goal, since many certification efforts are on large commercial plantations with limited biodiversity. Furthermore, certified forests represent less than 5 percent of the world's total forests, and the majority of certified forests are located in Europe, North America and other temperate, developed countries, with less than 10 percent of certified forests in tropical countries.[49]

Despite some growth in schemes in tropical regions, the majority of certified forests are still located in temperate countries.[50] Certification standards for both the process and performance of sustainable forest management must be adapted to local ecological and socioeconomic conditions and must include ecological, social, and economic elements.[51] The failure of such adaptation could be one reason why certification schemes have thus far failed to incorporate many tropical forests. Furthermore, while it is relatively easy to establish national standards for sustainable forest management, it is often difficult to make forest certification operational, due to the high transaction costs and detailed reporting requirements. Corruption and low government capacities for forest management and efficient, transparent data collection and dissemination could be additional reasons for the slow spread of initiatives in tropical countries. As Ramesteiner and Simula (2003) state, "It needs to be recognized that developed countries, countries in transition and developing countries are in quite different situations with regard to their needs, possibilities and resources to make use of certification."[52] Despite these limitations, Agrawal et al. (2008) note that certification initiatives are expanding into tropical countries as global awareness of the effects of local consumer choices on distant ecosystems grows.[53]

The cultivation and marketing of non-timber forest products (NTFPs), whether certified or non-certified, is a strategy that has been adopted by many governments and donor institutions to promote rural development that is compatible with conservation objectives by providing communities living in close proximity to natural forests with economic alternatives to more destructive livelihood and commercial activities, such as fuelwood collection and unsustainable (and often illegal) timber harvesting practices. Such initiatives have been particularly prominent in forest "buffer zones" adjacent to protected areas, and occasionally within the protected areas themselves. In many donor-funded conservation and development projects, NTFP promotion often goes hand-in-hand with micro-finance schemes to provide seed money to scale up the enterprises and/or make these enterprises financially viable in the long run.

Kusters et al. (2006) investigated the impacts of NTFP trade on both livelihood improvements of the producers and forest conservation outcomes, using fifty-five examples of NTFP trade from Asia, Latin America, and Africa.[54] They concluded that involvement in NTFP markets does in fact benefit peoples' livelihoods in specific ways, but that it may also increase inequality among beneficiaries, unless women are involved in the production-to-consumption process.[55] In terms of the management implications and ecological impacts, most of the operations were small scale and, in 80 percent of the cases, returns were not sufficient to allow investments for management measures to enhance the quality or quantity of production; commercial extraction from the wild has proven to be a source of resource depletion and degradation.[56] Although the study revealed that NTFP systems produced better ecological outcomes than most alternative land uses, except for natural forests, Kusters et al. (2006) also found a correlation between higher livelihood benefits and lower ecological quality, indicating that NTFP trade may not be conducive to the promotion of both conservation and development in natural forest ecosystems.[57] This must be taken into account by governments, as well as development and conservation organizations, in order to understand the potential trade-offs involved in promoting NTFP production and trade.[58]

Schreckenberg et al. (2006) argue that NTFP initiatives can be effective given the proper conditions and actions, such as integration into broader diversified livelihood strategies; supportive legal and regulatory frameworks for access, management, and commercialization of products; provision of credit to rural poor and small-scale entrepreneurs; policies that promote access to education and information that increase opportunities for entrepreneurship; and enhancements of transport and communications infrastructure to facilitate market access.[59] They also note three types of NTFP activities that contribute to poverty reduction: (1) "safety nets" that prevent people from falling into greater poverty by reducing their risk (e.g., products available year-round); (2) "gap-filling" activities that provide income to supplement more important farm and off-farm activities; and (3) "stepping stone" activities that help people move out of poverty.[60] Finally, they note four key factors that contribute to the success of NTFP initiatives: innovation, collaboration, entrepreneurs, and more conducive legislative and policy environments.[61] The same factors could be said to influence the success of timber certification efforts as well.

Perhaps one of the most obvious, yet most often overlooked, factors limiting the effectiveness of forest product certification and commercialization efforts—whether for NTFPs or sustainable timber production—is the fact that these efforts depend largely on market demand, and are thus vulnerable to changes in preferences and price sensitivity by distant consumers, and the resulting marketing advantages (or disadvantages) for producers.[62]

**Box 10.1. Impacts of Sustainable Forest Management Certification on Poverty Alleviation** [63]

Sustainable forest management (SFM) certification has become recognized as a potential tool for ensuring conservation while promoting the enhancement of local livelihoods, including poor and marginalized groups. Since 2004, Nepal has initiated one of the first SFM certification pilot programs involving twenty-two community forest user groups (CFUGs) in two districts of the country's Middle Hills region, under the auspices of the Forest Stewardship Council (FSC), an international certifying body.

This initiative aims to enhance sustainable forest management practices by encouraging the sustainable harvesting, sale, and processing of forest products by local households, communities, and enterprises. These products include handmade paper derived from different shrubs, especially *lokta* and *argheli*, as well as numerous essential oils from plants like wintergreen and juniper. Thus, one of the key objectives of the initiative is to enhance income-generating opportunities for local people, including poor and socially marginalized households. To facilitate this, smaller sub-groups have been formed within each CFUG involved in certification. Some of these sub-groups are comprised exclusively of marginalized households, while others are made up of households with varying socioeconomic status. Each sub-group has been provided with a forest plot, start-up funds, and sometimes training to grow various non-timber forest products (NTFPs).

Despite such support for these efforts, the economic benefits of SFM certification for local communities, particularly for poor households, have been limited. In fact, most sub-groups report that they have not yet been able to earn a significant profit from growing NTFPs, and many have also been excluded from receiving direct benefits from the sale of timber, which has not yet been certified but is being sold by CFUGs to local sawmills, and plywood and furniture factories. CFUGs do own a share in—and thus receive a percentage of the profits from—local cooperative enterprises. However, they do not typically receive a premium price for their products compared with other non-certified CFUGs, which also sell the same forest products to the local enterprises. Thus, there is no economic advantage to certification, especially considering the added costs associated with monitoring and verification. Beyond certification, CFUGs often have separate provisions for making loans or grants to socioeconomically marginalized households for various income-generating activities or for household construction and repairs.

Although the economic rewards from SFM certification have not yet been very substantial, members of certified CFUGs acknowledge that there have been some advantages to their participation in the certification scheme in terms of enhanced forest management and group governance outcomes. Many see a need to scale up certification efforts to incorporate other groups and enterprises, and thereby reduce the cost of monitoring the associated social, ecological and economic impacts. FSC is now piloting a new certification program in Nepal that combines sustainably produced forest products with the maintenance of ecosystem services, including carbon sequestration. There is some hope that expanding the scope of certification in this way could help increase benefits to local communities, but the capacity of SFM certification to make a significant impact on poverty alleviation remains in question.

## *Payments for Environmental Services*

Since the mid-1990s, a new paradigm called payments for environmental (or ecosystem) services (PES) has been gaining increasing popularity in efforts to promote the conservation and enhancement of forests, biodiversity, fresh water and other accompanying resources and services. The growing interest in PES has been driven partly by disenchantment with "command-and-control" approaches (i.e., fiscal and regulatory measures) and integrated conservation and development projects (ICDPs), along with reduced donor assistance for forest management and conservation initiatives.[64] PES schemes have been touted as a new model for combining conservation with development and poverty reduction goals. They confront the 'market failure' problem of tropical forestry—weak or absent markets for ecosystems services for carbon, water, or biodiversity—by formalizing economic relationships that recognize the value of forest conservation by compensating managers for their conservation efforts and, more specifically, the accompanying services they provide.[65]

Environmental services are typically provided by landholders or land managers, incorporating both private and community tenure arrangements. Those who purchase the services include national governments, municipalities, tourism companies, hydroelectric companies, agricultural users, fishing cooperatives, and other commercial users. In many countries, there are examples of informal PES arrangements at the local level, but no national level program or legal framework. Other countries have promoted formal PES programs at regional or national scales. PES schemes encompass a broad range of mechanisms for promoting the conservation and restoration of forest ecosystems and the many amenities that they provide, including drinking water, clean water for economic and household purposes, conservation of soil and biodiversity, and scenic beauty.[66] They could also include payments for the right to collect biological specimens, or "bioprospecting," in forests managed and/or owned by local and indigenous communities.

An environmental service can be described as any service that is provided through the maintenance and protection of a natural resource or ecosystem. Wunder (2007) provides a more specific definition of a PES based on five criteria: "(1) a voluntary transaction in which (2) a well-defined environmental service (or a land use likely to secure that service) (3) is being 'bought' by a (minimum of one) buyer (4) from a (minimum of one) provider (5) if and only if the provider continuously secures the provision of the service (conditionality). He points out that relatively few PES arrangements satisfy all of these criteria, and that the most difficult of these criteria to satisfy is "conditionality," due largely to the lack of effective monitoring and enforcement mechanisms.[67]

PES arrangements typically occur at the local, regional or national levels. There are very few examples of transboundary PES schemes. Although they do not typically involve the direct exchange of discrete quantities of a given product, PES schemes are like certification initiatives in that buyers of ecosystem services require some sort of assurance that their payments are actually promoting conservation and that the recipients are providing a real and measurable service, whether this is a concrete good, like water supply, or a more abstract good, like scenic beauty. In both cases, some sort of verification process is typically required. According to Richards and Jenkins (2007), there are four basic types of PES: (1) public (government) payments to forest owners or managers to protect ecosystems on their land; (2) trading between buyers and sellers based on a regulatory framework (including cap-and-trade systems); (3) private market arrangements whereby buyers contract directly with "upstream" buyers who provide a specific service; and (4) eco-labeling or certification schemes for natural resources, whereby consumers pay a premium to ensure the conservation and protection of natural resources and their supporting ecosystems (i.e., sustainable forest management).[68]

Like other market-based mechanisms, PES has its critics and its advocates. Critics argue that PES schemes will promote the return of strict command-and-control conservation measures, effectively de-linking conservation from development objectives; that the land-development rights and aspirations of communities could be threatened by these measures; and that market-oriented conservation could degrade culturally important not-for-profit conservation values.[69] Some are pessimistic about the potential of PES schemes to address both biodiversity conservation and poverty alleviation goals simultaneously, noting an inherent trade-off in these systems. Corbera et al. (2007) conclude that, "markets for ecosystem services are, in effect, limited in promoting more legitimate forms of decision making and a more equitable distribution of their outcomes."[70] They claim that such markets rely on political affiliation for their legitimacy and thus reinforce existing power structures and imbalances, socioeconomic inequities, and vulnerabilities; and that they do not inherently promote legitimate decision-making forums or the equitable distribution of benefits.[71] They also rely on effective project design. Furthermore, they stress, "All environmental decisions implicitly or explicitly involve questions, as well as trade-offs, regarding economic efficiency, environmental effectiveness, equity and political legitimacy."[72]

Advocates claim that PES promotes innovation in conservation to address the inability of current approaches to provide adequate economic incentives to conserve forests; that they can tap into new sources of funding from both the public and, especially, the private sector, and that they can enhance the livelihoods of participating poor communities.[73] Many have hailed the promise of PES schemes, noting their positive impacts on tenure security, community empowerment, and the development of organizational and social capital.[74] Governments and donors can play a key role in fostering equitable governance structures, secure tenure, and enabling policy, legal, and institutional frameworks; and environment-development trade-offs can be managed with adequate support from donors, governments, NGOs and the private sector.[75]

When devising PES mechanisms, it is important to consider the issue of "additionality." The operational question here is, would the same conservation outcome have occurred anyway, without the introduction of the financial incentives provided by PES? If the answer is yes, then the payment is not economically efficient and could actually be promoting perverse incentives. Wunder (2006) stresses that, in PES schemes, it is primarily a stakeholder's opportunity cost of conservation and their capacity to degrade the forest that should inform decisions about who should receive payments in order to have the greatest impact on forest conservation.[76] He notes that a large landowner who wants to clear his forest to plant commercially valuable crops has a much higher opportunity cost, as well as a high potential for degrading the forest.[77] Because the landowner's opportunity cost is so high, compensation schemes will only be able to preserve a fraction of the total forest and the remainder will likely be cleared for agriculture. Conversely, a remote indigenous group with a relatively low impact on the forest typically does not represent a big threat to the forest, so compensating them would not accomplish much "additionality," unless they are faced with a much better economic opportunity (such as a potential logging agreement with a commercial timber company) or they otherwise play a vital role in protecting the forest from other would-be destructive land users.[78] However, some claim that we have a moral responsibility to compensate the indigenous group for their forest conservation efforts.

Wunder (2006) concludes that it would be most efficient from a conservation standpoint to make PES investments to a stakeholder who has a moderate opportunity cost and a moderate-high (potential) impact on the forest, as the payments will buy a lot more conservation than they would with the large landowner.[79] He also stresses that it is extremely important to carefully consider the level of degradation, the opportunity costs of conservation, and possible perverse incentives of different actors, when devising PES schemes.[80]

The distributional implications of PES have been heavily debated in the literature. Engel et al. (2008) define PES as a "mechanism to improve the efficiency of natural resource management and not as a mechanism for poverty reduction."[81] While Pagiola et al. (2002) agree with this formulation that PES mechanisms are not meant to be poverty reduction tools, they also claim that there are important "synergies" that could be realized.[82] They go on to raise three important questions worth pondering about when designing PES programs bearing a heavy poverty reduction focus: (a) are the participants of the program poor? (b) are the poor households able to participate? and (c) how are poor people affected (indirectly) by PES programs? These questions encourage us to rethink the basics about program design, targeting and stakeholder engagement in PES programs.[83]

As most of the widely documented PES programs have involved watersheds, the general perception has been that upstream participants (i.e., environmental service providers) tend to be poorer. However, Miranda et al. (2003) found that participants in the PSA program in Costa Rica were from relatively wealthier backgrounds, though this claim has been contested by Munoz (2004).[84] User financed PES programs are typically more targeted compared to government financed ones, thereby allowing them to address poverty issues directly. Pagiola et al. (2002) employ a two-step structure while thinking about poverty and PES—how the PES program affects the participants and how and if the program alters conditions of local people indirectly.[85]

In an attempt to investigate the impacts carbon payments and contracts on poverty, Antle and Stoorvogel (2009) carry out empirical tests based on data from three case studies of PES schemes.[86] While these case studies are about carbon sequestration through interventions in agricultural soil management, they nonetheless bear important lessons for poverty alleviation and forest conservation. Antle and Stoorvogel (2009) find that the adoption rate of carbon sequestering practices depended mainly on the price of carbon, transaction costs, and ability to participate in the program.[87] Similarly, for severely degraded agricultural land, sustainable management practices were possible only under high carbon price scenarios. The impact on poverty was not clear. While the carbon payments did raise rural incomes, the findings of Antle and Stoorvogel (2009) suggest that the greatest beneficiaries of the additional income were not always the poorest people as they also tend to occupy the most severely degraded areas and face greater barriers to participation.[88]

Alix-Garcia et al. (2009) review Mexico's payment for environmental services program, particularly forest conservation for hydrological services.[89] The goal of this program is to develop an internal PES market and the experience thus far is telling on a few counts. First, payments went to forests that were not critical for hydrological services and ones that were not in danger of being deforested, raising questions on effectiveness and additionality. Second, the communities that chose to accept the payments were not necessarily those facing pressure to deforest. Finally, asymmetry of information between the service providers and the users of the service created accountability problems due to increased social distance.

Another way to look at the relationship between poverty and forest conservation is to study how landholders make decisions. Engel and Palmer (2009) apply a theoretical model of "community-firm interactions" based on negotiations that have taken place between communities and logging companies.[90] Their finding asserts that a trade off does in fact exists between poverty alleviation and maximum provisioning of environmental services. They argue that this is due to two factors: enforcing PES contracts and discount rates. Poorest communities do not find the payments high enough to reduce use of resources—in other words, the opportunity cost is high, thereby leading to non-compliance. Furthermore, since poorer communities tend to value short-term gains more than wealthier communities, immediate payments from logging companies, as opposed to future payments for environmental services at a certain level, induces poorer communities to opt for logging contracts instead.

PES mechanisms, while not a panacea, help respond to the market failure problem of forestry and are essential to an integrated approach to sustainable forest management and conservation. Richards and Jenkins (2007) conclude that, while PES schemes show promise for "raising the viability of sustainable forest management and conservation and delivering pro-poor benefits" by responding to the market-failure problem cited above, they should form only part of a diversified set of mechanisms that help reduce the opportunity costs of landowners for sustainable forest management and conservation.[91] Wunder (2006) concurs that, in order to be effective, PES must complement and function in tandem with other alternative livelihood strategies that promote the socioeconomic development of communities.[92] Richards and Jenkins (2007) further add that "avoided deforestation" schemes, such as Reduced Emissions for Deforestation and Degradation (REDD), have the most potential of all PES systems, but are also afflicted by a complex set of issues that challenge effective implementation. Such schemes are discussed in more detail below.[93]

## FOREST CARBON TRADING AND REDD PLUS: TOWARD THE GLOBALIZATION OF ECOSYSTEM SERVICES

Forest carbon trading, a relatively new market-based paradigm, strives to combine biodiversity conservation objectives with the enhancement of development and poverty alleviation outcomes and the mitigation of global climate change. This section discusses this new paradigm and its latest manifestation, REDD Plus, along with associated challenges for its implementation, and specific implications for poverty alleviation.

Housing over half of the global terrestrial carbon pool in biomass and soils, forests both contribute to and help to mitigate climate change through the continuous release, sequestration and storage of carbon dioxide.[94] Robledo et al. It is estimated that, since 1850, forests have been responsible for up to 90 percent of all greenhouse gas emissions from land use, land use change, and forestry (LULUCF); and that they currently contribute up to 17 percent of global annual greenhouse gas (GHG) emissions—through deforestation, forest degradation and forest fires—the second most significant source after the burning of fossil fuels for energy production.[95] There are signs that this contribution could increase in coming years due to further conversion of land for the growing of food crops and biofuels, in order to meet the world's growing demand for food and alternative energy sources.[96] In addition, a recent report to the Secretariat of the UN Framework Convention on Climate Change (UNFCCC) estimated that poverty, including the clearing of land for subsistence farming, is directly or indirectly responsible for nearly half (48 percent) of annual deforestation and degradation globally.[97] Thus, poverty alleviation, forest conservation, and carbon sequestration can be considered synergistic goals.[98]

Forests have played a part in voluntary carbon trading mechanisms for approximately two decades.[99] However, they first entered the international climate policy picture in 1997, as one type of project under the Kyoto Protocol's Clean Development Mechanism that could contribute to climate change mitigation.[100] The function of forests in sequestering carbon dioxide from the atmosphere has long been recognized. However, the significant contribution of deforestation to greenhouse gas emissions was formally acknowledged more recently,[101] leading to efforts to incorporate forest conservation efforts into the portfolio of climate change mitigation mechanisms. In fact, despite the existence of a policy provision for including reforestation and afforestation initiatives under the CDM for over a decade, only a handful of forestry projects have been approved under this mechanism to date.[102]

Following the publication of the *Stern Review*,[103] the cause of incorporating forests into international climate change policy mechanisms gained considerable momentum. The 13th Conference of Parties Meeting (COP 13), held in December 2007 in Bali, Indonesia, is considered a watershed moment in climate change negotiations for two main reasons: it elicited clear statements of support for a post-Kyoto climate policy regime from the United States and other Western countries; and it shifted the focus of carbon trading from an approach based purely on emissions reductions from fossil fuel consumption to one also involving forest carbon credits, through a newly proposed mechanism, reducing emissions from deforestation and forest degradation (REDD).

The global policy agenda and international funding institutions such as the World Bank and the United Nations have now embraced REDD and brought forests to the fore of international climate change negotiations and carbon trading schemes. However, not everyone is pleased by this development. Shamsuddoha and Chowdhury (2008), revealing their disdain for market-based approaches in general, claim: "In fact, [the] Bali climate conference sidelined its major agenda, emission reduction, and has focused on alternate ways of carbon capture through Reducing Emissions from Deforestation and Degradation (REDD) and emissions reduction through the CDM, both of which are basically....market mechanisms."[104] They note that, at the Bali meeting, Indonesia joined with ten other developing nations with large expanses of tropical forest (Brazil, Cameroon, Costa Rica, Colombia, Congo, the Democratic Republic of Congo, Gabon, Malaysia, Papua New Guinea, and Peru) to form a coalition to "demand that developed nations provide financial incentives to countries with tropical forests for their preservation."[105]

## *REDD, REDD Plus, and Beyond: Implications of Forest Carbon Trading for Poverty Alleviation*

In many ways, forest-carbon trading represents a synthesis of the three basic approaches discussed in the previous section—international transfer payments, transnational market-based mechanisms, and payments for environmental services. It resembles an international transfer payment scheme in that it would (in a national-level approach) facilitate payments between governments to finance conservation efforts via the production of carbon sequestration benefits—though in this case, these payments would be based on market prices for carbon sequestration. It is akin to transnational market-based mechanisms like forest product certification and commercialization schemes in that it relies on complex and systematic analysis and verification of forest management practices, standing carbon stocks, as well as mechanisms and institutions for sharing information and benefits in order to satisfy both the carbon "sellers" and the carbon "buyers" or investors. It is like a PES regime in that it represents a reimbursement by international consumers and companies and/or countries for the global environmental service of climate change mitigation via carbon sequestration in forest biomass and avoiding emissions from deforestation and forest degradation. Thus, carbon offsetting is an example of a global environmental service, for which national and international markets and policy regimes are currently evolving to compensate the producers of this service—those who regenerate or preserve forests—for their valuable contribution to addressing global climate change.

Despite these clear similarities, there are also important differences between forest carbon trading and other financial and market-based mechanisms. Perhaps the most significant difference is in its scale, since producers of carbon emissions reductions or sequestration services are selling carbon credits directly to governments, companies, and/or individuals in an expanding global marketplace, with internationally established prices and verification standards. Furthermore, these payments are not for direct services rendered to the buyer, but for the global public good of helping to reduce the amount of carbon dioxide in the atmosphere. They could be purchased voluntarily, or as an obligation by the buyer under some regional, national or international regulatory framework.

REDD+ is the most recent manifestation of forest carbon trading, proposing a global policy and institutional infrastructure for channeling investments from developed countries to developing countries for the conservation and sustainable management of forests. However, the socioeconomic implications of this evolving mechanism are still unknown. While REDD+ could, ostensibly, provide significant funds to forest-dependent communities for local development, there are a number of interrelated issues that could prevent the poor from receiving these benefits.

First, REDD+ is not like other market-based conservation mechanisms that provide households with a direct means of earning income through the sale of forest products or services (e.g., microenterprise, ecotourism, sustainable forest management certification). Therefore, poor and marginalized households would not have a direct economic stake or interest in the benefits. Rather, the income from REDD+ would be earned collectively at the community level or higher, so it would not necessarily benefit poor households. As a result, the ability of poor and marginalized groups to benefit from REDD+ would depend largely on existing governance and benefit-distribution systems at the local level. These systems are often lacking and/or corrupted by the interests of a few local elites and influential outsiders, who exploit their position or influence for personal gain. If, however, local governance systems are transparent, equitable, and pro-poor, then REDD+ could benefit the poor.

Second, there is also a risk of carbon payments being misused or captured by elite and/or corrupt community members, local politicians, administrators and private sector actors, and officials at the state/province or even national levels. It is no coincidence that many of the countries with the highest rates of deforestation are also the most corrupt.[106] In such contexts, providing a transparent, fair, and effective system for transferring the benefits of REDD+ from the national to the local level could prove to be a considerable challenge. To prevent this, strong institutional guidelines, safeguards, and governance mechanisms would have to be put in place.

Third, national and local imperatives to maximize carbon capture and storage could lead to the imposition of restrictions on forest management (official or unofficial—set either by the community or by external regulators and contracts). Such restrictions could adversely impact the ability of poor households, who are highly dependent on forests, to meet their subsistence needs and/or to earn a living by harvesting and selling various forest products. Again, strong safeguards must be built into any REDD+ mechanism at multiple scales—from the national to the community levels—in order to prevent such adverse effects.

Thus, carbon trading and REDD+ offer no guarantee of benefits for the poor. Rather, the ability of the poor and other marginalized groups to benefit depends on the robustness and responsiveness of existing and new institutional and governance systems and mechanisms.

**Box 10.2. Integrating Poverty Alleviation into Carbon Standards: The Plan Vivo Approach**

Amongst the many standards of voluntary carbon offsets, Plan Vivo is one of the few that addresses poverty directly. The standards are designed explicitly to work with smallholders to provide ecosystem services with the central goal of promoting sustainable rural livelihoods through community-based land use projects. Some of the distinguishing features of Plan Vivo are as follows:

**Scope.** Plan Vivo includes afforestation and reforestation, agroforestry, forest restoration, and avoided deforestation projects. This broad scope marks a major departure from other standards as it allows for comprehensive landscape-level interventions rather than focusing on specific types of land use change. This provides immense flexibility in project design as it can incorporate and bundle activities that would ordinarily be ineligible under other standards. With this flexibility, the standards are able to better reflect land use practices on the ground, thereby promoting the sustainability of the interventions. In addition, as project interventions can be very different across the board, Plan Vivo doesn't prescribe specific methods, but encourages projects to develop their own peer-reviewed specifications.

**Carbon ownership.** Plan Vivo recognizes the informal nature of land tenure in most rural settings by allowing user rights to be the legal basis for a project. Payments go directly to the target group, are provided at specific intervals, and are matched with a monitoring cycle.

**Capacity building.** Plan Vivo Standards require demonstration of capacity building of the target group over time. Capacity building plays a major role in maintaining good governance of the project. In addition, this "learning by doing" approach is expected to increase communities' sense of ownership over the project and, ultimately, to help diversify income streams to promote sustainable livelihoods.

**Livelihood co-benefits.** Community-led planning is carried out to ensure that project goals are consistent with a community's needs. An equitable benefit-sharing mechanism is created with full transparency. Recorded payments can be traced back to the target groups, including poor and marginalized, thus reducing permanence risks.

One of the most significant contributions of Plan Vivo standards has been to focus on the land use managers, that is, the rural communities, creating a framework that allows carbon credits to be generated according to prevailing land-use management options. This multi-pronged approach has added

significant "non-carbon" value to carbon credits as the interventions made under
Plan Vivo Standards help to increase local resilience to climate change, protect
and foster biodiversity, and provide a wealth of other ecosystem services beyond
the central goal of poverty alleviation.

More information is available at:
http://www.planvivo.org/documents/standards.pdf

## Challenges of Implementing Forest Carbon Trading and REDD Plus

In addition to the abovementioned challenges for realizing benefits from forest
carbon trading and REDD for poverty alleviation, there are significant barriers
to the effective implementation of such schemes in general. According to
Bumpus and Liverman (2008), "Carbon reductions as a resource show specific
spatial distribution patterns and practices that are mediated by their particular
environmental, socio-economic and political characteristics."[107] As such, the
potential for carbon trading is subject to four distinct types of challenges or
constraints: (1) biophysical and geographic; (2) technical and financial; (3)
social, economic, and institutional; and (4) policy and political. Each of these
types of constraints is elaborated upon below.

### Biophysical and Geographical Constraints

Certain biophysical constraints and uncertainties challenge the potential for the
success of forest carbon trading regimes. Different forests grow and sequester
carbon dioxide at diverse rates, according to such factors as their climate and
soil characteristics, local ecology and species composition, physical structure,
age (i.e., state of forest succession), type and frequency of disturbances,
management regime, and degree of anthropogenic pressure.[108]

In general, older forests serve to store carbon (striking a balance between
growth and decomposition) and younger forests tend to contribute to carbon
sequestration.[109] Furthermore, temperate forests are generally net carbon sinks,
due primarily to their reduced harvest levels, and to increased regeneration and
protection efforts; while tropical forests typically remain carbon emitters, as a
result of their rapid conversion to other land uses.[110] Therefore, temperate forests
lend themselves well to afforestation and reforestation initiatives, while
"avoided deforestation" schemes could be critical to preventing the further
destruction of tropical forests, which store up to half of their carbon balance in
vegetation.[111] Streck and Scholz (2006) also point out that, of the two general
approaches to forest carbon trading (afforestation/reforestation programs and
avoided deforestation), avoided deforestation holds the most physical promise

for reducing carbon losses from forest ecosystems in the short term, because it can take decades to restore carbon stocks that have been lost to land use conversion.[112]

## Technical and Financial Constraints

There are considerable technical and financial constraints to the effective implementation of forest carbon trading at the local, regional and national levels. Among the most prominent financial constraints is the inability of many governments and especially forest-dependent communities to pay the up-front costs associated with building the technical and institutional capacity required to accurately measure, monitor and record incremental growth in carbon stocks, and to pay external agents to verify these changes.[113] As a result, nearly all projects to date have targeted either larger private plantations or public forest lands, while excluding smaller private and community-managed forests.[114] According to Robledo et al. (2008), "At present, forestry activities in developing countries under the Kyoto Protocol's Clean Development Mechanism (CDM) are highly over-regulated.... A high level of expertise is required to get projects in motion, as well as heavy investments, thus discriminating against poor forest communities."[115] They further claim that, in order to engage in carbon trading, poor communities must either seek funding from outside investors, or obtain a subsidy to participate from the government or the NGO sector.[116] Neither of these is easy to secure.

Social and spatial scale can also pose significant constraints on effective and affordable participation in carbon trading. For instance, individual communities would face high transaction costs, struggling to come up with the resources or establish the connections to facilitate their successful participation. Conversely, if communities and forests are aggregated they could realize some economies of scale that might make it possible for them to afford the costs associated with carbon stock measurement, monitoring, verification and certification, whether under a project-based scheme or a national-level system. In terms of the scale of the forest itself, the critical question is: At what size or density does a forest become viable for carbon trading?

In addition to the above, there are some specific technical challenges associated with verifying and certifying forest carbon stocks, particularly for avoided deforestation schemes. In this regard, claim that three issues in particular stand out—"additionality", "permanence," and 'leakage':[117]

- Demonstrating *additionality* refers to the burden of proof that reductions in deforestation would not have occurred without a specific project intervention or financial incentive (i.e. carbon credit payment). In effect, the project or country must show that in the absence of the financial incentive, deforestation would have continued at historically projected rates.

- Establishing *permanence* means ensuring that any carbon sequestration and storage resulting from the intervention would remain in effect for the long term and would not be compromised by future anthropogenic or natural deforestation or degradation.

- *Leakage* implies an increase in deforestation and degradation (or displacement) outside of a specific project area, as a direct result of conservation efforts within the project area. In other words, leakage occurs when a community or project protects the project area at the expense of another adjacent or external area.

## Social, Economic and Institutional Constraints

A serious concern with forest carbon trading is that it could realize biophysical benefits, in terms of forest carbon accrual, at the expense of biodiversity conservation and/or the promotion of important social, economic and poverty-alleviation goals. Many feel that these so-called "co-benefits" should be an integral part of any carbon-trading scheme, and that appropriate socioeconomic monitoring systems should also be put in place.[118] A number of independent standards for ensuring co-benefits already exist. For instance, voluntary projects aiming to demonstrate their commitment to sustainable development outcomes, can adopt the Climate, Community and Biodiversity (CCB) Standards.[119] While many concur that some kind of co-benefit standards should be incorporated into formal forest carbon trading regimes, it remains unclear what the specifics of such standards would be.

While standards under the Kyoto Protocol mechanisms may be cumbersome for small producers and forest-dependent communities eager to engage in carbon trading, primarily due to the high transaction costs involved, voluntary markets present another type of challenge. Because of their lack of a unified standard methodology for measuring, reporting, and verifying carbon stocks, they can raise concerns among potential investors and carbon credit buyers as to their effectiveness in sequestering carbon dioxide (Peskett et al. 2007).[120] On the other hand, the voluntary markets can be more flexible and comprehensive in terms of the types of benefits offered (i.e. including not just carbon offsets, but also biodiversity conservation and sustainable development "co-benefits" such as employment, poverty alleviation, local capacity and legal status), and could present fewer barriers to entry to small producers (ibid).[121] They can also assess sustainable development outcomes at a variety of different levels, such as the socioeconomic impacts on communities, the potential for engagement by small producers, or the broader benefits to the host country in terms of technical and institutional capacity development (Peskett et al. 2007).[122]

In the interests of maintaining national sovereignty, the CDM allows the host government to decide whether a project activity contributes to sustainable development or not.[123] Peskett et al. (2007) sum up the main concern underlying this approach: "It is questionable whether delivering such benefits should fall within the remit of carbon offset standards, as they play no role in reducing [greenhouse gases] ... Adding to the number of objectives could possibly decrease the effectiveness of the standard in meeting this primary aim, especially if budgets are limited."[124] They also claim that, despite the confusion caused by the many divergent standards, there may be some advantage to multiple standards, in that competition between providers could actually increase accountability while keeping costs down.[125] However, there may be a trade-off between keeping costs down in order to involve more small producers and the effectiveness of carbon sequestration.[126]

One of the primary concerns with forest carbon trading is that it should provide economic benefits that are sufficient to compensate local people for their efforts and to make up for the loss of any other benefits or uses of the forest, whether subsistence or commercial, that they may have had to give up as a result of their involvement in carbon trading. Relative benefits will vary in different geographical contexts. The costs of mitigating deforestation and forest degradation in any particular context depend on: (a) the specific local drivers of deforestation (e.g., commercial agriculture, subsistence farming, wood extraction, etc.); (b) returns from alternative, non-forest land uses; (c) returns from alternative forest uses; and (d) any compensation paid directly to landholders. This implies that carbon trading could be more or less viable in some contexts according to the extent of pressure on the forest and the opportunity cost represented by other income-generating activities at both the household and community levels.[127]

The direct and indirect effects of REDD+ programs could be significant. Safeguards and standards are considered to be the tools that will allow the risks and harms of REDD+ programs to be identified and addressed, thereby ensuring the well-being of human and natural systems. This has become a contentious issue in the UNFCCC talks, as there is concern that the current design of REDD+ could significantly alter the access rights of indigenous peoples and local communities, despite some supportive language in the draft agreement. As implementation of REDD+ programs will be determined largely by formal tenure rights, communities that have been managing resources, but lack formal land titles, are in danger of losing their access. Another concern is that forest carbon trading could encourage plantations, and thereby lead to the loss of biodiversity. Though the Cancun Agreement has specified a list of safeguards, the text does not mention the means of making them operational.

As the UN process moves along, international agencies have gone forward and adopted safeguard policies. The multilateral Forest Investment Program (FIP) and the World Bank's Forest Carbon Partnership Facility (FCPF) have adopted the World Bank Policies and Procedures.[128] These stress the need to engage forest-dependent indigenous peoples and forest dwellers to assess who would be affected by programs such as REDD+, and how. The UN-REDD Programme, on the other hand, follows a rights-based approach and has applied the UN Declaration on Rights of Indigenous Peoples (UNDRIP), the stipulation for Free Prior and Informed Consent (FPIC) of the International Labor Organization (ILO) Convention 169, and the UN Development Group Guidelines on Indigenous Peoples.

Institutional constraints span the local to national levels and have to do primarily with both individual and collective (dis-)incentives and capacities for participation in carbon trading. While some estimates reveal that the potential economic earnings by local populations from forest carbon trading are substantial (Banskota and Karky 2007),[129] many are skeptical of the possible impacts these funds may have on governance and equity among both national and local level institutions (Lohmann 2006).[130] At the national level, there is concern that they may foment corruption among government officials charged with the accounting of carbon transactions and dispersing the funds to communities (ibid).[131] In addition, some fear that the funds will be used to bolster state-sponsored conservation by increasing security in protected areas and national forests and further limiting the access and use rights of forest-dependent communities.[132] According to Robledo et al. (2008) the carbon markets themselves also pose challenges with respect to the ability of poor forest-dependent communities to benefit: "As currently structured, carbon markets have been inequitable, thereby posing the risk of aggravating the growing economic gap between the forest dwellers and the rest of society. A lot needs to be done to make a post-2012 carbon sequestration approach obtainable for poor communities."[133]

There is also concern that carbon credit payments may present a challenge to local governance structures and social equity, as mentioned above. More specifically, some fear that such transactions will be characterized by a lack of transparency and that the funds may be co-opted by elite community members to use for their own personal enrichment and/or pet projects, without deliberative decision-making or adequate consideration for the interests of the broader community.[134] Involvement in carbon trading may also preclude some current forest uses and management activities by individuals or groups within a community. As a result, poor and disadvantaged groups may experience further marginalization through a loss of resource access and use rights. Corbera et al. (2007) contend that, "Equitable outcomes are more likely to be achieved when there is communal ownership of forest land and when economic power prior to the creation of the market scheme is fairly evenly distributed within a community."[135] However, they also note that "strong collective action does not guarantee procedural fairness, as women's interests in tree planting can still be ignored over a preference for fast-growing species in communally owned forests."[136] In summary, as Canadell and Raupach (2008) note:

> The challenges facing sustainable mitigation through forestry activities, anywhere but particularly in the tropics, are surmountable but large. They include the development of appropriate governance institutions to manage the transition to new sustainable development pathways.[137]

## Policy and Political Constraints

Policy constraints can be divided into incompatibilities and inconsistencies in both international and national-level policies concerning climate change, forest management, and biodiversity conservation. Internationally, neither the Kyoto Protocol nor any other binding international agreement currently incorporates mechanisms to reduce carbon emissions from deforestation and forest degradation. In fact, at the COP-9 meeting in Milan in 2003, the protection of existing carbon sinks (i.e. avoided deforestation) was declared ineligible under the CDM (though it is still eligible under the Joint Implementation mechanism),[138] effectively excluding many developing countries and local communities from participation in forest carbon trading mechanisms. However, the evolving REDD+ mechanism shows considerable promise for remedying this. Of course, alternative voluntary carbon markets exist, as described above, but they may remain limited in terms of their potential reach and returns, compared with an interna tional regulatory scheme such as REDD+. At the global level, there is also uncertainty about how REDD+ and other mechanisms fit with existing international protocols such as the UN Convention on Biological Diversity or the UN Convention to Combat Desertification.[139]

While constraints and oversights in international policy regimes are of great consequence, national level policy constraints and inconsistencies are perhaps of more immediate and lasting concern for the effective implementation of forest carbon trading schemes. While there seems to be some consensus on carbon trading among the world's governments—as evidenced by developments at the COP–13 and COP–14 meetings in Bali, Indonesia and Poznan, Poland, respectively—there is often less political consensus within countries as to what the appropriate means of implementing a given policy are, or who should carry it out. Drawing on the case of Nepal, Pokharel and Baral (2009) note that institutional and individual capacities and responsibilities for strategic action are often insufficient, unclear or dispersed among diverse bureaucracies or departments—or occasionally delegated to external consultants—in ways that elude effective coordination.[140] They stress the need to follow innovative policy processes that eschew the typical "blueprint" approach in favor of schemes that cater more closely to the national context and concerns; and point out that countries like Nepal should not pin all their hopes on the REDD model alone, but rather explore various carbon trading schemes simultaneously to determine which fits best with their national circumstances and capacities.[141] Karky and Banskota (2009) claim that there is also an urgent need to address inconsistencies among national policies concerning climate change, forestry, local governance, etc. and between national and international level policies on climate change.[142] They note, "Although [community-based forest management] has done fairly well in the mountain areas of Nepal, the challenge now is ... how to synchronize [national] policy with the emerging global climate policy so that the local communities that manage and conserve forests can benefit from the emerging global carbon markets under the UNFCCC."[143]

REDD has generated a lot of political controversy and discussion since it was officially put on the negotiating table at the Bali meeting in December 2007. The notion of "avoided deforestation" is a technically complicated one that invites criticism from many quarters. Some have moral qualms about paying countries for not cutting down their forests, stating that this introduces perverse incentives that could actually lead to the further destruction of forests before a baseline is set.[144] Others note that deforestation is a major source of global greenhouse gas emissions and that it is, therefore, imperative to create appropriate mechanisms to enlist diverse actors in forest conservation efforts.[145] Streck and Scholz (2006) argue for involving developing countries in carbon trading from a global social justice standpoint,

Developing countries administer the majority of the world's environmental resources... With increasing pressure on development and use of resource, they can hardly be expected to provide these services for free. By maintaining their rainforests, tropical countries provide an invaluable global service, one for which they have to be compensated.[146]

# CONCLUSION

This review of different financial and market-based mechanisms for forest conservation reveals that they offer no automatic gains for poverty alleviation. International transfer payments, such as debt-for-nature swaps and global and national conservation funds, provide no direct financial incentives to local communities and landholders to manage forests sustainably and their implementation is often plagued with political problems. While transnational market-based mechanisms such as sustainable forest management certification and NTFP production provide potential opportunities for the poor to earn direct economic benefits, these benefits are often limited due to their insufficient coverage, weak market linkages, high transaction costs, and/or their capture by more influential actors. Moreover, unless they are carefully managed, these market-based mechanisms could pose a threat to biodiversity. PES schemes have received a lot of attention in recent years for their potential to provide significant direct incentives to local landholders—thereby shaping their decisions regarding forest management and use—but the effective delivery of these incentives requires careful institutional design and paying heed to economic trade-offs, efficiency, potential perverse incentives, equity, and political legitimacy. There is no guarantee that PES schemes will reach the poor—they must be complemented by explicit efforts to ensure the effective political and economic participation of socioeconomically marginalized groups.

Market-based forest conservation programs, by themselves, do not necessarily provide benefits for poorer and socially marginalized groups. Institutional arrangements at multiple levels (international to local) are crucial in determining the impacts of these mechanisms on poverty alleviation and forest conservation. Emerging mechanisms like REDD offer an important opening to leverage global payments for environmental services like carbon sequestration toward poverty reduction. However, without the appropriate institutional and legal infrastructure that guarantees standards, social and ecological safeguards, and the tenure and access rights of local communities, these mechanisms will not be able to realize their economic and social goals to the maximum degree.

Carbon prices will largely determine the extent of changes in rural incomes. An international regulatory mechanism, for example, through a global climate change agreement, is needed to ensure that carbon prices are high enough to attract participation. The extent of mitigation commitments that developed countries make will directly impact the price. Combining carbon sequestration along with other environmental services could be a strategy to increase the total value of payments received. In addition, as we have seen in the case of degraded or threatened lands requiring substantial management alterations, carbon prices alone may not be able to shift management practices. Market-based mechanisms will need to recognize the diversity amongst poor peoples, particularly by paying attention to their different status of landholdings and their varied ability to participate.

In summary, achieving inclusive poverty reduction by leveraging market-based forest–conservation policies alone will be a challenge. As Engel and Palmer (2009) argue, realizing poverty reduction while achieving environmental gains requires additional policy tools and institutional structures, and expecting maximum environmental gains and poverty reduction simultaneously is unrealistic.[147] While instituting forest conservation policies, governments will need to continue to formulate targeted policies for poverty alleviation. The current body of literature on the nexus between poverty alleviation and achievement of environmental goals is an evolving area. More research needs to be carried out to understand how poverty reduction can be enhanced while pursuing forest conservation through market-based mechanisms. It has become increasingly evident that achieving gains from these financial and market-based mechanisms requires their further integration with other livelihood strategies that promote social and economic opportunity, empowerment, and equity at the community level, especially among the poor and other socially disadvantaged groups.

## NOTES

1. World Commission on Environment and Development, *Our Common Future* (Oxford, UK: Oxford University Press, 1987).

2. Helmut J. Geist and Eric F. Lambin, "Proximate Causes and Underlying Driving Forces of Tropical Deforestation," *BioScience* 52 (2002): 143-150; S. Sanderson, "Poverty and conservation: The new century's 'Peasant Question'?" *World Development* 33 (2005): 323-332; Ashwini Chhatre and Vasant Saberwal, "Political Incentives for Biodiversity Conservation," *Conservation Biology* 19 (2005): 310-317.

3. Michael Richards and Pedro Moura Costa, "Can tropical forestry be made profitable by internalizing the externalities?" *Natural Resource Perspectives*, No. 46 (October 1999); Stefano Pagioloa, ed., *Selling Forest Environmental Services: Market-based Mechanisms for Conservation and Development* (Earthscan Publications, 2002); B. Kelsey Jack, Carolyn Kousky, and Katharine R. E. Sims, "Ecosystem Services Special Feature: Designing payments for ecosystem services: Lessons from Previous Experience with Incentive-based Mechanisms," *PNAS* 2008 105:9465-9470.

4. Douglas J. McCauley, "Selling out on Nature," *Nature* 443 (2006): 27-28.

5. Ibid.

6. Ibid.

7. Nik Heynen, James McCarthy, Scott Prudham, and Paul Robbins, eds., *Neoliberal Environments: False Promises and Unnatural Consequences* (New Brunswick, Routledge, 2007).

8. Ibid.

9. Peter Kareiva, Amy Chang, and Michelle Marvier, "Environmental Economics: Development and Conservation Goals in World Bank Projects," *Science* 321, 5896 (19 September 2008): 1638-1639.

10. Ibid.

11. Ibid.

12. Charles E. Di Leva, "The Conservation of Nature and Natural Resources through Legal and Market-Based Instruments," *Review of European Community & International Environmental Law* 11, 1 (2002): 84-95.

13. Ibid.

14. Ibid.

15. C. Kremen, J. O. Niles, M. G. Dalton, G. C. Daily, P. R. Ehrlich, J. P. Fay, D. Grewal and R. P. Guillery, "Economic Incentives for Rain Forest Conservation across Scales," *Science* 288 (2000): 1828-1832.

16. Ibid.

17. Ibid.

18. Arian Spiteri and Sanjay K. Nepal, "Incentive-based Conservation Programs in Developing Countries: A Review of some Key Issues and Suggestions for Improvements," *Environmental Management* 37 (2006): 1-14.

19. Ibid.

20. Ibid.

21. Ibid.

22. Arun Agrawal, Ashwini Chhatre, and Rebecca Hardin, "Changing Governance of the World's Forests," *Science* (13 June 2008): 1460-1462.

23. Ibid.

24. Ibid.

25. Di Leva

26. Dal Didia, "Debt-for-Nature Swaps, Market Imperfections, and Policy Failures as Determinants of Sustainable Development and Environmental Quality," *Journal of Economic Issues* 35 (2001): 477-486; CRS (Pervaze Sheikh), *Debt-for-Nature Initiatives and the Tropical Forest Conservation Act: Status and Implementation*, CRS Report for Congress. Congressional Research Service (CRS). (Library of Congress: Washington DC, 2006).

27. Richards and Moura Costa

28. Ibid.

29. CRS

30. Ibid.

31. United States Tropical Forest Conservation Act of 1998. Public Law 105-214, 112 Stat. 885-894.

32. CRS.

33. Ibid.

34. Ibid.

35. Ibid.

36. Didia.

37. Ibid.

38. Ibid.

39. Ibid.

40. Ibid.

41. Richards and Moura Costa.

42. Ibid.

43. Ibid.

44. Ibid.

45. J. Christensen, "Win-Win Illusions," *Conservation in Practice* 5 (2004): 12-19. URL www.conbio.org?CIP/articles51wwi.cfm

46. Ibid.

47. Ibid.

48. Ewald Ramesteiner and Markku Simula, "Forest certification: An instrument to promote sustainable forest management?" *Journal of Environmental Management* 67 (2003): 87-98.

49. Ibid.

50. Auld, Graeme, Lars H. Gulbrandse and Constance McDermott. "Certification schemes and the impacts on forestry and forests." *Annual Review of Environment & Resources*. 33 (2008): 187–211.

51. Ibid.

52. Ibid.

53. Agrawal et al.

54. Koen Kusters, Ramadhani Achdiawan, Brian Belcher, and Manuel Ruiz Pérez, "Balancing Development and Conservation? An Assessment of Livelihood and Environmental Outcomes of Nontimber Forest Product Trade in Asia, Africa, and Latin America," *Ecology & Society* 11 (2006).

55. Ibid.

56. Ibid.

57. Ibid.

58. Ibid.

59. K. Schreckenberg, Elaine Marshall, Adrian Newton, Dir Willem te Velde, Jonathan Rushton, Fabrice Edouard. "Commercialisation of non-timber forest products: What determines success?" *Forestry Briefing* 10, (London: Overseas Development Institute (ODI), 2006).

60. Ibid.

61. Ibid.

62. Ramesteiner and Simula.

63. Acharya, B. P. "Practice and Implementation of Forest Certification in Nepal: A Case Study from Some CFUGs in Dolakha District." MS Thesis, University of Natural Resources and Applied Life Sciences, Vienna (October 2007); author's own research (2010-2011).

64. Richards and Jenkins.

65. Ibid.

66. Sven Wunder, "The Efficiency for Payments for Environmental Services in Tropical Conservation," *Conservation Biology* 21 (2007): 48-58; Sven Wunder, "Are Direct Payments for Environmental Services Spelling Doom for Sustainable Forest Management in the Tropics?" *Ecology and Society* 11 (2006).

67. Wunder, *Direct Payments for Environmental Services*.

68. Richards and Jenkins.

69. Wunder, *The Efficiency of Payments for Environmental Services*.

70. Esteve Corbera, Katrina Brown, and W. Neil Adger, "The Equity and Legitimacy of Markets for Ecosystem Services," *Development & Change* 38 (2007): 587-613.

71. Ibid.

72. Ibid.

73. Wunder, *The Efficiency of Payments for Environmental Services*.

74. Wunder, *Direct Payments for Environmental Services*; Richards and Jenkins.

75. Richards and Jenkins.

76. Wunder, *Direct Payments for Environmental Services.*

77. Ibid.

78. Ibid.

79. Ibid.

80. Ibid.

81. Stefanie Engel and Charles Palmer, "Designing Payment for Environmental Services with Weak Property Rights and External Interests," in Payment for Environmental Services in Agricultural Landscapes, ed. Leslie Lipper, Takumi Sakuyama, Randy Stringer and David Zilberman (Springer Science + Business Media, 2009): 35-57.

82. Pagiola, Stefano, ed., *Selling Forest Environmental Services: Market-based Mechanisms for Conservation and Development* (Earthscan Publications, 2002).

83. Pagiola.

84. M. I. Porras Miranda and M. L. Moreno, "The Social Impacts of Payments for Environmental Services in Costa Rica: a Quantitative Field Survey and Analysis of the Virilla Watershed." (IIED, London, 2003); R. Munoz, *Efectos del progama de pago por servicios ambientales en las condiciones de vida de los caminos de la Peninsula de Osa.* Masters thesis (San Jose: Universidad de Costa Rica, 2004).

85. Pagiola.

86. John M. Antle and Jetse J. Stoorvogel, "Payment for Ecosystem Services, Poverty and Sustainability: The Case of Agricultural Soil Carbon Sequestration," in *Payment for Environmental Services in Agricultural Landscapes*, eds., Leslie Lipper, Takumi Sakuyama, Randy Stringer and David Zilberman (Springer Science + Business Media, 2009): 133-161.

87. Antle and Stoorvogel.

88. Antle and Stoorvogel.

89. Jennifer Alix-Garcia, Alain de Janvry, Elisabeth Sadoulet, and Juan Manuel, "Lessons Learned from Mexico's Payment for Environmental Services Program," in *Payment for Environmental Services in Agricultural Landscapes*, edited by Leslie Lipper, Takumi Sakuyama, Randy Stringer and David Zilberman (Springer Science + Business Media): 163-188.

90. Engel and Palmer.

91. Richards and Jenkins.

92. Wunder, *Direct Payments for Environmental Services.*

93. Richards and Jenkins.

94. Charlotte Streck and Sebastian M. Scholz, "The Role of Forests in Global Climate Change: Whence We Come and Where We Go," *International Affairs* 82 (2006): 861-879; Carmenza Robledo, Jurgen Blaser, Sarah Byrne and Kaspar Schmidt. *Climate change and governance in the forest sector: An Overview of the Issues on Forests and Climate Change with Specific Consideration of Sector Governance, Tenure, and Access for Local Stakeholders.* (Washington, DC, Rights and Resources Initiative: 2008).

95. Nicholas H. Stern, S. Peters, V. Bakhshi, A. Bowen, C. Cameron, et al. *Stern Review: The Economics of Climate Change* (Cambridge, Cambridge University Press: 2006). http://www.hm-treasury.gov.uk/sternreview_index.htm; Streck and Scholz; Robledo et al.

96. Robledo et al.

97. J. Blaser and C. Robledo. Initial analysis on the mitigation potential in the forestry sector," report prepared for the Secretariat of the UNFCCC, Bonn, Germany                              (August                              2007). (http://unfccc.int/files/cooperation_and_support/financial_mechanism/applicatio n/pdf/blaser.pdf)

98. Blaser and Robledo.

99. A. G. Bumpus and D. Liverman, "Accumulation by decarbonization and the governance of carbon offsets," *Economic Geography* 84 (2008): 127-55. (http://www.opencarbonworld.com/carbon-library/goverance%20of%20offsetts.pdf)

100. Ibid.

101. Stern et al.

102. Robledo et al.

103. Stern N. H., Peters S., Bakhshi V., Bowen A., Cameron C., et al. *Stern Review: The Economics of Climate Change.* Cambridge, UK: Cambridge University Press, 2006.

104. M.D. Shamsuddoha and Karim Chowdhury, "The Political Economy of UNFCCC's Bali Climate Conference: A Roadmap to Climate Commercialization," *Development* 51 (Rome: 2008): 397-402.

105. Ibid.

106. Michael R. Brown, "Limiting Corrupt Incentives in a Global REDD Regime," *Ecology Law Quarterly* 37 (237-267): 2010.

107. Bumpus and Liverman.

108. Streck and Scholz.

109. Streck and Scholz.

110. Ibid.

111. Ibid.

112. Ibid.

113. Robledo et al.

114. Streck and Scholz; Robledo et al.

115. Robledo et al.

116. Robledo et al.

117. Streck and Scholz; Bumpus and Liverman.

118. Bhaskar R. Karky and Kamal Banskota, "Reducing Emissions from Nepal's Community Managed Forests: Discussion for COP 14 in Poznan," *Journal of Forest and Livelihood* 8 (2009).

119. Streck and Scholz.

120. Leo Peskett, Cecilia Luttrell and Mari Iwata. "Can standards for voluntary carbon offsets ensure development benefits?" *Institute of Forestry Briefing* 13 (London, Overseas Development Institute: July 2007).

121. Peskett et al.

122. Peskett et al.

123. Peskett et al.

124. Peskett et al.

125. Peskett et al.

126. Peskett et al.

127. Robledo et al.

128. Kristen Hite, "Safeguards and REDD," Paper presented at RRI Workshop on Standards, Safeguards, and Recourse Mechanisms for Forests and Climate, (Washington, DC.: May 12, 2010).

129. Banskota and Karky.

130. Larry Lohmann, ed., "Carbon Trading: A Critical Conversation on Climate Change, Privatization and Power," *Development Dialogue* No. 48 (September 2006).

131. Ibid.

132. H. Bachram, "Climate Fraud and Carbon Colonialism: The New Trade in Greenhouse Gases," *Capitalism Nature Socialism* 15 (2004).

133. Robledo et al.

134. Lohmann.

135. Esteve Corbera, Katrina Brown, and W. Neil Adger, "The Equity and Legitimacy of Markets for Ecosystem Services," *Development & Change* 38 (2007): 587-613.

136. Corbera et al.

137. Joseph G. Canadell and Michael R. Raupach. 2008. Managing Forests for Climate Change Mitigation. *Science* 13 (2008): 1456-1457.

138. Streck and Scholz.

139. Streck and Scholz.

140. Bharat Pokharel and Jagadish Baral, "From Green to REDD, from Aid to Trade: Translating the Forest Carbon Concept into Practice," *Journal of Forest and Livelihood* 8 (2009).
141. Ibid.
142. Ibid.
143. Ibid.
144. Shamsuddoha and Chowdhury.
145. Stern et al.
146. Streck and Scholz.
147. Engel and Palmer.

## BIBLIOGRAPHY

Acharya, Bishnu Prasad. "Practice and implementation of forest certification in Nepal: A Case Study from some CFUGs in Dolakha District." MS Thesis, University of Natural Resources and Applied Life Sciences, Vienna (October 2007).

Agrawal, Arun, Ashwini Chhatre and Rebecca Hardin. "Changing Governance of the World's Forests." *Science* (13 June 2008): 1460-1462.

Alix-Garcia, Jennifer, Alain de Janvry, Elisabeth Sadoulet, and Juan Manuel. "Lessons Learned from Mexico's Payment for Environmental Services Program." In *Payment for Environmental Services in Agricultural Landscapes*, eds. Leslie Lipper, Takumi Sakuyama, Randy Stringer and David Zilberman. Springer Science + Business Media, 163-188.

Antle, M. John and Jetse J. Stoorvogel. "Payment for Ecosystem Services, Poverty and Sustainability: The Case of Agricultural Soil Carbon Sequestration." In *Payment for Environmental Services in Agricultural Landscapes*, eds. Leslie Lipper, Takumi Sakuyama, Randy Stringer and David Zilberman. Springer Science + Business Media, 2009: 133-161.

Auld, Graeme, Lars H. Gulbrandse and Constance McDermott. "Certification schemes and the impacts on forestry and forests." *Annual Review of Environment & Resources* 33, 2008: 187–211.

Bachram, Heidi. "Climate Fraud and Carbon Colonialism: The New Trade in Greenhouse Gases." *Capitalism Nature Socialism* 15 (2004).

Banskota, Kamal and Baskar Karky (ICIMOD). "Reducing Carbon Emissions Through Community Forestry in the Himalaya". Kathmandu: International Center for Integrated Mountain Development (ICIMOD), 2007.

Blaser, Jurgen and Carmenza Robledo. "Initial analysis on the mitigation potential in the forestry sector." Report prepared for the Secretariat of the UNFCCC, Bonn, Germany, August 2007. http://unfccc.int/files/cooperation_and_support/financial_mechanism/application/pdf/blaser.pdf

Brown, Michael R. "Limiting Corrupt Incentives in a Global REDD Regime." *Ecology Law Quarterly* 37 (2010): 237-267

Bumpus, G. Adam and Diana M. Liverman. "Accumulation by Decarbonization and the Governance of Carbon Offsets." *Economic Geography* 84(2008): 127-55. http://www.opencarbonworld.com/carbonlibrary/governance%20of%20offsetts.pdf

Canadell, G. Joseph and Michael R. Raupach. 2008. "Managing Forests for Climate Change Mitigation." *Science* 13 (2008): 1456-1457.

Chhatre, Ashwini and Vasant Saberwal. "Political Incentives for Biodiversity Conservation." *Conservation Biology* 19(2005): 310-317.

Christensen, Jon. "Win-Win illusions." *Conservation in Practice* 5 (2004): 12-19. http://www.conbio.org?CIP/articles51wwi.cfm

CRS (Pervaze Sheikh). "Debt-for-Nature Initiatives and the Tropical Forest Conservation Act: Status and Implementation." CRS Report for Congress. Congressional Research Service (CRS), Library of Congress: Washington DC, 2006.

Corbera, Esteve, Katrina Brown, and W. Neil Adger. "The Equity and Legitimacy of Markets for Ecosystem Services." *Development & Change* 38(2007): 587-613.

Di Leva, Charles E. "The Conservation of Nature and Natural Resources through Legal and Market-Based Instruments." *Review of European Community & International Environmental Law* 11 no. 1 (2002): 84-95.

Didia, Dal. "Debt-for-Nature Swaps, Market Imperfections, and Policy Failures as Determinants of Sustainable Development and Environmental Quality." *Journal of Economic Issues* 35(2001): 477-486.

Engel, Stefanie and Charles Palmer. "Designing Payment for Environmental Services with Weak Property Rights and External Interests." pp. 35-57. In *Payment for Environmental Services in Agricultural Landscapes*, eds. Leslie Lipper, Takumi Sakuyama, Randy Stringer and David Zilberman, Springer Science + Business Media, 2009.

Geist J. Helmut and Eric F. Lambin. "Proximate Causes and Underlying Driving Forces of Tropical Deforestation." *BioScience* 52(2002): 143-150.

Heynen, Nik, James McCarthy, Scott Prudham, and Paul Robbins (Eds). *Neoliberal environments: False promises and unnatural consequences."* New Brunswick: Routledge, 2007.

Hite, Kristen. "Safeguards and REDD." Paper presented at RRI Workshop on Standards, Safeguards, and Recourse Mechanisms for Forests and Climate. Washington, DC, May 12, 2010.

Jack, B. Kelsey, Carolyn Kousky, and Katharine R. E. Sims. "Ecosystem Services Special Feature: Designing Payments for Ecosystem Services: Lessons from Previous Experience with Incentive-based Mechanisms". *PNAS* 2008 105:9465-9470.

Kareiva, Peter, Amy Chang, and Michelle Marvier. "Environmental Economics: Development and Conservation Goals in World Bank Projects." *Science* 321 no. 5896 (19 September 2008): 1638 - 1639.

Karky, Bhaskar and Kamal Banskota "Reducing Emissions from Nepal's Community Managed Forests: Discussion for COP 14 in Poznan," *Journal of Forest and Livelihood* 8 (2009).

Kremen, Claire, J. O. Niles, M. G. Dalton, G. C. Daily, P. R. Ehrlich, J. P. Fay, D. Grewal and R. P. Guillery. "Economic Incentives for Rain Forest Conservation across Scales." *Science* 288 (2000): 1828-1832.

Kusters, Koen, Ramadhani Achdiawan, Brian Belcher, and Manuel Ruiz Pérez. "Balancing Development and Conservation? An Assessment of Livelihood and Environmental Outcomes of Nontimber Forest Product Trade in Asia, Africa, and Latin America." *Ecology & Society* 11 (2006).

Lohmann, Larry (Ed). "Carbon trading: A Critical Conversation on Climate Change, Privatization and Power". *Development Dialogue* no. 48 (September 2006).

McCauley, Douglas J. "Selling Out on Nature." *Nature* 443(2006): 27-28.

Miranda, Miriam, Ina T. Porras and Mary Luz Moreno. "The Social Impacts of Payments for Environmental Services in Costa Rica: A Quantitative Field Survey and Analysis of the Virilla Watershed." London: IIED, 2003.

Munoz, Rodrigo. "Efectos del progama de pago por servicios ambientales en las condiciones de vida de los caminos de la Peninsula de Osa." Masters thesis. Universidad de Costa Rica, San Jose, 2004.

Pagiola, Stefano (ed). *Selling Forest Environmental Services: Market-based Mechanisms for Conservation and Development.* Earthscan Publications, 2002.

Peskett, Leo, Cecilia Luttrell and Mari Iwata. "Can standards for voluntary carbon offsets ensure development benefits"? *Institute of Forestry Briefing* 13 (July 2007). London: Overseas Development Institute, 2007.

Pokharel, Bharat and Jagadish Baral. "From Green to REDD, from Aid to Trade: Translating the Forest Carbon Concept into Practice," *Journal of Forest and Livelihood* 8(2009).

Ramesteiner, Ewald and Markku Simula. "Forest Certification: An Instrument to Promote Sustainable Forest Management?" *Journal of Environmental Management* 67(2003): 87-98.

Richards, Michael and Michael Jenkins (ODI). "Potential and Challenges of Payments for Ecosystem Services from Tropical Forests." *Forestry Briefing 16*, Forest Policy and Environment Programme, Overseas Development Institute (ODI): London, 2007.

Richards, Michael and Pedro Moura Costa (ODI). "Can Tropical Forestry be Made Profitable by Internalizing the Externalities?" *Natural Resource Perspectives*, No. 46 (October 1999). Overseas Development Institute (ODI): London, 1999.

Robledo, Carmenza, Jurgen Blaser, Sarah Byrne and Kaspar Schmidt (RRI). "Climate Change and Governance in the Forest Sector: An Overview of the Issues on Forests and Climate Change with Specific Consideration of Sector Governance, Tenure, and Access for Local Stakeholders." Washington, DC: Rights and Resources Initiative (RRI), 2008.

Sanderson, Steven. "Poverty and Conservation: The New Century's 'Peasant Question'?" *World Development* 33(2005): 323-332.

Schreckenberg, Kate, Elaine Marshall, Adrian Newton, Dir Willem te Velde, Jonathan Rushton, Fabrice Edouard. "Commercialisation of Non-timber Forest Products: What Determines Success"? *Forestry Briefing* 10. London: Overseas Development Institute (ODI), 2006.

Shamsuddoha, M. D. and Karim Chowdhury. "The Political Economy of UNFCCC's Bali Climate Conference: A Roadmap to Climate Commercialization." *Development* (Rome) 51 (2008): 397-402.

Spiteri, Arian and Sanjay K. Nepal. "Incentive-based Conservation Programs in Developing Countries: A Review of Some Key Issues and Suggestions for Improvements." *Environmental Management* 37(2006): 1-14.

Stern N. H., Peters S., Bakhshi V., Bowen A., Cameron C., et al. Stern Review: The Economics of Climate Change. Cambridge, UK: Cambridge University Press, 2006. http://www.hm-treasury.gov.uk/sternreview_index.htm

Streck, Charlotte, and Sebastian M. Scholz. "The Role of Forests in Global Climate Change: Whence We Come and Where We Go." *International Affairs* 82 (2006): 861-879.

United States Tropical Forest Conservation Act of 1998. Public Law 105-214, 112 Stat. 885-894.

WCED (World Commission on Environment and Development). *Our Common Future*. Oxford, UK: Oxford University Press, 1987.

Wunder, Sven. "Are Direct Payments for Environmental Services Spelling Doom for Sustainable Forest Management in the Tropics?" *Ecology and Society* 11 (2006).

Wunder, Sven. "The Efficiency of Payments for Environmental Services in Tropical Conservation." *Conservation Biology* 21(2007): 48-58.

# 11

## BANKING ON SOCIAL INSTITUTIONS FOR POVERTY REDUCTION

*Hari B. Dulal and Roberto Foa*

In recent years there has been growing interest in including estimates of "intangible" capital, such as knowledge, skills, and institutions, in national asset accounting. In accordance with these efforts, this chapter attempts to provide the first worldwide evaluations of "social" capital, understood as the norms and networks that reduce transaction costs and enable collective action, as a proportion of national wealth. Using a new dataset that combines some 200 items from twenty-five sources, a composite of indices—measuring intergroup cohesion, gender equity, the strength of local community, the extent of crime and interpersonal trust, and levels of civic engagement—is formed and used to explain variance in the intangible capital residual, the proportion of national income that is left over after physical and natural capital have been accounted for. We show that social capital is the main component of national wealth and a major productive asset for societies and their constituent communities around the world.

Intangible capital encompasses human capital, which includes the sum of the knowledge, skills, and know-how possessed by a population as well as the level of trust in a society and the quality of its formal and informal social institutions (World Bank 2006). The majority of wealth in the world lays in the form of intangible capital—an amalgam including human and social capital, which reflects the quality of formal and informal institutions. Intangible wealth—human, institutional and social capital contributes 59–80 percent of social welfare, and the relative contribution rises with income across all regions and income classes (Hamilton and Ruta 2006). Rich countries are largely rich because of the skills of their populations and the quality of the institutions supporting economic activity. This is well reflected in the relationship between natural capital and income. The share of natural capital in total wealth tends to fall with income, while the share of intangible capital rises (World Bank 2006). Using a panel dataset with observations for 115 countries for the years 1995, 2000, and 2005, Ferreira and Hamilton (2010) show that the shares of produced, natural, and intangible capital in production are 32 percent, 7 percent and 18 percent, respectively. However, when the sample was limited to OECD countries, the only statistically significant factor of production was intangible capital, with a 50 percent share. The finding reinforces the fact that intangible factors, rather than produced or natural capital, are the principal sources of consumption growth in high-income countries. Labor productivity, which is comparatively high in the Western developed countries, can be attributed to higher intangible capital. Investment on intangible capital has shown to have a positive impact on labor productivity. Using international comparable data on intangible capital investment by business within a panel analysis from 1995–2005 in EU–15 country, Roth and Thum (2010), show a positive and significant relationship between intangible capital investment by business and labor productivity growth.

As well as having a major impact upon labor productivity and rates of economic growth, intangible capital has an important role in reducing levels of absolute and relative poverty. Minh Quang Dao (2008) shows that the fraction of the population below the poverty line is linearly dependent upon a range of human capital variables, including the gender parity ratio in primary and secondary schools and the maternal mortality rate. Using another sample of thirty-five developing countries, he also finds that income inequality linearly depends on the same explanatory variables plus the infant mortality rate and the primary school completion rate, which suggests that both relative as well as absolute poverty depend upon human capital accumulation. Specific country and regional studies have suggested similar conclusions, for example in Nigeria (Okunmadewa 2005), Pakistan (Kurosaki and Khan 2001), the Philippines (Asian Development Bank 2005), or Ghana (Rolleston 2009). Further studies have also shown that not only human capital, but also other aspects of social norms and behavior commonly referred to as "social" or "cultural" capital are critical in explaining why certain groups are able to exit poverty more rapidly than others. Narayan and Pritchett (1997) report findings from a study of 6,000 people living in eighty-seven villages in Tanzania, showing that a one standard deviation increase in village-level social capital predicts a 20 to 30 percent increase in expenditure per person, for each household in the village. In a follow-up publication, Narayan and Pritchett (1999) explain this statistical relationship by arguing that higher group membership rates imply more enjoyment of public services, the use of more advanced agricultural practices, joining in communal activities, and participation in credit programs. Other scholars have suggested further mechanisms linking social institutions and poverty reduction, including norms of gender equity (Van Staveren 2003), tolerance and trust (Knack and Keefer 1997; Tabellini 2010), or attitudes to work and saving (Inglehart 1997).

## THEORETICAL OVERVIEW

Intangible capital is the difference between total wealth and the sum of produced and natural capital and is calculated as a residual. Solow's residual and Tobin's Q are the most famous examples used. Solow's Residual measures the actual productivity which cannot be attributed to labor or capital accumulation and has to be attributed to technology. Tobin's Q is the ratio of the market value of a company versus the recorded asset value with the difference attributed to intangibles. The intangible capital residual includes all assets that are neither natural nor produced, and includes both human and social capital that are not accounted in wealth estimates. In other words, it consists of all forms of capital not immediately manifested in tangible matter (Webster and Jensen 2006). Webster and Jensen argue that, since the stock of available tangible matter is fixed, the sole source of productivity growth and the only way to enhance the (material) quality of life must be through the growth and deployment of intangible capital in the production process. According to Cummins (2004), although intangible capital is not a distinct factor as physical capital or labor, it is the "glue," that creates value from other factor inputs. Goldfinger (1997) acknowledges the role of intangible assets in wealth creation and argues that the creation and manipulation of the intangible assets to be the source of economic value and wealth rather than the production of material goods. As the sustainable competitive advantage of a nation is to a large extent dependent on the possession of relevant capability differentials, having a larger stock of capability differentials in the form of intangible capital—better human resources and social institutions—could in part help answer the question as to why some countries have more sustained economic growth and development, whereas others do not.

## THE VALUATION OF INTANGIBLE CAPITAL

We distinguish between two kinds of intangibles that contribute to the residual: human capital and social capital. For the purpose of this study, we define human capital as the knowledge, skills, and attributes possessed by individuals that contribute to personal and societal well-being and social capital as the informal institutions which reduce transaction costs and facilitate collective action. Although human capital has often been defined and measured with reference to educational attainment, it is a broader concept that includes attributes that reflect how various non-cognitive skills and attributes contribute to the greater well-being under different socioeconomic and cultural conditions. Meanwhile, social capital has variously been defined as general social trust (Fukuyama 1995), networks of civic engagement (Putnam 1993), non-discrimination, rule of law (Knack and Keefer 1997), or communal cooperation (Ostrom 2000).

Human capital has been most extensively studied in the form of education (Cohen and Soto 2007; World Bank 1999; World Bank 2000, Psacharopoulos and Patrinos 2004; Kimenyi et al. 2006; Wigley and Akkoyunlu-Wigley 2006; Vinod and Kaushik 2007). An investment in improving the skills and knowledge of the labor force determines the stock of human capital and returns to investment in education. Psacharopoulos and Patrinos (2004) empirically demonstrate the positive returns of investment in education. They show that the average rate of return to an additional year of schooling is 10 percent and that education produces the highest returns in low- and middle-income countries. Using time series and panel regressions for data on a group of eighteen large developing countries, Vinod and Kaushik (2007) empirically demonstrate that human capital has a statistically significant impact on economic growth. They cite India as an example where knowledge-based industries, notably computer software, telecommunications, pharmaceuticals, chemicals, and biotechnology have emerged as a result of sustained investment in institutes of higher education. Finally, Anderson and Keys (2007) demonstrate that if introduced at a young age education does contribute to future earning capacity (Anderson and Keys 2007), while other studies show that education also plays role in economic growth (Vinod and Kausik 2007), and promoting social cohesion by reducing ethnic tensions (Gradstein and Justman 2000).

Following Barro and Lee (2000), we measure human capital using the average number of years of schooling. This follows an established precedent whereby years of schooling has been used as a measure of educational achievement (Cohen and Soto 2001; World Bank 1999; World Bank 2000; Scarpetta and Visco 2000; Psacharopoulos and Patrinos 2004; Kimenyi 2006; Wigley and Akkoyunlu-Wigley 2006; Mamuneas et al. 2006; Vinod and Kaushik 2007). Following Woolcock and Narayan (2000), we define social capital as the norms and networks that enable collective action (Woolcock and Narayan 2000). A range of studies have examined the contribution of social capital to economic outcomes. Based on their study of Tsimane', a native Amazonian society of foragers and farmers in Bolivia, Godoy et al. (2007) report that social capital is positively and significantly associated with private investments in social capital, even after controlling for individual-level variables from an optimal investment model, spillovers from group social capital, village income inequality, and market openness. They also found that village income inequality and market openness were negatively associated with private investments in social capital. Similar results have been reported by Narayan and Pritchett (2002), who show that community level social capital has high and significant effects upon per capita income, based upon their sample of eighty-seven villages in Tanzania.

Multi-regional and cross-country studies have replicated the results cited based on household and village surveys. Using a sample of twenty regions in Italy, Helliwell and Putnam (1995) demonstrate that social capital has strong and significant relationship with economic growth, reconfirming earlier findings by Putnam, Leonardi and Nanetti (1993). Knack and Keefer (1997) have expanded the study of social institutions and growth to a cross-country sample, using data from twenty-nine nations empirically demonstrate the role of generalized trust and strength of civic norms on average annual growth in per capita income from 1980 to 1992, with the finding that countries with a higher starting level of "social trust" saw greater subsequent economic expansion. Similar studies have been conducted in the growing literature on the institutional determinants of growth; Easterly and Levine (1997), for example, use a sample of ninety-six nations to demonstrate that low level of social capital resulting from ethno-linguistic fractionalization impacts average annual growth in per capita, both directly and indirectly.

## *The Data and Model Estimates*

In our study, we have chosen human capital and social capital as intangibles because the two-part taxonomy suits our empirical model and data. As human capital has been most widely analyzed in the economics literature among the components of intangible capital, it is also included in our study to see what percentage of the variation in the IC residual is explained by human capital (in form of number of years of schooling). Although the human capital variable does not take into account the quality of education of those trained, it is highly unlikely to bias the coefficients, given the high correlation between this measure and direct assessments of educational outcomes (Cohen and Soto 2001; World Bank 1999; World Bank 2000; Scarpetta and Visco 2000; Psacharopoulos and Patrinos 2004; Kimenyi 2006; Wigley and Akkoyunlu-Wigley 2006; Mamuneas et. al. 2006; Vinod and Kaushik 2007). Following the precedent of World Bank (2006), in our study, immigrants settled abroad are considered as a special form of human capital. As emigrant workers send money to their families in the form of remittances, they are a part of total national wealth even though they are not physically present in the country. We include remittances in our model because the workers who choose to immigrate to foreign shores in order to find better employment opportunities do send money back to the country of their origin to support their families and in some cases own businesses. The variable *Inworkrempc* in the model is the percentage of migrant remittances and compensation of employees working abroad as a share of GDP, which includes all the legal economic flows generated by migrants to their country of origin. The role of remittances in influencing development through increase in the investment level in the source country has been widely acknowledged (Taylor 1999; Nyberg-Sørenson et al. 2002; León-Ledesma and Piracha 2004; Tewolde 2005). Emigrants have been known to raise incomes of their families at home significantly through remittances (Grubel and Scott 1966).

Despite the enormous interest in immigrant remittances on economic development and income growth in the emigrants' country of origin, empirical investigations have been limited by the availability of good quality data. It is extremely difficult to gather accurate data on remittances because many remittances are not channeled through the proper payment system and hence do not appear in the official statistics on remittances (Chami et al. 2005). Due to the unavailability of good-quality data for cross-country comparisons, most study on effects of remittances has been limited to the particular immigrant group. As data collected to study the impact of remittances are gathered from various sources, conflicting results are reported, which has made extremely difficult to draw general inference. For the purpose this study, we use a panel of aggregate data on remittances and years of schooling from the World Bank's World Development Indicators (WDI) database.

## Social Capital

The major innovation of our analysis is to include a composite index of social capital in our decomposition of the intangible capital residual. This measure of the "stock" of social capital is an index compiled from six sub-indices, each measuring an aspect of social institutional quality. The source of these measures is the Indices of Social Development project hosted at the Institute for Social Studies (ISS), which combines 200 items from some twenty-five sources into six estimates of how social institutions vary across countries (Foa 2011; Foa and Tanner 2011). The project adopts a working definition of social institutions as the informal norms and conventions that pattern human interaction (North 1991), and among the universe of all social institutions, those constitute *social capital* which serve to reduce transaction costs and enable collective action.

Reflecting the definitions of social institutions and social capital, the 200 items are siloed into five subareas where social norms serve to reduce transaction costs and facilitate collective action: crime and social trust, intergroup cohesion, gender equity, civic engagement, and strength of local community. The first area, gender equity, specifically estimates the level of discrimination against women. Included in this subindex are data on gender health, educational, and wage disparities, as well as data on the norms of discrimination that sustain these over time, such as the proportion of managers who believe men have more right to a job than women, or the proportion of parents who believe that boys should be prioritised in access to education. The second area, inter-group cohesion, reflects the extent of social conflict among ethnic, religious, or other social identity groups. It is measured by data such as ratings on the level of ethnic and religious tensions as well as the number of riots, assassinations, and acts of terrorism. The third area is crime and personal trust. Included in this subindex are data on citizens' trust in their society, neighbors, and community; data on crime victimization; and data on homicide and other acts of interpersonal aggression. The fourth area, strength of community, and measures the level of engagement in local associations and networks. Strength of community is measured using data on levels of engagement in local voluntary associations, time spent socializing in community groups, and membership of developmental organizations. Finally, the fifth area is the level of civic engagement, which measures the extent to which social practices encourage more active and critical engagement with political authorities. The strength of civil society is measured using survey data on participation in civic activities such as petitions or marches, access to media through newspaper and radio, and the density of international civil society organizations.

Why are these measures of social institutions a form of "capital?" They do so, because they facilitate the exchange of goods and information through their effect in reducing transaction costs. Norms of inclusion and non-discrimination, for example, serve to reduce the distortions in the labor market that are introduced by arbitrary exclusion of ethnic or other minorities. Likewise, norms of gender equity also enhance allocative efficiency, allowing women to translate human and economic capital into returns in the workplace. Crime and low social trust impose additional monitoring and enforcement costs in economic life, potentially inhibiting exchanges that might otherwise have occurred, while the presence of systemic intergroup violence, even below the level of open civil conflict, can have a similar effect.

Finally, the norms that enable collective action, whether in the local community or in the nation as a whole, facilitate public goods provision, the exchange of information among individuals, and between citizens and the public service providers. The general effect of social capital is therefore to reduce transaction costs, whether between transacting parties in the private sector or between public service providers and citizens.

For the purposes of empirical testing, the six subindices are combined into a single variable reflective of the "stock" of social capital in that society. This is done by summing the standardized score for each subindex, weighted by the average correlation between that subindex and the others (pairwise correlations used in the weighting schema are shown in table 11.1). The full six subindices were available for eighty-five countries, yielding an initial eighty-five scores, and to ensure the efficient use of information, we also impute social capital scores for an additional seventy-four countries for which between three and five subindices were available. This yields a total of 159 country observations for the composite social capital variable.

**Table 11.1. Correlation Between Sub-Indices Used in the Social Capital Index**

|  | Civic Engagement | Crime and Social Trust | Intergroup Cohesion | Local Community | Gender Equity |
|---|---|---|---|---|---|
| **Civic Engagement** | 1 |  |  |  |  |
| **Crime and Social Trust** | 0.479 | 1 |  |  |  |
| **Intergroup Cohesion** | 0.5451 | 0.397 | 1 |  |  |
| **Local Community** | 0.1311 | 0.0998 | -0.002 | 1 |  |
| **Gender Equity** | 0.6713 | 0.229 | 0.4986 | -0.2381 | 1 |

The model basically represents the residual as a function of human capital within the nation and abroad and social capital expressed as institutional quality. The per capita years of schooling of the working population is to capture domestic human capital and remittances by those outside the country.

We use Cobb-Douglas function:

$$R = AS^{\alpha}{}_S F^{\alpha}{}_F P^{\alpha}{}_I$$

R = Intangible residual
A = Constant
S = Years of Schooling per worker
F = Remittances from Abroad
P = Social Capital
$\alpha_s$ = Elasticity of the residual

## Empirical Results

The result of the regression analysis shown in table 11.2 reveals that our independent variables explain 88 percent variation in the residual. The result indicates goodness of fit of the model evaluated by means of the adjusted $r^2$ (0.88).The coefficients for all three independent variables are positive and different from zero at the 95 percent confidence interval. The coefficient of *school years* suggests that a 1 percent increase in school years will increase the intangible capital by 0.47 percent, whereas 1 percent increase in remittance from abroad will increase intangible capital by 0.14 percent. Out of three variables selected, social capital however shows the most significant association with the intangible capital residual. The estimation shows that a 1 percent increase in the stock of social capital results in a 1.10 percent increase in intangible capital. However, coefficients lower than one indicates decreasing marginal returns. The negative dummy coefficients support the general assumption that low-income, middle-income, and upper-middle income countries possess lower levels of intangible residual capital.

**Table 11.2. Elasticities of IC with Schooling, Remittances, and Social Capital**

| | |
|---|---|
| LIC | -2.34605 (0.4450128)*** |
| LMI | -1.69097 (0.3422432)*** |
| UMI | -1.316136 (0.3150706)*** |
| log social capital | 16.18202 (6.056749)** |
| log schooling | 0.4704271 (0.2155866)* |
| log remittances | 0.1392153 (0.0484213)** |
| Constant | -22.46867 (12.43407) |
| | |
| N | 74 |
| adj. $r^2$ | 0.88 |

Dependent variable: Intangible Capital Residual


**Marginal Returns**

| | Marginal Returns to Schooling | Marginal Returns to Social Capital | Marginal Returns to Remittances |
|---|---|---|---|
| Low-income countries | 903 | 127 | 17 |
| Lower-middle income countries | 1,723 | 412 | 15 |
| Upper-middle income countries | 2,537 | 547 | 84 |
| High-income OECD | 15,805 | 3,274 | 153 |

# COUNTRY CASE STUDIES

Country examples can increase our intuitive understanding of the decomposition of intangible wealth. In this section, we illustrate the relative contributions of human capital, social capital, and remittances to national wealth by comparing three landlocked, low-income countries in sub-Saharan Africa: Mali, Rwanda, and Lesotho. While all three share important characteristics and have substantial intangible capital residuals, these residuals can be explained by the different relative endowments of the three countries. This can be seen from table 11.3, which decomposes the intangible capital residual for each of the three cases.

**Figure11.1. Decomposition of intangible capital**

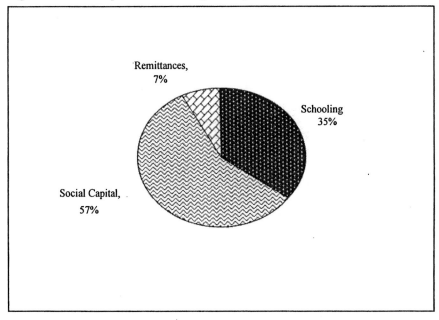

**Table 11.3. Decomposition of the Intangible Capital Residual**

| Country | Income per capita ($) | Intangible residual (% GNI) | Shares of residual (%) | | | Levels | | |
|---|---|---|---|---|---|---|---|---|
| | | | Schooling | Social capital | Remittances | Schooling (yrs) | Social capital | Remittances ($ per capita) |
| Rwanda | 226 | 54 | 54 | 45 | 1 | 3 | 15 | 2 |
| Mali | 208 | 47 | 13 | 82 | 5 | 1 | 40 | 18 |
| Lesotho | 480 | 76 | 25 | 33 | 42 | 4 | 40 | 393 |

Mali, a landlocked Sahelian country in West Africa, is the poorest among the three countries with a per capita income of just $208. Nonetheless, the country has an intangible capital residual that accounts for about half of per capita wealth. Of this, over four-fifths (82 percent) rests in the country's stock of social capital, and one-eighth (13 percent) in the country's human capital. From among the six social development subindices, Mali scores strongly on local community and on intergroup cohesion, and has a strong civil society relative to other countries of a similar economic level. Intergroup cohesion has helped avert the kind of civil conflict experienced by neighboring states such as Niger or Cote d'Ivoire, while civic engagement has since 1991 supported a successful process of democratization. Stability and openness may in part explain the country's strong 5 percent annual rate of growth since 1994. The estimates here suggest that growth might be yet stronger still, were there a higher degree of human capital; at present, the average citizen experiences just one year of schooling.

Rwanda, a small landlocked country in central Africa, has a similar GDP per capita to Mali, estimated at $226. Yet the country has a very different distribution between social and human capital. Social capital accounts for only 45 percent of the residual, while schooling account for 54 percent. A country with a historically centralized but very effective state, Rwanda's human capital is result of a comprehensive system of primary education that has raised average years of schooling to three years per capita—about treble the level found in Mali. Yet Rwanda has a very low stock of social capital. On the index used in this paper, Rwanda has a score of just fifteen, compared to forty in Mali. Seventeen years after a genocide that killed 15 percent of the country's population, Rwanda remains a country plagued by low social trust, disputes over justice and land rights, and disengagement from civil society (Human Rights Watch 2007). The primary medium-term growth risk in Rwanda remains political instability from ethnic insurgency. Efforts to enhance national unity, equitable growth, and justice and reconciliation will remain central in bolstering the country's long-run growth prospects.

Lesotho—also a small, landlocked country in sub-Saharan Africa—is by far the richest of the three countries, with a per capita GDP of $480 in 2000. However, as our decomposition of the intangible capital residual shows, this is not due to higher levels of social or human capital. Levels of average schooling, at four years per capita, are similar to Rwanda, while the stock of social capital is comparable to Mali. By contrast, the country's relatively greater wealth is entirely a product of the large inflow of remittances from abroad: the average Mosotho receives $393 per capita a year in remittances, primarily from Basotho working as miners in neighboring South Africa. Consequently, remittances account for 42 percent of Lesotho's stock of intangible capital, compared to less than 5 percent in Mali or Rwanda.

These findings suggest very different possibilities for raising the stock of intangible capital in developing countries that face geographical restraints. Mali has clearly achieved nascent growth through its stock of social capital. Ethnic cohesion, civic engagement, and community development have in turn ensured political stability, accountability, and the better management of collective resources, and these factors have sustained productivity increases in agriculture, fishing, and mining over the course of the last decade. The example suggests that a similar level of cohesion in a country such as Rwanda would yield a large marginal gain, by creating the expectations of stability and transparency that enable sustained (physical) capital accumulation. On the other hand, Rwanda, unlike Mali, has attained a level of state organization that is unusual for a low-income country, and this has enabled sustained investment in primary education and skills. Provided the country can maintain its current political stability, this stock of human capital will constitute a vital resource that can be leveraged to support growth through technology transfer and improved quality in management and administration.

Finally, Lesotho has achieved important secondary growth by relying on a migrant labor force that brings substantial remittances home every year to their dependents within the country. The key challenge for such a country is to channel this inflow into expenditures that will support long-run sustainable growth within the country; this in turn will in no small degree involve solving domestic political and social conflicts which inhibit inward investment and domestic job creation. The outbreak of violence after disputed elections in 1998, for example, led to widespread destruction and disinvestment, and in part this instability is the consequence of a weak civil society, which if stronger would help build political consensus, reduce elite capture, and foster transparency. The social development indices give a very low civil society score for Lesotho, and strengthening the conduits between citizens and the state would help approximate the peaceful political transition seen in a country such as Mali.

## CONCLUSION

This chapter represents an addition to the relatively sparse literature on intangible capital. We argue that cross-country differences in intangible capital stock may be explained by differences in human and social capital. The evidence provided by this study helps policymakers by showing that intangible capital is among the fundamental determinants of the developmental status of countries, and consequent success in poverty reduction. The major hurdle that developing nations face today is their lower stock of intangible capital and the inability to convert human and social capital into revenue, cost savings, and other forms of tangible benefits. The intangible residual obtained from the wealth estimates has provided researchers with an opportunity to carry out cross-country valuation of human and social capital. In our study, by decomposing the intangible wealth residual, we have tried to highlight the importance of social capital, remittances, and years of schooling. Developing countries with lower stock of human and social capital can produce high levels of output per worker in the long run and work to reduce poverty by investing in education and improvement of institutional quality.

As a result of our analysis, we find that among the most important aspects of intangible wealth is social capital—the norms and networks that reduce transaction costs and enable collective action. Social trusts, non-discrimination, cohesion across social and ethnic groups, and engagement in civic and community associations are all important dimensions of this stock, and together account for 57 percent of the world's intangible capital. By serving to reduce transaction costs, these institutions facilitate processes of production, exchange, and the provision of public goods. Social trust, for example, ensures the maintenance of informal, non-legally binding agreements; non-discrimination enables individuals to enter into contracts of work or exchange with those of differing backgrounds or beliefs; and intergroup cohesion ensures the continuity of institutions and therefore the expectations upon which investment and negotiation decisions are made. We are a long way from knowing the policies that serve to build up or deplete social capital stocks over time, yet by recognizing the place of social capital in contributing to societal wealth, we make an important first step toward a research paradigm in which the effects of policies upon a society's social capital are factored into cost–benefit analyses of programs and projects.

### BILIOGRAPHY

Anderson, A. Gary and James D. Keys. "Building Human Capital through Education," *Journal of Legal Economics* 14, no. 1 (2007): 49-74.

Asian Development Bank (ADB). "Poverty in the Philippines: Income, Assets, and Access." ADB: Manila, 2005.

Barro, J. Robert and J. W. Lee. "International Data on Educational Attainment: Updates and Implications," *Oxford Economic Papers* 53, no. 3 (2001): 541-63.

Chami, Ralph, Connel  Fullenkamp and Samir Jahjah. "Are Immigrant Remittance Flows a Source of Capital for Development?" IMF Working Paper WP/03/189, 2005.

Cohen, Daniel and Soto Marcelo. "Growth and Human Capital: Good data, Good results." *Journal of Economic Growth*" 12, 2007:51–76.

Cummins, G. Jason. "A New Approach to the Valuation of Intangible Capital." NBER Working paper no. 9924, 2004.

Easterly, William and Levine Ross. "Africa's Growth Tragedy: Policies and Ethnic Divisions," *Quarterly Journal of Economics* 112, 4: 1997.

Ferreira, Susana  and Hamilton Kirk.  "Comprehensive Wealth, Intangible Capital, and Development." The World Bank Policy Research Working Paper no. 5452. Washington DC: The World Bank, 2010.

Foa, Roberto. "Indices of Social Development–Research Applications." Institute for Social Studies  Working Paper Series. The Hague, 2011. www.indsocdev.org

Foa, Roberto and Tanner C. Jeffery.  "Methodology of the Indices of Social Development." Institute for Social Studies Working Paper Series. The Hague, 2011. www.indsocdev.org

Goldfinger, Charles. *Understanding and Measuring the Intangible Economy: Current Status and Suggestions for Future Research.* Helsinki: CIRET seminar, 1997.

Gradstein, Mark and Justman Moshe. "Education, Social Cohesion, and Economic Growth". *American Economic Review* 92, no. 4 (2002): 1192-1204.

Grubel, G. Herbert and Anthony Scott. "The International Flow of Human Capital," *American Economic Review* 56 (1966): 268–74.

Hamilton, Kirk and Ruta Gianni. "Measuring Social Welfare and Sustainability". Washington DC: The World Bank, 2006. http://www.iaos2006conf.ca/pdf/Kirk%20Hamilton.pdf

Helliwell, F. John and Robert D. Putnam. "Economic Growth and Social Capital in Italy," *Eastern Economics. J.* 21 (1995): 295–307.

Inglehart, Ronald. "The Impact of Culture on Economic Development: Theory, Hypotheses, and Some Empirical Tests." In ed. Ronald Inglehart, *Modernization and Postmodernization.* Princeton, NJ: Princeton University, 1997.

Kimenyi, S. Mwangi, Germano Mwabu and Damiano Kulundu Manda. "Human Capital Externalities and Private Returns to Education in Kenya." *Eastern Economic Journal* 32, no. 3 (2006): 493-13.

Knack, Stephen and Keefer Philip. "Does Social Capital Have an Economic Payoff? A Cross-Country Investigation." *The Quarterly Journal of Economics* 112, no. 4, (1997):1251-88.

Kurosaki, Takashi and Humayun Khan. "Human Capital and Elimination of Rural Poverty: A Case Study of the North-West Frontier Province, Pakistan." IER Discussion Paper Series B no. 25, Hitotsubashi University, January 2001.

León-Ledesma, Miguel and Piracha Matloob. "International Migration and the Role of Remittances in Eastern Europe," *International Migration* 42, no. 4(2004): 66–83.

Mamuneas, P. Theofanis, Andreas, Savvides and Thanasis Stengos. "Economic Development and the Return to Human Capital: A Smooth Coefficient Semiparametric Approach," *Journal of Applied Econometrics* 21, 2006: 111-32.

Minh Quang Dao. "Human Capital, Poverty, and Income Distribution in Developing Countries," *Journal of Economic Studies* 35, no. 4 (2008):294-03.

Narayan, Deepa. "Voices of the Poor: Poverty and Social Capital in Tanzania." Environmentally and Socially Sustainable Development Network, Studies and Monograph Series no. 20. Washington, DC: World Bank, 1997.

Narayan, Deepa and Lant Pritchett. "Cents and Sociability: Household Income and Social Capital in Rural Tanzania," *Economic Development and Cultural Change* 47, no. 4(1999): 871-97.

Nyberg-Sørensen, Ninna, Nicholas Van Hear and Poul Engberg-Pedersen. "The Migration-Development Nexus: Evidence and Policy Options." *International Migration*, 40, no. 5 (2002): 49-71.

OECD. "The Well-Being of Nations: The Role of Human and Social Capital." OECD, 2001.

Okunmadewa, F. Y, S. A. Yusuf and B. T. Omonona. "Effects of Social Capital on Rural Poverty in Nigeria," *Pakistan Journal of Social Sciences* 4, no. 3 (2007): 331-39.

Okunmadewa, F. Y. O. Olaniyan, S. A. Yusuf, A. S. Bankole, O. A. Oyeranti, B. T. Omonona, T.T. Awoyemi and K. Kolawole. "Human Capital, Institutions and Poverty in Rural Nigeria," Research Report Submitted to *African Economic Research Journal of Consortium (AERC) Kenya*, for the second Phase of Collaborative Poverty Research Project, 2005.

Psacharopoulos, George and Harry Anthony Patrinos. "Returns to Investment in Education: A further update," *Education Economics* 12, no. 2 (2004): 111-34.

Putnam, Robert, Robert Leonardi and Raffaella Nanetti. *Making Democracy Work: Civic Traditions in Modern Italy*. Princeton, NJ: Princeton University Press, 1993.

Putnam, Robert. *Bowling Alone: The Collapse and Revival of American Community*. New York: Simon and Schuster, 2000.

Rolleston, Caine. "Human Capital, Poverty, Educational Access and Exclusion: The Case of Ghana 1991-2006." Project Report Consortium for Research on Educational Access, Transitions and Equity (CREATE). UK: Falmer, 2009.

Roth, Felix and Thum Anna-Elisabeth. "Does Intangible Capital Affect Economic Growth?" Centre for European Policy Studies Working Document no. 335. Brussels: The Centre for European Policy Studies, 2010.

Taylor, Edward. "The New Economics of Labor Migration and the Role of Remittances in the Migration Process," *International Migration* 37, no. 1 (1999): 63-87.

Tabellini, Guido. "Culture and Institutions: Economic Development in the Regions of Europe." *Journal of the European Economic Association* 8 (2010): 677-16.

Tewolde, Berhane. "Remittances as a Tool for Development and Reconstruction in Eritrea: An Economic Analysis," *Journal of Middle Eastern Geopolitics* 1, no. 2 (2005): 21-32.

Van Staveren, Irene. "Beyond Social Capital in Poverty Research," *Journal of Economic Issue* 37, 2003.

Vinod, Hrishikesh D. and Surendra K. Kaushik. "Human Capital and Economic Growth: Evidence from Developing Countries." *American Economics* 51, no. 1 (2007): 29-39.

Webster, Elizabeth M. and Paul H. Jensen. "Investment in Intangible Capital: An Enterprise Perspective," *The Economic Record* 82, no. 256 (2006): 82–96.

Wigley, Simon and Arzu Akkoyunlu-Wigley. "Human Capabilities versus Human Capital: Gauging the Value of Education in Developing Countries," *Social Indicators Research* 78, no. 2 (2006): 287-304.

Woolcock, Michael and Deepa Narayan. "Social Capital: Implications for Development Theory, Research, and Policy," *World Bank Research Observer* 15, no. 2 (2000): 225-50.

World Bank. "Knowledge for Development." World Development Report 1998/1999. New York: Oxford University Press, 1999.

World Bank. "Entering the 21st Century." World Development Report 1999/2000. New York: Oxford University Press, 2000.

World Bank. "Where is the Wealth of Nations? Measuring capital in the 21st century." Washington DC: The World Bank, 2006.

## 12

CLIMATE CHANGE ADAPTATION AND THE INTERNATIONAL AID
REGIME PROSPECTS FOR POLICY CONVERGENCE

*Craig Johnson*

## INTRODUCTION

Incorporating climate change adaptation into contemporary development policy
has quickly become one of the most pressing and controversial challenges in the
international aid regime (Ayers and Huq 2009; Boyd et al. 2009; Brooks et al.
2009). There are a number of reasons for this. First, in keeping with the aims of
the 1992 Earth Summit, any effort to support adaptation must be "new and
additional" to existing official development assistance, which in practice further
bureaucratizes what is already a highly bureaucratic system (Ayers and Huq
2009; and below). Second, climate change competes with other aid priorities,
such as primary education, poverty reduction and human health, reinforcing the
separation of "climate" and "non-climate change" issues and agendas (cf. Klein
et al. 2007). Third, adaptation implies an inter-generational perspective that is
often at odds with an aid paradigm that uses time-bound targets and indicators to
achieve quantifiable development outcomes, highlighting the challenge of
changing historical policies, practices and cultures in ways that lead to
sustainable reductions in poverty, insecurity (of various kinds) and vulnerability
to climate change (Ayers and Huq 2009; Boyd et al. 2009; Brooks et al. 2009).

The global food shocks of 2008 and 2010 illustrate the vulnerability of
globalized food production systems to sudden macro-economic and
environmental change. They also provide some insight about the ways in which
climate change will affect the livelihoods of poor and primarily rural
populations in the developing world. According to one estimate, the "mega-
heatwave" of 2010 reduced the Russian grain harvest by 25 percent, causing
global shortages and economic losses in the range of US$15 billion (Carrington
2011). The World Bank estimates that 40 percent of all official assistance and
concessional loans are at risk of losses due to climate change (Ayers and Huq
2009, 681, citing Burton). Moreover, the IPCC (2007) projects that crop yields
in South Asia and sub-Saharan Africa will fall by 50 and 30 percent as a result
of changing rainfall patterns and temperatures by 2020. Another study (by Cline,
cited by UNDP 2007: 18) predicts that declining yields in South Asia alone

could increase the number of people affected by malnutrition to 600 million by 2080. Incorporating the possible impacts of saltwater intrusion, coral bleaching and ocean acidification, the possibility of large-scale shocks to the global food system looms large (IPCC 2007; UNDP 2008).

In its broadest sense, climate change implies transformations in rainfall, temperature and seasonality, whose impact will be inconsistent with the strategies farmers, resource managers and policy makers have traditionally used to sustain livelihoods in environmentally sensitive regions and sectors, such as farming, forestry and fishing (Perch-Nielson et al. 2008). Some of these will be sudden, cataclysmic and deeply destructive to human life, health and property. Others will be gradual, variable and complex, forcing long-term transformations (e.g. changes in land use from rain-fed agriculture to grazing) that will create new vulnerabilities as well as opportunities for the poor. Under "normal" conditions, one can identify a range of strategies (e.g. crop diversification, borrowing, consumption smoothing, migration, public works programs, food security measures) that can be used to manage and adapt to changing climatic conditions. However, when changes in climate entail major transformations in rainfall, temperature and seasonality, the idea of using "traditional" adaptation measures becomes far less likely (McLeman and Smit 2006; Smit and Wandel 2006; Burton 2004 [2009]; 2009; Perch-Nielson et al. 2008; Bogardi and Warner 2009).

Successful adaptation implies an ability to learn from past experiences and to coordinate actions in ways that challenge or change existing behaviours, thereby improving resilience to future shocks and stresses (Holling 2001; Folke et al. 2005; Folke 2006; Olsson et al. 2006; Smit and Wandel 2006; IPCC 2007; Pahl-Wostl 2009). "Maladaptation," by contrast, implies that social actors are unable to change their behaviour to manage changing circumstances and events (Pahl-Westl 2009: 358). In Pahl-Wostl's words, it implies an "inability...to develop expectations, coordinate collective action and improve routines and practices," (Pahl-Westl 2009: 358).

### *To what extent is the international aid regime able to support the kinds of adaptation that will be necessary to reduce vulnerability to climate change? And what are the main policy priorities?*

This chapter considers these questions by exploring the challenge of incorporating adaptation into the international aid regime, a term I use to describe the rules, norms, and practices that structure international aid giving and assistance.[2] It starts from the premise that international efforts to support adaptation are rooted in a regime of institutions, interests and ideas that frame and constrain the ability of international actors to cooperate in ways that reduce long-term vulnerability to climate change. The following section first explores the historical and institutional limitations of fostering adaptation within the international aid regime. The next section then develops a typology that may be used to conceptualize different forms of climate risk: "catastrophic shocks," livelihood shocks, systemic shocks and cultural shocks. The fourth section identifies a number of ways in which adaptation and development policy may be combined to reduce these risks. The final section concludes the chapter.

## ADAPTATION AND THE INTERNATIONAL AID REGIME

There is now a growing consensus that human emissions of greenhouse gases are leading to unprecedented transformations in the Earth's climate, creating new forms of vulnerability to rapid-onset disasters and long-term environmental change. The IPCC (2007) predicts that "hot extremes, heat waves and heavy precipitation will continue to become more frequent," during the next century (IPCC WGI 2007: 15), projecting that the effects of climate change will be worst in Africa and Asia, where agriculture, public health systems, food supplies, and human settlements are least able to adapt to extreme climate events, such as flooding, windstorms, disease and drought (IPCC WGII 2007). Notwithstanding substantial improvements in public infrastructure, it is projected that climate change will disrupt food supplies, health systems, human settlements and major livelihoods, especially ones deriving from agriculture, forestry and coastal fishing (Biermann and Boas 2007; IPCC WG-II 2007).

Recent efforts to support climate change adaptation have called upon the international community to provide new and predictable funding sources that can be used to reduce environmental vulnerability over the long term (Klein et al. 2007; UNDP 2007; Ayers and Huq 2009; Boyd et al. 2009; Brooks et al. 2009). The UNDP's HDR (2007: 30), for instance, calls for "at least US$ 86 billion in 'new and additional' finance for adaptation" by 2016, as well as an additional $US2 billion to for the UN's Central Emergency Response Fund by 2016 and the World Bank's Global Facility for Disaster Reduction and Recovery. Others, such as Peskett et al. (2009) and Ayers and Huq (2009) have identified new funding sources, such as an international levy on air travel, which could be used to support future adaptation. The World Bank (2010) recommends that 0.7 percent of the stimulus packages being used to manage the current economic crisis be directed towards a "vulnerability fund" that would be used to support adaptation efforts in the developing world.

Arguably, the most visible regime governing international adaptation financing is the UN Framework Convention on Climate Change (UNFCCC) and the Kyoto Protocol, which provide three main channels of funding for adaptation in low-income countries:

- The Least Developed Country Fund (LDCF), which is intended to help with the identification of adaptation priorities in the National Adaptation Programmes of Action (NAPAs);
- The Special Climate Challenge Fund (SCCF), which is designed to support adaptation, as well as mitigation and technology transfer; and
- The Adaptation Fund, which is supported under the Kyoto protocol and aimed at supporting "concrete adaptation" actions (Huq and Reid 2004 [2009: 317])

Funding for the LDCF is provided by "Annex I" countries and administered by the Global Environment Facility (the GEF). As of October 2007, the fund had received US$163m, US$9.4m of which had gone to the preparation of individual NAPAs and US$28.5m going to eight priority implementation projects (Hedger et al. 2008: 12). Over the same period, US$59m had been allocated to the SCCF. The Adaptation Fund was created through the Kyoto Protocol and is funded by a 2 percent levy on all carbon emissions reductions certified under the Clean Development Mechanism (CDM). Depending on the price of carbon, estimates project that the fund will be able to generate anywhere between US$400m and 1.2 billion between 2008 and 2012.[3]

However, the ability of the UNFCCC to fund adaptation efforts faces a number of challenges. One concerns the gap between funding commitments and the estimated costs of adapting to climate change. According to the World Bank (2006, cited in Adger et al. 2007), the cost of "climate proofing" existing development investments would be as high as US$40 billion dollars per year, which (at 2006 levels) would be the equivalent of almost half of the entire ODA budget for 2006.[4] However, estimates reported by Ayers and Huq (2009: 678-9) suggest that the UNFCCC was US$100 million short of what it had been promised in 2009, reflecting donor concerns about the "adequate and accountable mechanisms in developing countries for receiving and disbursing money," (Ayers and Huq 2009: 678).

In principle, the agreement in Copenhagen to provide US$100 billion in additional adaptation financing by 2020 will provide an important source of funding.[5] However, a second and related challenge concerns the stipulation that adaptation financing remains "new and additional" to official aid commitments. Under the terms of the UN Framework Convention on Climate Change (UNFCCC), financing for climate change adaptation must be "new and additional" to existing ODA, which in 2009 stood at US$119.6bn (OECD 2010). However, establishing whether donors have in fact adhered to the stipulations outlined in the UNFCCC presents a significant challenge. In 2010, for instance, donors claimed they had already exceeded their pledge to provide US$30bn for the period 2010–2012, when in fact the total amount that could be counted as new and additional was $US8.2bn (Fallasch and de Marez 2010).

Third, there is a more general lack of consensus about what adaptation means, and how it should be applied to existing decisions about future vulnerability and risk (cf. Burton  2004 [2009]; Huq and Reid 2004 [2009]; Moench 2007 [2009]; Peskett et al. 2009; Ayers and Huq 2009; Brooks et al. 2009; Boyd et al. 2009). Although the UNFCCC provides a number of articles (4.1, 4.4, 4.8 and 4.9) that identify the ways in which parties to the Convention may be expected to support adaptation, the Convention lacks a precise definition that can be operationalized through the UNFCCC (Burton, 2004 [2009]). Similarly, although the GEF has responsibility for assisting developing countries in the preparation of NAPA documents, neither its website nor its brochure "Linking Adaptation to Development" (GEF 2007) provides a precise definition of what adaptation may entail.

Fourth, concerns have been raised about the disbursement of adaptation financing. According to research recently conducted by the Overseas Development Institute in London, nearly a third of the US$760m allocated by the Global Environment Facility between 2006 and 2009 had gone to India, China and Brazil (Vidal, 2009). In contrast, less than $US100m had gone to the poorest forty-nine countries, highlighting the importance of geopolitics in the definition and implementation of adaptation programming. Although some donors (e.g. JICA and the World Bank) are now developing allocation criteria for adaptation programming, most donor-supported adaptation programs still operate on an *ad hoc* basis (cf. Ayers and Huq 2009), reflecting the commercial and geopolitical patterns of historical aid giving practices (cf. Riddell 2006; de Haan 2008).

A fifth challenge relates to the institutional structure of the UNFCCC. According to Ayers and Huq (2009), the rules and procedures governing the disbursement of GEF-managed funds are confusing and burdensome, posing serious obstacles for low-income countries. Particularly counterproductive, they argue, is the requirement that the LDCF and the SSCF can only be used to support "additional" adaptation needs that arise as a result of climate change, which in effect rules out an entire class of "baseline" development needs, i.e. ones "that would occur anyway in the absence of climate change" (Ayers and Huq 2009: 678). Reflecting upon the *ad hoc* nature of adaptation financing, Ayers and Huq (2009) argue in favour of a funding mechanism that provides a predictable stream of revenue generation outside of the GEF, in particular the Adaptation Fund. Unlike the SSCF and the LDCF, the Adaptation Fund is managed by an independent board, which has "representation from the five UN regions as well as special seats for the Least Developed Countries and Small Island Developing States," (Ayers and Huq 2009: 679).

Finally, it has been argued that the National Adaptation Plans of Action (the NAPAs) are exceedingly top-down and technocratic, promoting plans that reflect the priorities of donors, international agencies (especially the GEF) and national governments, as opposed to vulnerable populations (Ayers and Huq 2009). Like the international negotiations that gave them force, the UNFCCC and the Kyoto Protocol are products of an international system whose principal aim is to uphold the jurisdictional and territorial sovereignty of individual nation states. Correspondingly most of the adaptation programming that has so far been supported under the Kyoto Protocol and the UNFCCC has transpired on a country-by-country basis (Huq and Reid 2004 [2009]). Although national governments should of course play a leading role in the formulation and implementation of adaptation policy, a principal concern here is that funding for *some* national governments has come at the expense of populations living within the territorial confines of the nation state.

An over-riding concern is therefore that international efforts to promote adaptation lack the sustainability and coherence to build long-term resilience to global climate change.

## *Mainstreaming Adaptation*

Reflecting upon these challenges, a number of observers have highlighted the need to "mainstream" climate change adaptation into donor policy and practice (e.g. Klein et al. 2007; UNDP 2007; Ayers and Huq 2009). The UNDP Human Development Report (2007: 30), for instance, highlights the need to "integrate adaptation into poverty reduction strategies that address vulnerabilities linked to inequalities based on wealth, gender, location and other markers for disadvantage" (UNDP 2007: 30). Arguably the most systematic effort to integrate poverty reduction and environmental vulnerability at the international level is the MDG framework, which provides an important mechanism through which bilateral and multilateral donors now design and evaluate their programs and policies. Most relevant to the present discussion is MDG 7, which seeks to "ensure environmental sustainability," by setting out four specific targets aimed at (7a) integrating "the principles of sustainable development into country policies and programmes and reverse the loss of environmental resources," (7b) reducing biodiversity loss, (7c) halving the proportion of people without sustainable access to safe drinking water and basic sanitation by 2015 and (7d) improving the lives of at least 100 million slum dwellers by 2020.[6] Although none of these targets is specifically concerned with climate change, targets 7a and 7b do have indicators measuring progress in relation to the conservation of land area covered by forests and CO2 emissions per capita and per $1 GDP (PPP).

However, the MDGs have been criticized for fostering a "log-frame" culture that is pre-occupied with time-bound goals and targets, as opposed to long-term systematic change (e.g. Saith 2006; Eyben 2006). Although the goals of cutting in half world poverty, achieving universal primary education, promoting gender equality etc. are not on their own unworthy of our attention, the concern is that the MDGs are framed so broadly and ambiguously that they fail to articulate with sufficient clarity or purpose the ways in which governments, NGOs and people may act to address the underlying factors and processes that perpetuate poverty, hunger, inequality, etc. (Saith 2006; Eyben 2006). As Saith (2006) has argued, they are also conspicuously silent about the historical and causal relationship between the poverty reduction agenda and the (neo-liberal) conditions under which poverty reduction strategies are now being pursued, highlighting the idea that policies aimed at reducing vulnerability operate in a wider political economy of resource access and entitlement that frames and constrains the ability of resource users and managers to learn and adapt in ways that would strengthen resilience to future environmental shocks and stresses (Oliver-Smith 2004; Wisner et al. 2004; Adger 2006).

Reflecting upon the institutional challenges of taking forward a systematic approach to climate change adaptation, a number of authors identify specific steps that need to be taken in order to mainstream climate change into development policy and practice. Klein et al. (2007), for instance, argue that a comprehensive approach to mainstreaming entails the institutionalization of norms and practices that make climate change considerations a central part of development planning. Portfolio screening can play a part in this process, but it also requires "a more sophisticated understanding of the complex relationships that determine people's vulnerability to climate change," which takes into consideration the long term implications of development interventions and climate vulnerability (Klein et al. 2007: 41). Similarly, Ayers and Huq (2009) identify a "four-step program" that can be used to better integrate climate change considerations into development planning and administration.

More critical interpretations question the ability of donors and the development paradigm more generally to support policies and practices that actually reduce vulnerability to future environmental shocks and stresses. In a recent article, Nick Brooks, Natasha Grist and Katrina Brown (2009: 753) argue that contemporary approaches to adaptation "run the risk of locking societies into unsustainable and maladaptive patterns of the development," highlighting the powerful ways in which development policies often create and exacerbate human vulnerability to rapid-onset disasters and long-term environmental change. At the heart of their analysis is the idea that long-term adaptation requires an ability to learn from past experiences and coordinate actions in ways that challenge assumptions, change behaviours, reduce vulnerability and improve resilience to future shocks and stresses (cf. Pahl–Wostl 2009; Armitage et al. 2009). Most development agencies, by contrast, focus on models that are time bound by short-term goals and targets. Framed in this way,

> Adaptation is presented essentially as a means of "neutralising," or at least minimising, the impacts of climate change, in pursuit of predetermined development goals and desirable development outcomes, via processes that are manageable and, by implication, predictable. (Brooks et al. 2009: 752)

Broader "cultural" critiques question the carbon–based model of economic growth and development. The World Bank (2010), for instance, calls for a "fundamental transformation" that would challenge the "carbon-intensive growth paths of developing countries." The report highlights the challenge of overcoming behavioural and institutional inertias, as well as using "green stimulus" funds to support low carbon technologies, energy efficiency and improved water and waste management. More fundamentally, Brooks et al. (2009: 741) highlight the need to move away from growth and "yield maximization models towards alternatives encouraging resilience and risk spreading." Among the measures they identify are changes from low to high diversity food systems, support for mobile livelihoods (e.g. pastoralism) and, echoing the spirit of the Brundtland Report (WCED 1987), policies that integrate finance, energy and agriculture into fields typically limited to the "environmental ministries" (Brooks et al. 2009: 755).

A final and arguably more pragmatic perspective identifies what Hetberg and colleagues (2009) have called a "no-regrets" approach to poverty and vulnerability reduction. By this they envision a wide range of policies that can be used to "generate net social benefits under all future scenarios of climate change and impacts," (Hetburg et al. 2009: 89). Among the most viable, they argue, are ones that promote social protection, including primarily social funds for "community-based adaptation," safety nets for climate-induced disasters, livelihood programs, microfinance for building livelihoods and assets and new forms of weather-based index insurance. Similarly, the UNDP (2007: 30) identifies the need to "empower and enable vulnerable people to adapt to climate change by building resilience through investments in social protection, health, education and other measures."

In principle, the idea of identifying "no-regrets" approaches that "generate net social benefits under all future scenarios" provides an important guide for future policy. In practice, utilitarian approaches present a number of challenges, including especially the methodological challenge of assigning values to net social benefits and the related political challenge of governing preferences about competing policy options, outcomes and trade-offs. The inter-generational aspect of climate change further complicates matters in the sense that there is no guarantee that perceptions of what constitutes benefits and trade-offs will remain consistent over time, highlighting the need to specify the kinds of policies that may be used to address different climate change scenarios.

The following section identifies a range of policy interventions that may be used to address poverty and vulnerability over the long term.

## ADAPTING TO WHAT? A TYPOLOGY OF CLIMATE VULNERABILITY AND RISK

Table 12.1 lays out a typology that may be used to conceptualize different forms of climate vulnerability and risk. The categories are neither exhaustive nor mutually exclusive, but provide a heuristic that helps to identify the kinds of adaptation that may be used to reduce different forms of climate vulnerability.

**Table 12.1. A Typology of Climate Vulnerability and Risk**

| Type | Examples | Possible Impacts | Adaptation strategies |
|---|---|---|---|
| **Catastrophic risk** | Windstorms, flash floods | Major loss of life, major casualties, forced displacement | Early warning systems, evacuation plans and procedures, sea wall construction, provision of temporary shelter, medical supplies |
| **Livelihood risk** | Failed harvests due to flooding, soil erosion and drought | Loss of income, assets, destructive coping strategies | Cash transfers, asset transfers, food aid |
| **Systemic risk** | Price inflation due to failed harvests, transmission of water-borne diseases | Loss of income, loss of purchasing power, distress sales, distress migration | Food aid, monetary policy interventions; index-based insurance |
| **Cultural risk** | Long -term changes in rainfall, temperature; sea level rise; river and coastal erosion | Forced displacement; unplanned economic diversification | Planned economic diversification; permanent resettlement |

*Catastrophic risk* implies exposure to rapid-onset events whose speed, scale and magnitude overwhelm the ability of public, private and voluntary institutions to prevent substantial losses of life and injuries requiring complex and extended medical care. Depending upon national and local adaptive capacity, catastrophic shocks can entail the following characteristics (Coppola 2006:129):

- Widespread loss of life;
- Large numbers of injuries requiring hospitalization;
- Populations displaced for extended periods of time;
- Disruption of transportation, communication, energy distribution and relevant health and safety networks;
- Significant impacts on environment and/or permanent damage.

*Livelihood risks implies exposure to rapid- or slow-*onset environmental changes that degrade the assets and incomes of vulnerable populations in environmentally sensitive regions and sectors, such as farming, forestry and fishing. Depending on their speed, scale and intensity, livelihood shocks can lead to wider *systemic* shocks that force vulnerable populations into destructive and potentially irreversible coping strategies (see below). Livelihood shocks can also lead to "poverty spirals" where affected populations are forced into selling land, livestock and other valuable assets at prices below market value, reducing household consumption to the extent that stunting, wasting and other forms of malnutrition begin to occur and re-locating to areas in which economic opportunities, environmental vulnerability, personal safety and access to social welfare programs are all poor (Chambers 1983; Davies 1993; Scoones 1998; Sen 1981, 1999; Ericksen 2008; de Sherbinin 2008; Johnson and Krishnamurthy 2010).

*Systemic risk* arises when markets, social safety nets and other distribution systems are unable to provide basic entitlements, such as food and clean drinking water, during times of market instability and environmental stress.[7] During the global food crises of 2008 and 2010, for instance, a combination of environmental (e.g. failed harvests in grain producing regions of Russia and Australia) and socio-economic (e.g. land conversions for bio-fuel production) factors contributed to spiralling food prices. Multiple feedback mechanisms, inter-connected sub-systems, multiple and interacting controls, indirectly obtainable information and an incomplete understanding of the system make for high levels of unpredictability, which heighten the possibility of "cascading effects," events in which failures in one part of the system lead to unpredictable and uncontrollable failures in other parts of the system (Perrow 1984).

Finally, *cultural risk* implies environmental changes that alter fundamentally and negatively the ways in which people lead their lives. Empirical studies of drought in the Sahel and the dustbowl of the 1930s suggest that prolonged periods of climatic changes can bring about fundamental changes in land use, human settlement patterns, livelihoods and migration, creating new socio-economic cultures that render pre-existing norms and practices obsolete (cf. McLeman and Smit 2006). Whether environmental changes will have wider cultural impacts will depend on numerous factors, including:

- the *assets, networks and various forms of capital* that are available to vulnerable households and populations (McLeman and Smit 2006; Perch-Nielson et al. 2008);[8]
- *demographic and socio-cultural factors*, including age, gender, life cycle status, health and existing livelihood options (McLeman and Smit 2006; Perch-Nielson et al. 2008)

- the *speed and intensity* of change—i.e. whether the onset is unexpected and rapid, constraining efforts to coordinate mobilization and migration;
- the *scale* of change—i.e. whether it changes affecting large areas, potentially re-shaping major agro-ecological systems and national boundaries;
- the *scope* of change—i.e. whether populations have the option of returning to the life and livelihood they knew before the event or whether the change is permanent;
- the *sensitivity and resilience* of social and ecological systems to absorb, respond and recover from external environmental shocks (Smit and Wandel 2006).

Within ecological systems, rigidities manifest themselves when a particular species becomes dominant, thereby preventing other competitors from utilizing resources within the system (Holling 2001). Within socio–ecological systems, they may take the form of rules and path dependencies that help to sustain the existing structure, but at the same time make it less able to adapt and change in response to new pressures and events (Holling 2001; Folke et al. 2005; Folke 2006; Olsson et al. 2006). Periods of sudden collapse may be triggered by an external disturbance (e.g. wind, fire, disease, inflation, economic crisis), which exposes the underlying rigidities of the system. But it is also during these periods of crisis and destruction that new opportunities for innovation and re-organization may be expected to occur. Holling (2001), for instance, introduces the idea of an "adaptive cycle" whereby human–ecological systems undergo long periods of accumulation and transformation (e.g. the growth of a forest or a bureaucracy; conversion of forest for agriculture), punctuated by shorter period that create new opportunities for innovation and change (Holling 2001: 394).

The ability of vulnerable populations to anticipate and adapt to rapid and prolonged periods of environmental change is therefore dependent upon the ways in which and extent to which social actors are able to challenge or change existing behaviours, thereby improving resilience to future shocks and stresses. To what extent is the international aid regime able to support the kinds of adaptation that will be necessary to reduce vulnerability to climate change? And what are the main policy priorities? The following section next identifies a number of micro-economic interventions that can be used to reduce vulnerability to different forms of risk.

# ADAPTATION AND THE AID REGIME: PROSPECTS FOR POLICY CONVERGENCE

Formulating policies that can accommodate the adaptation needs of future generations poses particular challenges for the international aid regime. First, the timelines affecting (for instance) global concentrations and impacts far exceed the average life expectancy of a typical development policy or project (cf. Brooks et al. 2009). Second, the regional distribution of climate change impacts will vary, and in some cases, the impacts will not be universally negative (IPCC 2007). Third, the scale and magnitude of climate-induced hazards will vary, highlighting the need for regional and international priorities (Smit et al. 2000 [2009]; Smit and Wandel 2006).

## *Governing Catastrophic Risk: The Role of Disaster Risk Reduction*

Arguably the most immediate risk of catastrophic losses resulting from climate change is the threat of sea level rise. The IPCC (Nicholls et al. 2007) projects with "very high confidence" that coastal areas will become increasingly vulnerable to "an accelerated rise" in eustatic sea levels, ranging from 0.2 to 0.6 metres or more by the year 2100. Others (e.g. Young and Pilkey 2010) contend that if the Greenland and West Antarctic icesheets are factored into the equation, global sea levels could rise by a range of 4 to 6 metres. The IPCC identifies three "key hot spots of societal vulnerability," in which a combination of land use and population pressure may exacerbate exposure to windstorms, erosion and floods: river deltas, especially the seven Asian "mega-deltas"; low elevation coastal urban areas, especially ones prone to subsidence; and small islands, especially coral atolls.

Even a modest change in sea levels could have a devastating impact on populations living in low elevation coastal zones, including especially small island states such as Kiribati and Tuvalu and the so-called "mega-deltas" of the Nile, the Ganges–Brahmaputra and the Mekong (Erikson et al. 2006, cited in IPCC 2007C). According to a study recently published by the Feinstein International Center at Tufts University (Webster et al. 2009: 17–18), the cost of financing humanitarian responses to climate induced disasters could increase from US\$ 89 to 135 million dollars per year (an increase of 52 percent) with a 15 percent rise in the intensity and impact of cyclonic storms. If tropical storms become 10 percent more frequent than they are today, the study estimates that humanitarian costs would jump from US$89 million to 149 million per year, a rise of 67 percent. Moreover, if the past provides a reliable indication of what we can expect in the future, hydro-climatic events will have a heavy toll on property, livelihoods and human life, especially in Asia. Between 2000 and 2006, 1,886 storm and flood disasters killed more than 57,000 people, affecting almost a billion people and causing damage estimated at 390 billion US$ (CRED 2006, cited in WGBU 2008: 69). Between 2000 and 2004, climate-related disasters affected roughly 262 million people, 98 percent of whom were in the developing world UNDP (2007: 16).

In theory, disaster risk reduction or DRR can help to manage catastrophic risk by (1) identifying the risk factors that would lead to an unacceptable loss of life, livelihood, health and property; and (2) developing strategies that would reduce exposure and vulnerability to these and related natural hazards, including, conceivably laws that prohibit the construction of homes and buildings in high risk areas (e.g. low lying coastal zones, floodplains), the construction of cyclone shelters, improvements in the availability and delivery of drugs, water purification, and other medical supplies, and strategies for evacuation and communication during times of crisis (Wisner et al. 2004; Coppola 2006; ISDR 2007).

In recent years, humanitarian policies have articulated the need to establish a continuum of "relief to development," developing specific policies that may be used during the immediate relief period, in which the need for medical assistance, food aid, cash provisioning and shelter is particularly crucial; and the post-relief rehabilitation phase, in which resettlement of affected communities, reconstruction of homes and restoration of affected livelihoods (e.g. crop re-planting, repairing equipment, etc.) typically occurs; and, ultimately investment in assets and income-generating activities that will build resilience and reduce poverty over the long-term.

However, trade-offs exist between the policies that donors, governments and other humanitarian agencies use to rehabilitate lives and livelihoods and the factors affecting human exposure to extreme climate events, such as windstorms, flooding and drought (Wisner et al. 2004; Moench 2007 [2009]; O'Brien et al. 2008). First, efforts to promote "relief to development" (e.g. public insurance, subsidies, compensation and reconstruction) create strong incentives to *relocate* and to *re-engage* in geographical areas whose exposure to natural disasters is chronically high (Wisner et al. 2004; Moench 2007 [2009]; O'Brien et al. 2008). Second, if they are indeed truly effective, technological approaches to DRR will entail the disruption and possible displacement of populations living and working in vulnerable coastal areas, highlighting the importance of recognizing, protecting and ideally expanding the entitlements and capabilities of populations most directly exposed to the risks of forced displacement and climate change (Wisner et al. 2004; Moench 2007 [2009]).

Such insights highlight the factors influencing settlement and livelihood in marginal ecological areas (cf. Wisner et al. 2004; Hulme 2008). They also highlight the importance of developing policies that recognize and protect the needs and capabilities of affected populations.

## *Governing Livelihood Risk: The Role of Asset and Cash Transfers*

Whether households and economies more generally will cope and therefore recover from livelihood shocks is dependent upon the scale, intensity and magnitude of the disturbance in question and the extent to which the strategies employed during times of crisis exacerbate or reduce the highly destructive spirals that lead to deeper and potentially irreversible forms of poverty and destitution. Recent debates in the development policy literature have highlighted the idea of using cash transfers, asset transfers and other forms of "social protection" to assist vulnerable populations affected by the loss or reduction of incomes and assets during periods of market volatility, social instability and severe environmental stress (Doocy et al. 2006; Farrington and Slater 2006; Prowse 2008; Davies et al. 2008; Slater et al. 2008; Tanner et al. 2008; Teichman 2008; Todd et al. 2010; Maluccio 2010). Although there is disagreement about the perceived merits of using cash versus asset transfers (Farrington and Slater 2006), the basic idea is that the inability to meet basic needs (of nutrition, healthcare, etc.) during droughts, famines and other volatile periods is the result of a loss or devaluation in skills, assets and incomes in relation to the cost of food, shelter and other basic entitlements. The perceived solution is therefore to provide affected populations with cash or asset transfers, which can theoretically improve their ability to "command" scarce resources in volatile market settings.

Cash transfers have been used to address short-term consumption needs in the context of rapid-onset disasters, such as earthquakes, flooding and entitlement failures during droughts. Asset and conditional cash transfers (CCTs) on the other hand aim to influence longer-term behaviour by conditioning the transfer of assets or cash in relation to a particular policy outcome, such as nutrition, education and healthcare. Beyond the idea of meeting basic needs, social protection programs can also support vulnerable households in ways that prevent destructive poverty spirals while at the same time providing assets that can build resilience to future environmental and economic shocks (Deshingkar et al. 2005; Dreze and Sen, 1989; Matin et al. 2008; Sen 1981; 1999; Slater et al. 2008).

Whether it makes sense to support cash transfers instead of asset or conditional cash transfers in the context of forced displacement and climate change depends in very large part on the speed, scale and magnitude of the disaster in question and the extent to which the program is aiming to effect long-term behavioural change. In the short term, cash and food transfers can provide a vital safety net for households whose incomes, assets and livelihoods are negatively affected as a result of environmental shocks and stresses, such as windstorms, flooding and drought (Doocy et al. 2006; Tanner et al. 2009). When it has worked effectively, for instance, India's food-for-work program has provided a vital safety net during food shortages, entitlement failures and famine (Deshingkar et al. 2005). Similarly, the DFID-funded Chars Livelihoods Programme (CLP) has used cash for work programmes to support cash poor households during the "monga" lean season (http://www.clp.-bangladesh.org).

The challenge, however, is to ensure that micro-economic policy interventions are able to build livelihoods, capacity and resilience over the long-term. Cash transfers, for instance, have been shown to provide an important means of preventing disaster-induced poverty spirals, but they are less effective at addressing the inter-generational factors contributing to risk, poverty and vulnerability. Over the long term, the challenge is to identify programs (including conceivably asset transfers) that can reduce vulnerability to future climate change (cf. Hetberg et al. 2009; Johnson and Krishnamurthy 2010).

Table 12.2 identifies a number of scalable and sustainable technologies (e.g. sand dams, Lasage et al. 2008; and "scuba rice" Barclay 2009) that can be used in areas affected by water shortages, salinity, flooding and coastal and riverbank erosion. Conceivably and pending careful cost–benefit evaluations, asset transfers supporting the acquisition of flood resistant rice and water harvesting technologies can help build the resilience of vulnerable populations to potentially catastrophic forms of environmental change.

**Table 12.2. Assets for Adaptation**

| Hazard type | Asset | Description |
|---|---|---|
| Flooding | Scuba Rice | Scuba rice is a new breed of high yielding, flood resistant rice that can withstand 2 weeks of complete submergence and recover adequately for a reasonable harvest. Scuba rice works by suppressing the elongation process, effectively stunting the rice plant into a dormant state during flooding. As a result, the plant is able to conserve energy for a post-flood recovery (Barclay 2009). |
| Flooding | Snorkel Rice | Snorkel rice is a new breed of low yielding, flood resistant rice that can significantly elongate its internodes to function as a snorkel to allow gas exchange with the atmosphere during long flooding periods. This trait prevents the rice plants from drowning. (Hattori, Nagai, Furukawa et al. 2009) |
| Drought | Sand Dams | Sand dams are small structures built in ephemeral rivers to block seasonal water flows to create reservoir to store excess water to overcome periods of drought. The sand dams increase water availability, food security, and household incomes. It is a community based adaptation approach with low systems cost and easy maintenance. (Lasage, Aerts, Mutiso et al. 2008) |

When targeted effectively, asset transfers can provide an important means of building assets, diversifying incomes and spreading risk. However, as Teichman (2008) points out, asset and cash transfers are in the final analysis micro-economic interventions, which cannot take the place of systemic and structural change.

## Governing Systemic Risk: The Role of Stocks, Diversification and Insurance

One of the challenges of managing systemic risk is that policies aimed at mitigating systemic shocks can also lead to unforeseen cascading events, which in turn affect the ability of policies and institutions to manage and pool risk across wider scales. During the global food shocks of 2008 and 2010, for instance, many governments used export restrictions to shore up domestic food supplies, which in turn led to further price shocks in the world market. Others used price controls to stabilize food prices, which in effect exacerbated food inflation for rural food producers, especially agricultural labour. Writing in the midst of the crisis, World Bank president Robert Zoellick (2011) argued that export bans represent a particularly destructive form of adaptation, highlighting the need to develop alternative ways of managing threats to the global food system. Specifically he identified a number of ways in which donors and international institutions, such as the World Food Program, can help to manage systemic food shocks:

1. Improve the quality of information about the quality and quantity of grainstocks;
2. Improve long-range weather forecasting and monitoring, especially in Africa, where institutional capacity remains weak;
3. Improve knowledge about the relationship between international prices and local prices in poor countries;
4. Ring-fence humanitarian food aid, ensuring "that food for humanitarian purposes be allowed to move freely";
5. Target the most vulnerable populations, "such as pregnant and lactating women and children under two";
6. Provide "fast-disbursing" grants, loans and credit, which can be used "as an alternative to export bans or price fixing";
7. "Establish small regional humanitarian reserves," that can be used to ensure food provisioning "in disaster-prone, infrastructure-poor areas," such as the Horn of Africa;
8. "Develop a robust menu of other risk management products", including weather insurance and rainfall index insurance;
9. Incorporate smallholder farmers into food sourcing purchases during humanitarian crises.

Zoellick's editorial helps to illustrate some of the factors that have long plagued humanitarian food aid policy, including especially, aid tying, price fixing and the use of large buffer stocks. Although it doesn't necessarily question the forces driving commercial monoculture, the emphasis on local provisioning through smallholder agriculture provides some interesting ways forward.

Recent work on the viability of smallholder agriculture has identified a number of ways in which local non-traditional value chains can improve the food security of the poor (Henson 2006). First, diversification into horticulture and other non-traditional commodities has been shown to improve and diversify agricultural incomes. Second, non-traditional commodities have been shown to create new and more equitable forms of labour. Third, they can improve the variety and availability of nutritious food, thereby yielding health and nutrition benefits for the poor. However, promoting local food production systems also poses challenges. For one, traditional production systems are often dispersed with "multi-layered and fragmented supply chains" that limit the prospects for linking smallholder farmers with high value markets (Henson 2006). Second, the costs of establishing processing facilities that meet health, environmental and other regulatory norms and standards can be prohibitively expensive for low-income producers and economies (Henson 2006). Third, the establishment of supply chains can exacerbate resource intensity by incorporating intensive use of inputs including ground water, crop diversification, and the application of agro-industrial fertilizers and pesticides. Fourth, the restructuring of agricultural and food markets has been shown to create new vulnerabilities to market and price volatility while at the same time creating new and more profitable income streams (Henson 2006). Fifth, in some cases, agricultural restructuring has led to the marginalization and devaluation of female labour (Shah 2006).

The empirical literature therefore yields a number of ambiguities about the social, economic and environmental implications of promoting non-traditional food systems. At the same time, studies have revealed that "even in very poor countries with low levels of overall economic development and where supply chains are predominantly traditional in nature," it is possible to find "enclaves" in which dynamic supply chains have emerged (Henson 2006: 2). The challenge is therefore to identify alternative food production and distribution systems that can improve incomes and nutritional standards while at the same time ensuring the environmental sustainability, price stability and food security of the poor.

Another crucial area of intervention entails the provision of insurance that can cover the risks of income and asset losses resulting from climate change. Drawing upon the very large range of micro-insurance products now provided by organizations like BRAC, Grameen and Basix, micro-insurance can be tailored to meet the specific needs of populations whose labour is devalued (either permanently or temporarily) by the impacts of climate change.

For populations living in vulnerable areas (e.g. floodplains, low-elevation coastal zones), micro-insurance programming can document and potentially insure people's existing assets (including inter alia land, housing, livestock, food stocks, etc.) and incomes affected by disabilities resulting from injury and infectious disease.

A final form of social protection is index-based insurance (Hetberg et al. 2009). Unlike crop insurance, which insures against actual crop losses, index-based insurance extends coverage in relation to a pre-established index of rainfall, soil moisture, etc. below which compensation payments are made (Hetberg et al. 2009: 97–98). The advantage of using index-based insurance is it reduces the transaction costs of establishing actual losses, thereby improving the ability of insurers to establish eligibility for compensation. However, index-based models also have their limitations. First, they require a reliable means of establishing minimum thresholds, which may exceed the capacity of public and private providers in low-income countries. Second, they fail to cover losses resulting from environmental and other exogenous disturbances that do not meet the minimum threshold (cf. Hetberg et al. 2009: 97). Third, index-based instruments do not necessarily cover losses due to lost or devalued labour. Finally, they create (or at least fail to remove) incentives to live and work in hazardous regions and sectors, which may exacerbate vulnerability over the long term (Hetberg et al. 2009).

## *Governing Cultural Risk: The Case of Human Resettlement*

This final issue highlights the wider, "cultural" factors affecting livelihood and settlement in vulnerable environments. According to a study recently commissioned by the US National Science Foundation, an estimated 600 million people (or 10 percent of the Earth's population) live in areas that are within 100 km of the coast and at elevations less than 100 m above sea level (McGranahan et al. 2007). Of these, 360 million people are in urban areas, reflecting the impact of urbanization and Third industrial growth strategies on patterns of economic development and human settlement. A central trend in this process has been the development of large urban centres whose ability to organise and absorb pools of labour facilitated both the import substitution policies of the 1960s and '70s (primarily in Latin America) and then the export led strategies of '80s and '90s.

Concerns about urbanization, coastal vulnerability and rising sea levels have raised new fears that a changing climate could lead to new patterns of large-scale population displacement (Brown et al. 2007; Reuveny 2007; Christian Aid 2007; WBGU 2008). According to the UN (2007) 93 percent of urban growth is expected to occur in developing nations, the vast majority (80 percent) taking place in Asia and Africa. In a widely cited article, Norman Myers (1993) warns that climate change could displace as many as 250 million people by the year 2100. More recently, the British charity, Christian Aid (2007), projects that as many as a billion people may be displaced as a result of climate change by the year 2050 (cf. Biermann and Boas 2007, 2008a; IPCC WG-II 2007; Bogardi and Warner 2009).

Such estimates are of course speculative and controversial,[9] but they also raise important questions about the real and ideal relationship between the freedom to decide basic questions about mobility, settlement and livelihood on the one hand and social responsibility (very broadly defined) on the other.[10] At one end of the spectrum is the idea that the international community must act quickly and proactively, using new protocols (defining for instance new rights for "climate refugees") and planned resettlement to protect human populations displaced by the effects of climate change (Biermann and Boas 2007, 2008a, 2008b; Bogardi and Warner 2009). At the other is a concern that efforts to protect and possibly relocate human populations overstate the probability of environmentally induced migration, and understate the impact of preventative action (such as resettlement) on basic human rights and freedoms (Black 2001; Castles 2001; Boano et al. 2008; Hulme 2008; Hartmann 2010).

In principle, the idea of acting early to support voluntary resettlement could provide an important means of diversifying and strengthening poor people's assets, livelihoods and income sources, thereby reducing vulnerability to rapid onset disasters and long term environmental change. In practice, it faces a number of logistical and ethical challenges. First, notwithstanding safeguards that would ensure the rights of individuals, the emphasis on groups may exclude important sub-groups, such as widows, landless labourers and people with disabilities, raising questions about the terms on which groups would be recognized and defined for the purposes of international climate refugee protection. Second, it raises questions about the specific ways in which and terms on which national governments and international institutions may manage and encourage voluntary resettlement (cf. Hulme 2008).

By far the most difficult issue facing any discussion about relocation concerns the immediate and long-term impact of human resettlement on livelihood, economic security and material well-being (Cernea 1997; Cernea and Schmidt-Soltau 2006). If the history of "development-induced displacement" is anything to go by, resettlement policies which fail to provide adequate compensation and social protection will almost certainly create new vulnerabilities whose impact will be greatest on the poor. Of particular concern here are policies that fail to recognise or address:

- The loss or devaluation of historical patterns of settlement, mobility and livelihood
- The loss or devaluation of rights, assets and entitlement
- The loss or devaluation of social security and status, creating new risks associated with social discrimination, social exclusion and violence

A related argument stems from political ecology (e.g. Blaikie and Brookfield 1987; Wisner et al. 2004; Davis 2002), and suggests that efforts to relocate poor and vulnerable populations from hazard-prone areas will counter-act the very strong factors (including perceptions of what constitutes acceptable vulnerability and risk) that induce very poor people to live and work in vulnerable areas in the first place. When land-use values (and therefore property values) are low, environmentally vulnerable areas (e.g. floodplains, drylands, coastal areas) tend to attract very large numbers of poor people, creating social conditions under which the quality of housing, roads, public services, etc. also tends to be low.

For populations displaced (either permanently or temporarily) by climate hazards and disasters, the ability to recognize and protect existing land rights and other household assets will be crucial. Developing ways of documenting formal and informal (i.e. customary) rights of access and entitlement will provide an important means by which national governments and international institutions may recognize and ideally compensate entitlements lost or devalued during periods of migration and resettlement (cf. Cernea 1997; Cernea and Schmidt-Soltau 2006; Raleigh et al. 2008). Future programs will also need to recognize the impact of gender and gender relations on access and entitlement (both within and beyond the household), targeting specifically the assets and income streams that are particularly important to women and female-headed households.

If the history of "successful" land reform is anything to go by (see, for instance, Putzel 1992), the ability to challenge existing inequalities in society will require an authority (i.e. a state) that has the capacity and autonomy to challenge (and potentially compensate) powerful groups disenfranchised by social redistribution. It will also require active and participatory engagement between the affected populations (i.e. the beneficiaries and potential losers) and the state (Putzel 1992). For states that lack the necessary authority and autonomy to carry out ambitious redistributive reform efforts, adaptation programming can help to facilitate (at the very least) the documentation of existing (formal and customary) entitlements (see, for instance, Sen's concept of "entitlement mapping"), which would provide an important baseline for future forms of insurance and social protection.

# CONCLUSION

Efforts to incorporate climate change adaptation into development policy have fostered important discussions about the viability and sustainability of using foreign aid to reduce poverty and vulnerability to environmental change. They have also highlighted the limitations of taking forward a long-range adaptation strategy within confines of the international aid regime. If we accept the premise that resilient systems are ones that have the capacity and the flexibility to anticipate vulnerability, to pool risk and to learn from past events, it remains unclear whether the international aid regime in its current form will be able to achieve these ends. First, the norms and expectations surrounding "new and additional" adaptation financing have created new forms of bureaucracy that impede the ability of vulnerable countries and communities to access adaptation financing. Second, adaptation and climate change more generally remain separate administrative categories with little or no substantive linkages with other aid priorities, such as primary education, poverty reduction and human health. Finally, there is a notable gap between the timelines typically used to design, implement and evaluate foreign aid policies and the inter-generational challenges of climate change.

Overcoming these challenges will entail an ability to identify the sectors and regions (or "hot spots") most exposed to high risk, high probability losses, and to develop strategies that will reduce the risk that these eventualities will occur. Assuming that low-income countries will bear the greatest burden of adapting to climate change, international policies need to start thinking strategically about the challenge of strengthening adaptive capacity. The preceding analysis has identified a number of short- and long-range policies that may be used to build assets, reduce vulnerability and pool risk. At the heart of this framework are policies aimed at reducing climate risk and vulnerability; building assets, incomes, and resilience; and preparing the international system for future "systemic" and "cultural" shocks, such as global food shortages and population displacement.

Over the longer term, logistical issues of this kind are arguably less important than ethical and geopolitical concerns about international and inter-generational burden sharing. Some (such as Biermann and Boas 2008a; Byravan and Rajan 2006) have suggested that the burden of accommodating future "climate exiles" be determined in relation to a country's historical greenhouse gas emissions. Although it certainly appeals to the idea of "restoring" justice by requiring the biggest polluters to bear the burdens of their behavior, the notion that immigration policy may be determined only or primarily on the basis of historical emissions would need to resolve a number of difficult ethical and political questions, including especially the period over which the principle in question would be meant to apply and the political conditions under which it would be expected to happen. How long, for instance, would future generations be expected to bear the burden of their forebears? [11] How long would these and other "rules of allocation" be expected to apply? Could the principles be revoked if the environmental conditions change? [12] At what level and through which institutions would these issues need to be addressed? Would the existing international regime provide an adequate and appropriate institutional forum?

Beyond the challenge of establishing who will bear the burden of future mitigation efforts (Caney 2005, 2008) therefore lies an equally and possibly more difficult challenge of establishing who will bear the burden of protecting or accommodating populations physically and permanently displaced by climate change. Although scholars are now beginning to explore the normative and geopolitical dimensions of these questions[13] future research will need to advance this scholarship by identifying the empirical conditions under which successful forms of adaptation may be expected to work and by theorizing much more extensively the principles of fairness and justice that would guide national and international responses to future climate disasters.

## NOTES

1. The seconf and fourth sections of the chapter expand upon material previously published in Johnson and Krishnamurthy (2010). The author would like to thank Yvonne Su for research assistance in relation to material presented in the fourth section.

2. International regimes can be usefully defined as "rules agreed to by states ...concerning their conduct in specific issue areas (trade, monetary exchange, navigation on the high seas or in the air, non proliferation .... etc.) and often associated with international and non-governmental organizations linked to these regimes" (Viotti and Kaupi, 2010: 131).

3. Annett Möhner, UNFCCC Secretariat, *"Needs of Developing Countries with Regard to Adaptation to Climate Change – A View from the Adaptation Fund and Current Negotiations"* paper presented at the 7th Open Meeting of the International Human Dimensions of Global Change Program, United Nations University, Bonn, Germany, 30 April 2009.

4. Calculating the costs of adapting to climate change is of course a difficult undertaking (cf. Oxfam 2007; Ayers and Huq 2009). The Bank's estimates are based on the proportion of government, private, ODA and foreign direct investments that are in theory sensitive to climate risk and the additional cost of climate-proofing these investments (Oxfam 2007: 17). Oxfam on the other hand calculates the cost of "scaling up" (1) community-based projects of NGOs; (2) "the most urgent and immediate national adaptation needs"; and (3) the "costs that are not adequately taken into account in any of the above" (Oxfam 2007: 19).

5. Outside of the UNFCCC, many donors are now actively promoting adaptation programs of their own. The Overseas Development Institute (ODI 2010) in London, for instance, estimates that alongside the UN Framework Convention on Climate Change and the Kyoto Protocol are another eighteen private, multilateral and bilateral channels through which adaptation is currently being supported.

6. Accessed February 28, 2011 at http://www.undp.org/mdg/goal7.shtml

7. Systems can be usefully understood as sets of relationships whose component parts yield properties that would not be produced by any single part (Perrow 1984). *Complex systems*, on the other hand, are systems in which interactions are multiple, unpredictable and generally unintended.

8. In their study of out-migration from eastern Oklahoma during the 1930s, for instance, McLeman and Smit (2006) report that renters of land were more likely to migrate than were landowners, highlighting important linkages between land-holdings, agrarian class structures and migration.

9. Space restrictions preclude an extended treatment of the ways in which models of environmental migration are theorizing the factors that motivate household decisions to move (either temporarily or permanently) in the first place. Suffice to say however that some of the more alarmist predictions have been criticised for understating the factors that are well known to motivate household and individual decision making, the epistemological and normative terms on which adaptation may be defined and the extent to which "environmental migration" has been happening or—on the basis of future projections—may be expected to happen in the future. For others (e.g. Biermann and Boas 2008a, 2008b), the assertion that migration is multi-dimensional and complex is important, but it does not obviate the need to take action on the issue.

10. See, for instance, the very interesting exchange between Mike Hulme (2008) and Frank Biermann and Ingrid Boas (2008, 2008b).

11. For an excellent treatment of the distribution of burdens of GHG mitigation, see Caney (2005).

12. See Hulme (2008) for a brief critique along these lines.

13. See, especially, Bates (2002), Byravan and Rajan (2006), Biermann and Boas (2007, 2008), and Kolmannskog (2008).

## BIBLIOGRAPHY

AWG–LCA (Ad Hoc Working Group on Long-Term Cooperative Action under the Convention). *Report of the Ad Hoc Working Group on Long-term Cooperative Action under the Convention on its eighth session.* Held in Copenhagen, December 7–15, 2009 FCCC/AWGLCA/2009/17. AWG-LCA, 2010. http://unfccc.int/resource/docs/2009/awglca8/eng/17.pdf

Adger, N. et al. "Assessment of adaptation practices, options, constraints and capacity." In *Climate Change 2007: Impacts, Adaptation and Vulnerability: Contribution of Working Group II to the Fourth Assessment Report of the Intergovernmental Panel on Climate Change,* edited by M.L. Parry et al., 717-43. Cambridge: CUP, 2007.

Adger, W. N. "Vulnerability" *Global Environmental Change* 16 (2006): 268-81

Anthes, R.A, R. W. Corell, G. Holland, J. W. Hurrell, M. C. MacCracken, K. E. Trenberth. "Hurricanes and global Warming–Potential Linkages and Consequences." *Bulletin of the American Meteorology Society* 87 (2006): 623-28.

Armitage, D., M. Marschke, and R. Plummer. "Adaptive Co–Management and the Paradox of Learning." *Global Environmental Change* 18, 2008: 86-98.

Ayers, J, and Huq, S. "Supporting Adaptation to Climate Change: What Role for Development Assistance?" *Development Policy Review* 27, no.6 (2009): 657-92.

Barclay, A. "Scuba Rice: Stemming the Tide in Flood-Prone South Asia". *Rice Today,* 2009.

Barnett, J., and S. O'Neill. "Maladaptation." Global Environmental Change 20 (2010): 211-13.

Barnett, J., and Adger W. N. "Climate Change, Human Security and Violent Conflict." *Political Geography* 26(2007): 639-55.

Berkes, F., and Folke, C. "Linking Social and Ecological Systems for Resilience and Sustainability," In *Linking Social and Ecological Systems* ed. Berkes, F. Folke, C. Cambridge: Cambridge University Press, 1998.

Biermann, F., and Boas, I. "Preparing for a Warmer World: Towards a Global Governance System to Protect Environmental Refugees." *Global Governance* Working Paper no. 33, November 2007. Accessed from glogov.org

Biermann, F., and Boas, I. "Protecting Climate Refugees: The Case for a Global Protocol." *Environment* 50, no.6 (2008a): 8-16.

———. "Response to Hulme." *Environment* 50, no.6 (2008b).

Black, R. "Environmental Refugees: Myth or Reality"? UNHCR Working Papers 34 (2001): 1-19.

Boano, C., Zetter, R. and Morris, T. (2008) "Environmentally Displaced People: Understanding the Linkages between Environmental Change, Livelihoods and Forced Migration." *Forced Migration Policy Briefing 1,* Refugees Studies Centre. UK: Oxford University, 2008. www.rsc.ox.ac.uk

Bogardi, J., and K. Warner. "Here comes the flood." *Nature Reports Climate Change* 3 (2009): 9-11.

Boyd, E., N. Grist, S. Juhola, and V. Nelson. "Exploring Development Futures in a Changing Climate: Frontiers for Development Policy and Practice." *Development Policy Review* 27, no. 6 (2009): 659-74.

Brooks, N., N. Grist and K. Brown. (2009) "Development Futures in the Context of Climate Change: Challenging the Present and Learning from the Past." *Development Policy Review* 27, no. 6 (2009): 741-65.

Brown, O. "Migration and Climate Change IOM Migration" Research Series No. 31. Geneva: International Organization for Migration, 2008.

Burton, I. "Climate change and the adaptation deficit." Pp. 89-98 in *The Earthscan Reader on Adaptation to Climate Change,* ed. E.L.F. Schipper and I. Burton. London: Earthscan, (2004 [2009]).

Burton, I. "Beyond borders: The need for strategic global adaptation," *IIED Opinion,* December 2008. www.iied.org/pubs

Byravan, S., and Rajan, S. C. "Providing New Homes for Climate Change Exiles." *Climate Policy* 6 (2006): 247-52.

Caney, S. "Cosmopolitan Justice, Responsibility, and Global Climate Change." *Leiden Journal of International Law* 18 (2005): 747-75.

Caney, S. "Human Rights, Climate Change, and Discounting." *Environmental Politics* 17, no. 4 (2008): 536-55.

Carrington, D. "Deadly Heatwaves will be more Frequent in Coming Decades, Say Scientists," *The Guardian,* 2011, accessed March 21, 2010, http://www.guardian.co.uk/environment/2011/mar/17/deadly-heatwaves-europe/print.

Castles, S. "Environmental Change and Forced Migration: Making Sense of the Debate." *Refugees Studies Centre Working Paper No. 70,* UK: University of Oxford, 2002.

Chambers R. *Rural Development: Putting the Last First.* Longman: London, 1983.

Cernea, M. M. "The Risks and Reconstruction Model for Resettling Displaced Populations." *World Development* 25, no. 10 (1997): 1569-89.

Cernea, M. M., and Schmidt-Soltau, K. "Poverty Risks and National Parks: Policy Issues in Conservation and Resettlement," *World Development* 34, no. 100(2006): 1808-30.

Coppola, D. P. *Introduction to Disaster Management.* Oxford: Elsevier Press, 2006.

Davies, M., Oswald, K., Mitchell, T., and Tanner, T. "Climate Change Adaptation, Disaster Risk Reduction and Social Protection," Briefing Note. Sussex: Institute of Development Studies Brighton, 2008.

Davies, S. "Are Coping Strategies a Cop Out?" Pp. 99-116 in *The Earthscan Reader on Adaptation to Climate Change,* ed. E.L.F. Schipper and I. Burton. London: Earthscan (1993 [2009]).

De Haan, A. "Livelihoods and Poverty: The Role of Migration—a Critical Review of the Migration Literature." *The Journal of Development Studies* 36, no. 2 (1998): 1-4

De Sherbinin, A., VanWey, L. K., McSweeney, K., Aggarwal, R., Barbieri, A., Henry, S., Hunter, L. M., Twine, W. and Walker, R. "Rural Household Demographics, Livelihoods and the Environment." *Global Environmental Change* 18 (2008): 38-53.

Deshingkar, Priya. "Maximizing the Benefits of Internal Migration for Development," Pp. 21-63 in *Migration and Poverty Reduction in Asia,* ed. F. Laczko. Department for International Development (UK), and Ministry of Foreign Affairs, Peoples Republic of China: International Organization for Development, 2005. Accessed June 29, 2009, http://www.preventtraffickingchina.org/english/Website_Files/Deshingkar.pdf.

Deshingkar, Priya, Craig Johnson, and John Farrington. "State Transfers to the Poor and Back: The Case of the Food for Work Programme in India." *World Development* 33, no. 4 (2005): 575-91.

Doocy, Shannon, Michael Gabriel, Sean Collins, Courtland Robinson, Peter, Stevenson. "Implementing Cash for Work Programmes in Post-Tsunami Aceh: Experiences and Lessons Learned." *Disasters* 30, no.3 (2006): 277-96.

Dreze, Jean, and Amartya, Sen. *Hunger and Public Action*. Oxford: Clarendon Press, 1989.

Eakin, H, and L.A. Bojorquez-Tapia. "Insights into the Composition of Household Vulnerability from Multicriteria Decision Analysis." *Global Environmental Change – Human and Policy Dimensions* 18, 2008: 112-27.

Ericksen, P. "Conceptualizing Food Systems for Global Environmental Change Research." *Global Environmental Change* 18 (2008): 234-45.

Eyben, R. "The Road not taken: International Aid's Choice of Copenhagen over Beijing." *Third World Quarterly* 27, no. 4(2006): 595-608.

Fallasch, Felix, and Laetitia de Marez. "New and Additional? An Assessment of Fast-Start Finance Commitments of the Copenhagen Accord." Climate Analytics, 2010, accessed March 17, 2011, www.climateanalytics.org.

Farrington, J. and Slater, R. "Cash Transfers: Panacea for Poverty Reduction or Money down the Drain?" *Development Policy Review* 24, no. 5 (2006): 499-11.

Folke, C. "Resilience: The Emergence of a Perspective for Social-Ecological Systems Analysis." *Global Environmental Change*16 (2006): 253-67.

Folke, C., T. Hahn, P. Olsson, and J. Norberg. "Adaptive Governance of Social-Ecological Systems," *Annual Review of Environmental Research* 30 (2005): 441-73.

Fraser, E., and A. Rimas. "The Psychology of Food Riots: When Do Price Spikes Lead to Unrest." *Foreign Affairs,* 30 January, 2011.

Fussel, H-M. "Vulnerability: A Generally Applicable Conceptual Framework for Climate Change Research." *Global Environmental Change* 17 (2007): 155-67.

Hattori, Y., Nagai, K., Furukawa, S., et al. "The Ethylene Response Factors SNORKEL1 and SNORKEL2 Allow Rice to Adapt to Deep Water." *Nature* 460, 2009: 1026-1031.

Hedger, M., T. Mitchell, J. Leavy, M. Greeley, A. Downie and L. Horrocks. "Adaptation to Climate Change from a Development Perspective," a desk review commissioned by the Evaluation Office of the Global Environment Facility, Brighton: Institute of Development Studies, 2008, accessed January 15, 2009, www.ids.ac.uk.

Helmer, M., and Hilhorst, D. "Natural Disasters and Climate Change." *Disasters* 30, no.1 (2006): 1-4.

Heltberg, R., Siegel P. B. and Jorgensen S. L. "Addressing Human Vulnerability to Climate Change: Toward a 'no–regrets' Approach." *Global Environmental Change* 19 (2009): 89–9.

Holling, C. S. "Understanding the Complexity of Economic, Ecological, and Social systems," *Ecosystems* 4, 2001:390–405.

Hulme, M. "Commentary—Climate Refugees: Cause for a New Agreement?" *Environment* 50, no. 6, 2008.

Huq, S., and H. Reid. "Mainstreaming Adaptation in Development," pp. 313-22 in *The Earthscan Reader on Adaptation to Climate Change,* ed. E.L.F. Schipper and I. Burton. London: Earthscan, (2004 [2009]).

IOM (International Organization for Migration). "Migration, Climate Change and the Environment," *IOM Policy Brief,* May 2009, accessed June 29, 2009. www.iom.int/envmig

IPCC WG-I. "IPCC, 2007: Summary for Policymakers," in S. Solomon et al. (eds.) The Physical Science Basis: *Contribution of Working Group I to the Fourth Assessment Report of the Intergovernmental Panel on Climate Change.* Cambridge: Cambridge University Press, 2007.

IPCC WG-II. "IPCC, 2007: Summary for Policymakers," Pp. 7-22 in *Climate Change 2007: Impacts, Adaptation and Vulnerability: Contribution of Working Group II to the Fourth Assessment Report of the Intergovernmental Panel on Climate Change,* ed. M.L. Parry et al. Cambridge: Cambridge University Press, 2007.

ISDR. "Words into action: A Guide for Implementing the Hyogo Framework". Geneva: United Nations, 2007, accessed May 12, 2010, http://www.unisdr.org/eng/hfa/docs/Words-into-action/Words-Into-Action.pdf

Johnson, C., and K. Krishnamurthy. "Dealing with Displacement: Can 'Social Protection' Facilitate Long-Term Adaptation to Climate Change?" *Global Environmental Change* 20 (2010): 648-55.

Klein, R. et al. "Portfolio Screening to Support the Mainstreaming of Adaptation to Climate Change into Development Assistance." *Climatic Change* 84 (2007): 23-44

Kolmannskog, V.O. "Future Floods of Refugees: A Comment on Climate Change, Conflict and Forced Migration." Oslo: Norwegian Refugee Council, 2008.

Laczko, F., and Aghazarm, C. (eds.) *Migration, environment and climate change: Assessing the evidence.* Geneva: International Organization for Migration, 2009.

Lasage, R., Aerts, J., Mutiso, G., et al. "Potential for Community Based Adaptation to Droughts: Sand Dams in Kitui, Kenya." *Physics and Chemistry of the Earth* 33, 2008: 67-73.

Maluccio, J. A. "The Impact of Conditional Cash Transfers on Consumption and Investment in Nicaragua." *Journal of Development Studies* 46, no. 1 (2010): 14-38.

Matin, I. et al. "Crafting a Graduation Pathway for the Ultra Poor: Lessons and Evidence from a BRAC Programme." *Chronic Poverty Research Centre Working Paper 109,* 2008, accessed May 5, 2010, www.chronicpoverty.org

McGranahan, G., Balk, D., Anderson, B. "The Rising Tide: Assessing the Risks of Climate Change and Human Settlements in Low Elevation Coastal Zones." *Environment and Urbanization* 19, no. 1 (2007): 17-37

McLeman, R., and Smit, B. "Migration as Adaptation to Climate Change." *Climatic Change* 76(2006): 31-53.

Moench, M. "Adapting to Climate Change and the Risks Associated with other Natural Hazards: Methods for Moving from Concepts to Action," Pp. 249-82 in *The Earthscan Reader on Adaptation to Climate Change,* ed. E.L.F. Schipper and I. Burton. London: Earthscan (2007 [2009]).

Myers, N. "Environmental Refugees in a Globally Warmed World," *BioScience* 43, 1993: 752–61.

Nelson, D. R., Adger, W. N. and Brown, K. "Adaptation to environmental change: Contributions of a resilience framework." *Annual Review of Environmental Resources* 32, 2007: 395–19.

Nicholls, R. J, P. P. Wong, V. R. Burkett, J. O. Codignotto, J. E. Hay, R. F. McLean, S. Ragoonaden and C. D. Woodroffe."Coastal Systems and Low-Lying Areas," *Climate Change 2007: Impacts, Adaptation and Vulnerability. Contribution of Working Group II to the Fourth Assessment Report of the Intergovernmental Panel on Climate Change,* ed. M. L. Parry, O. F. Canziani, J. P. Palutikof, P. J. van der Linden and C. E. Hanson. Cambridge, UK: Cambridge University Press, 2007: 315-356.

O'Brien, K. et al. "Disaster Risk Reduction, Climate Change Adaptation and Human Security," A Report commissioned for the Norwegian Ministry of Foreign Affairs, GECHS Report 2008: 3, accessed   May 12, 2010. http://www.crid.or.cr/digitalizacion/pdf/eng/doc17943/doc17943-a.pdf

OECD (Organization for Economic Cooperation and Development). "Development Aid Rose in 2009 and Most Donors will meet 2010 Aid Targets." Press release by OECD, April 14, 2010, Accessed May 12, 2010. http://www.oecd.org/document/11/0,3343,en_2649_34487_44981579_1_1_1_1,00.html

Oliver-Smith, A. "Theorizing Vulnerability in a Globalized World: A Political Ecological Perspective." pp. 10-24 in *Mapping Vulnerability: Disasters, Development and People*, ed. Greg Bankoff, Georg Frerks and Dorothea Hilhorst. London: Earthscan, 2004.

Olsson, P., L. H. Gunderson, S. R. Carpenter, P. Ryan, L. Lebel, C. Folke and C. S. Holling. "Shooting the Rapids: Navigating Transitions to Adaptive Governance of Social-Ecological Systems." *Ecology and Society* 11, no. 1 (2006). http://www.ecologyandsociety.org/vol11/iss1/art18/

Osbahr, H., C. Twyman, W. N. Adger, and D. S. G. Thomas. "Effective Livelihood Adaptation to Climate Change Disturbance: Scale Dimensions of Practice in Mozambique." *Geoforum* 39, 2008: 1851-64.

Pahl-Wostl, C. "A Conceptual Framework for Analyzing Adaptive Capacity and Multi-level Learning Processes in Resource Governance Regimes." *Global Environmental Change* 19, 2009: 354-65.

Perrow, C. "Normal Accidents: Living with High-risk Technologies". New York: Basic Books, 1984.

Perch-Nielson, S. L., Battig, M. B. and Imboden, D. "Exploring the Link Between Climate Change and Migration." *Climatic Change* 91, 2008: 375-93.

Peskett, L., N. Grist, M., Hedger, T. Lennartz-Walker, and I. Scholz. "Climate Change Challenges for EU Development Co-operation: Emerging Issues," *EDC 2020 Working Paper no. 3*. Bonn: European Association of Development Research and Training Institutes, 2009.

Prowse, M. "Pro-Poor Adaptation: The Role of Assets," ODI Opinion No. 117. London: Overseas Development Institute, 2008.

Putzel, J. *A Captive Land: The Politics of Agrarian Reform in the Philippines*. London: Catholic Institute for International Relations, 1992.

Raleigh, C, Jordan, L. and Salehyan, I. *Assessing the Impact of Climate Change on Migration and Conflict*. Washington DC: World Bank, 2008.

Reuveny, R. "Climate Change-Induced Migration and Violent Conflict," *Political Geography* 26 (2007): 656-673.

Roger, Riddell. *Does Foreign Aid Really Work?* Oxford: Oxford University Press, 2008.

Saith, A. "From Universal Values to Millennium Development Goals: Lost in Translation," *Development and Change* 37, no. 6(2006): 1167-99.

Scoones, I. "Sustainable Rural Livelihoods: A Framework for Analysis", IDS Working Paper 72. Brighton: Institute of Development Studies, 1998.

Sen, A. *Poverty and Famines*. New York and Oxford: Oxford University Press, 1981.

Sen, A. *Development as Freedom*. New York and Oxford: Oxford University Press, (1999 [2001]).

Slater, R., J. Farrington, and R. Holmes. "A Conceptual Framework for Understanding the Role of Cash Transfers in Social Protection," ODI Project Briefing No. 5. London: Overseas Development Institute, 2008.

Smit B., and J. Wandel. "Adaptation, Adaptive Capacity and Vulnerability." *Global Environmental Change* 16, 2006: 282–92.

Stal, M. Mozambique: Case Study Report. "EACH-FOR Environmental Change and Forced Migration Scenarios." Project, 2009. http://www.each-for.eu/

Swift, J. "Understanding and Preventing Famine and Famine Mortality," *IDS Bulletin* 24, no. 4(1993): 1-16.

Tanner, T., T. Mitchell, E. Polack and B. Guenther. "Urban Governance for Adaptation: Assessing Climate Change Resilience in Ten Asian Cities," IDS Working Paper 315. Brighton: Institute of Development Studies, 2009.

Teichman, J. "Redistributive Conflict and Social Policy in Latin America," *World Development* 36, no. 3 (2008): 446-60.

Todd, J. E. et al. "Conditional Cash Transfers and Agricultural Production: Lessons from the *Oportunidades* experience in Mexico." *Journal of Development Studies* 46, no.1 (2010): 39-67.

Traynor, I. "EU puts €100bn-a-year Price on Tackling Climate Change." *The Guardian,*      30      October,      2009. http://www.guardian.co.uk/environment/2009/oct/30/eu-climate-change-funding-deal

Tschakert, P., and K. A. Dietrich. "Anticipatory Learning for Climate Change Adaptation and Resilience," *Ecology and Society 15,* no. 2 (2010). http://www.ecologyandsociety.org/vol15/iss2/art11/

UNDP. Human Development Report 2008: "Fighting Climate Change: Human Solidarity in a Divided World." New York: United Nations Development Program (UNDP), 2008.

UNDP. Human Development Report 2009: "Overcoming Barriers: Human Mobility and Development." New York: United Nations Development Program, 2009.

UNFCCC. "Climate Change: Impacts, Vulnerabilities and Adaptation in Developing Countries." United Nations Framework Convention on Climate Change,      2007.      Accessed      June      10,      2009. http://unfccc.int/adaptation/items/4159.php 10 June 2009.

Vidal, J. "Poorest Nations Shortchanged by Billions in Climate Change Pledges." *The Guardian*, 21 February, 2009.

Viotti, P., and M. Kaupi. *International Relations Theory Fourth Edition.* Toronto: Longman, 2010.

Warner, K., C. Ehrhart, A. de Sherbinin, S. Adamo and T. Chai-Onn. "In Search of Shelter: Mapping the Effects of Climate Change on Human Migration and Displacement." CARE International and CIESEN, 2009. Accessed March 17, 2010, http://ciesen.columbia.edu/publications.html

Watts, J. "China Bids to Ease Drought with $1bn Emergency Water Aid." *The Guardian,* 11 February 2011. Accessed February 14, 2011. http://www.guardian.co.uk/environment/2011/feb/11/china-drought-emergency-water-aid

WBGU (German Advisory Council on Global Change). *Climate Change as a Security Risk.* London: Earthscan, 2008.

Webster, M., J. Ginnetti, P. Walker, D. Coppard, and R. Kent. *The Humanitarian Costs of Climate Change.* Cambridge MA: Feinstein International Center, Tufts University, 2009.

Wisner, B., P. Blaikie, T. Cannon, and I. Davis. *At risk: Natural hazards, people's vulnerability and disasters.* London and New York: Routledge, 2004.

World Bank. "World Development Report 2010: Development and Climate Change." Washington DC: The World Bank, 2010.

Young, R., and Pilkey, O. "How High Will Sea Levels Rise? Get Ready for Seven Feet." Yale Environment 360, 2010, accessed October 29, 2010. http://e360.yale.edu/content/feature.msp?id=2230

Zoellick, Robert. "Free markets can still feed the world." Financial Times, January 5 2011. Accessed March 14, 2011 http://www.ft.com/cms/s/0/64ccfdae-1904-11e0-9c12-00144feab49a.html#axzz1GapbNvNc.

# 13

## Lessons on Equitable Growth: Stories of Progress in Vietnam, Mauritius and Malawi

*Milo Vandemoortele and Kate Bird[1]*

The world has seen notable progress in terms of economic growth and human development in the past two decades. In most countries, however, this progress has been inequitable (see Jaumotte, Lall, and Papageorgiou 2008). Wealthier and more powerful groups have benefited more than the middle and low income groups, and marginalised groups. Rising inequality undermines a country's potential for sustained economic and human development. For example, many argue that rising inequality in the US contributed to its financial crisis in 2008 (Stiglitz et al. 2009). On the other hand, more equity can contribute to more sustained and broad-based development (Wilkinson and Picket 2009).

How have some countries grappled with the challenges to achieve equitable growth? This chapter identifies and unpacks key lessons on equitable growth from three case studies. These case studies examine the links among economic growth, equity and poverty reduction. These three case studies have been carefully selected as part of a larger study that includes twenty-four "stories of progress" on development across eight different areas of development (agriculture and rural development, economic conditions, water and sanitation, health, education, governance, social protection and environmental conditions). The case studies sought to illustrate national-level progress that is equitable and sustainable, and where a significant share of the population has benefited. Selection was based on qualitative[2] and quantitative data (i.e. DHS, MICS, MDG, WDI databases and more). The quantitative indicators[3] to select the economic conditions case studies included: (i) annual GDP growth rates and household final consumption expenditure per capita; (ii) the proportion of the population below the international and the national poverty lines; and (iii) the Gini coefficient. This quantitative analysis was triangulated with qualitative research. We also sought to select case studies where progress met the following criteria: *scale*, with progress experienced by a significant share of the population; equity and sustainability. We also sought to select case studies which would avoid country duplication, would provide regional variation, highlight progress which has not previously been widely recognized, contribute to the MDG review process and provide lessons for cross-national learning.

A first case study, Mauritius, provides examples of reconciling high levels of growth with reductions in inequality and poverty. A second case study, Viet Nam, illustrates that initial levels of equity can contribute to high economic performance. A third case study, Malawi, showcases a recent example of economic growth, poverty education and falling inequality. The second section two briefly describes the three case studies. The third section three identifies four key factors that contribute to inclusive growth, namely: policy continuity coupled with government commitment; equitable investment to improve the quality of human capital; macroeconomic stability to contribute to national and international investor confidence; and pragmatic trade liberalization that enabled the country to develop competitive advantages. The overarching lesson is that there is no single policy or intervention that will lead to equitable growth. Rather, it is a package of policies that harness and develop the country's competitive advantage, often pioneered by a pragmatic, dedicated and inclusive state.

The next section provides an overview of the progress in economic conditions that the three case studies have experienced. The final section concludes by outlining the main lessons learnt and by highlighting some key policy recommendations.

## PROGESS IN EQUITABLE GROWTH

All three case study countries have seen economic growth, coupled with poverty reduction. Mauritius and Malawi experienced a reduction in inequality and Viet Nam initially focused on equity when it implemented the fundamental economic reforms that triggered its economic progress. Of late, Viet Nam has depriorisised equity, and has seen rising inequity (both in terms of income and human development).

Since independence, Mauritius' economy has developed equitably and beyond expectations. In 1968 the odds were stacked against the country. It was remote from world markets, commodity dependent and had extreme cultural diversity and racial inequality. It was a mono-crop economy with power concentrated among a small elite group, had high unemployment, high population growth, declining terms of trade, faced economic crisis and had low levels of human development. Since then it has developed into the leading performer in sub-Saharan Africa in terms of economic and human development. It has succeeded where other countries with comparable initial conditions and similar trade advantages have failed. It has also exhibited notable resilience to the global financial crisis.

Viet Nam's economy was transformed in the 1990s through a series of economic, social and political reforms—the Doi Moi policies. These contributed to an average annual growth rate of 7.4 percent per year between 1990 and 2008, coupled with a dramatic reduction in poverty from 58 percent to 14.5 percent between 1993 and 2008, based on the national poverty line. The proportion of people living on less than a dollar a day fell from 63 percent to 21.5 percent between 1993 and 2006. Exports rose by an average 21 percent per year between 1991 and 2007. From having been a marginal player on international markets, Viet Nam has become one of the leading exporters of a number of agricultural commodities—such as rice and coffee. This pattern has been repeated in the manufacturing sector, which saw a fourteen-fold increase in export value from 1995 to 2007.

Income growth has been concentrated mainly around the large cities and in areas with export-oriented economic activities. Growth has been slower in the Northwest and Central Highlands regions and among particular ethnic minorities. This has exacerbated income inequalities. The Gini coefficient rose from 33 in 1993 to 43 in 2008: disparities between urban and rural populations, between ethnic minorities and the majority of the population and between the rich and poor have widened.

Malawi is increasingly recognised as an example of progress in economic growth, poverty and inequality reduction. The country has enjoyed seven years of uninterrupted economic growth. Between 2003 and 2009 Malawi averaged 7 percent annual GDP growth rates, performing well above the sub-Saharan average of 5 percent per year. This progress has been paralleled with notable poverty reduction (NSO 2009). Nationally reported figures show a decline in the rate of poverty from 52 to 39 percent, between 2004 and 2009. WIDER's data base on inequality (WIID2C) points to a notable drop in inequality, with the Gini coefficient falling from 62 in 1993 to 39 in 2004. Progress in several aspects of human development is also evident: using MDG indicators, the country ranks among the top twenty performers in both absolute and relative progress (ODI 2010). Malawi has also been lauded by the IMF and the UN Secretary General as a success story. Nevertheless, Malawi's story is complicated in that progress has been underpinned by a set of precarious factors that raise valid concerns about its sustainability.

## KEY FACTORS CONTRIBUTING TO PROGRESS

Four factors emerge from analysis of the case studies on equitable growth. They include policy continuity, equitable investment in human capital, macroeconomic stability and pragmatic and sequenced liberalisation. It is important to note that these factors are not independent of each other—rather their complementarity is key to explaining progress in economic conditions in the three countries. It is also important to note that these four factors do not represent a comprehensive list of factors leading to equitable growth. Rather they highlight key factors shared across the case studies. The case studies spellout additional and country-specific factors that have contributed to equitable growth.

## *Policy Coherence and Continuity*

Policy coherence and continuity are key factors contributing to progress in Viet Nam, Mauritius and Malawi. Policy coherence occurs when a country's policy framework aims towards a specific set of development objectives and does not contain significant internal contradictions. Evidence from both developed and developing countries indicates that progress in development occurs slowly and in a non-linear manner. Thus it is important that policy coherence is maintained over time. Good policies grow over time. Underpinning policy coherence and continuity are government commitment and proper institutional arrangements.

Since independence, Mauritius' government engaged in a concerted "nation building" strategy that was maintained over the subsequent four decades. This strategy was followed in the early decades following independence and set the foundations for Mauritius' sustained progress. Behind this nation building strategy are several factors. Fearing an autocratic government dominated by the Hindu majority, 44 percent of the population (virtually the entire non-Indian population) voted against independence, voting to remain a British colony. Many were also sceptical that Mauritius could maintain peace or achieve development, citing the challenges outlined above. There were also riots in the mid-1960s, including one in which twenty-five people were killed (Sriskandarajah 2005: 69). To ensure legitimacy and to assuage the misgivings the portion of the population against independence, a nation building strategy in the post independence era was considered necessary.

The "nation building" strategy involved a strategic partnership between the major ethnic groups, negotiated economic redistribution, a better balance of economic power (previously concentrated within the minority Franco-Mauritian group) and political power and the building of strong institutions to ensure economic redistribution and political representation. Sensitive discussions on how to avoid the embedding of ethnic politics and the possibility of ethnic conflict were had in the early years post-independence (ibid.: 68). These resulted in institutional arrangements that bolstered the emergence of a democratic developmental state that ensured economic redistribution and inclusive political representation.

This strategy enabled a long-termist approach to policy making. Thus policy continuity enabled effective structural changes to the Mauritian economy, which allowed the country to maintain its competitiveness. This policy continuity, with the long-term aim of diversifying, liberalizing and modernizing the economy, is said to have "played an important role in building strong business confidence among investors—domestic and foreign" according to a key informant.

A key foundation for Viet Nam's progress in economic conditions was consistent and pragmatic leadership by the Party, which did not necessarily rely on strong individual leaders. Within the party, there was an overarching objective to work towards development. Thus, internal debate and policy innovation were encouraged and a trial-and-error approach was accepted. In addition, once a decision was made, full support was expected from all party members, with (partial) accountability mechanisms arising from decentralized power structures and an educated and politically aware population. These "democratic forces" enabled the adoption and implementation of an effective set of policies, the *Doi Moi* policies, which were crucial to progress in Viet Nam's economic conditions.

A strong sense of social solidarity and equity, by both the population and the government, contributed to a coherent policy framework in the late 1980s, which laid the foundations for subsequent progress. Rama (2008) writes that "Vietnam's success in adopting comprehensive reforms is also associated with the determination to avoid creating losers from a material point of view." This was reflected in policy designs that aimed to improve living standards and avoid rising inequality. The former objective has, however, gained precedence and inequality has been allowed to rise. Economic growth and price stability have remained central objectives of Viet Nam's macroeconomic management (Kokko 2008).

Malawi's development planning was mostly disjointed and largely driven by external actors. Policy coherence has become stronger since 2004/05.[4] Increasing national ownership has enabled policy makers to adhere to a more coherent and long-term policy agenda for Malawi. Experts interviewed by the authors provided examples of increased assertiveness of government officials vis-à-vis donors and international actors. Examples range from the design and effective implementation the Farm Input Subsidy Programme[5] to the monitoring of the Millennium Development Goals (MDGs). This has contributed to progress in economic conditions by reducing inefficiencies that resulted from disjointed policy making, prioritising the allocation of government funds according to nationally agreed objectives and supporting sustained coherence of development policy (i.e. Malawi Development Growth Strategy).

## *Human Capital*

Investing in human capital (i.e. education, skills, and health) is recognised as important for economic development. Two important components of human capital that have contributed to equitable economic development in the case study countries are the quality and equity of human capital. High-quality human capital means that the labour force is sufficiently flexible to be able to take advantage of national and international market opportunities, move up the value chain, adapt to changing opportunities and diversify and attract international investors. When human capital is equitably shared within a population (usually a result of equitable investment in human capital) the broader population can contribute to and benefit from economic growth—thus enabling equitable growth.

In Viet Nam initial equitable investments in education and health enabled it to take advantage of national and international market opportunities. According to Nguyen and Nguyen (2007), among the key country-specific advantages that enabled Viet Nam to attract Foreign Direct Investment (FDI) was its relatively high level of education and labour quality. Initial equitable investment in education set the foundation for future improvements in education outcomes. Improvements in education outcomes are one of the key factors contributing to sustained progress in economic conditions and in Viet Nam's ability to continue attracting foreign investment (ibid.). The focus on equity in the initial years of reforms in the late 1980s, has however dwindled. In recent years, progress on economic and human development indicators has been inequitable. The wealthiest groups, urban populations and non-minorities are benefiting significantly more than the poorer groups, the rural population and the ethnic minorities.

Viet Nam's rapid growth has been as impressive as that in China or India, but it has also been inequitable. Perhaps less inequitable than in China, because of a combination of geography, initial equitable provision of services (especially education), equitable distribution of land, a high degree of female participation in the workforce and a decentralised power structure. Rising inequality in Viet Nam is worrying, however, and threatens the sustainability of progress in economic conditions (Kokko 2008; Rama 2008). This suggests that the challenge now is to better understand what is needed to reverse widening disparities in income, education quality and health services and so avoid undermining future growth. Experiences in Indonesia may offer some clues on this topic—as it has enjoyed economic growth and equalising improvements in under-five morality rates (see data annex in ODI [2010] for equity adjusted indicators on U5MR).

Mauritius, at independence, inherited a system of education and health services[6] free at the point of delivery. The government, engaged in a nation building strategy, determined to avoid social and political tensions and to support solidarity and equity by investing in social services. It has continued to provide these services, funded through taxation. Contrasting with Viet Nam, Mauritius has been able to maintain relatively equitable investments in human capital and has expanded services, with an aim to "broaden the circle of opportunities" for its population and ensure "inclusive growth" according to a key informant.

The pool of unemployed but educated and easily adaptable workers was an essential input in the success of export-oriented strategy of the 1980s (Subramanian 2009 and MoF Respondent). For example, when an Export Processing Zone (EPZ) was set up, Mauritians responded to opportunities with vigour and, sometime after their establishment, around 90 percent of entrepreneurs in the EPZ and the manufacturing sectors were Mauritian nationals (MoF Respondent). Indeed, businesses faced no major shortages in human capital, know-how and education to exploit market opportunities. This contributed to average real output growing at 7 percent between 1984 and 1988 (MoF Respondent). The social infrastructure underpinning the flexible labour force in the 1980s was crucial and consisted of free education, health services, a non-contributory basic retirement pension and an extensive set of social security schemes. More recently, education and skill levels have remained very much the same and have no longer been adequate to take advantages of emerging opportunities. Actions to stimulate competitions recommended by an IMF report include: "make labor markets more flexible, raise the average skill level and further reduce the cost of doing business" (Imam and Minoiu 2008: 20).

Malawi, unlike Mauritius and Viet Nam, has a low human capital base. Government efforts to invest in human capital have been fragmented and weak. Low-quality human capital is one reason for low rates of foreign direct investment into the country. It has also inhibited Malawians' ability to take advantage of rising market opportunities (particularly at the international level). A lesson for Malawi from Viet Nam and Mauritius is that in order to be able to reduce dependence on agriculture and to move up the value chain, the country needs to work towards an equitable improvement in the quality of human capital. Reinvesting rents from recent growth in the agricultural sector into human development would mimic the successful Mauritian strategy of development.

Underpinning the high quality and equitable investment in human capital in Mauritius and Viet Nam is a wide network of infrastructure, with the institutional arrangement and government commitment that enabled equitable and high quality human capital.

## *Macroeconomic Stability*

The stabilisation of the economy through a balancing of fiscal deficits, stabilising inflation rates and interest rates, among other policies, set the scene for equitable growth across the three case study countries. Achieving macroeconomic stability has been a process of mutually reinforcing factors. Fundamental shifts in the structure of the economy contributed to it becoming more stable, and this stability has in turn contributed to economic growth, and poverty and inequality reduction, enabling further structural changes to the economy. Underpinning this macroeconomic stability is a concerted government effort, national ownership of development policy and enabling institutional arrangements.

In Malawi, between 2004 and 2010, macroeconomic policy has focused on fiscal discipline and stabilising economic fundamentals. Increased agricultural production has contributed to and benefited from macroeconomic stability, low interest rates and controlled inflation. There has been a two-way causality, a mutual reinforcement. For example, Malawi has reduced its inflation rate, reported at 7 percent per annum at the end of July 2010 (NSO), after having reached levels above 70 percent in the 1990s and early 2000s. A large component of the consumer price index (CPI) basket is food, representing 58 percent and the food component is largely maize. Thus adequate maize supplies in recent years have contributed to stabilising inflation.

The person credited with leading this change, President Mutharika, understands the link between macroeconomic stability and agricultural growth. His development policy emphasized infrastructure and agricultural development, while maintaining fiscal discipline, tackling corruption and implementing public sector reform.[7]

In the 1970s, Mauritius was suffering from economic instability, experiencing a decline in its terms of trade, a worsening of its balance of payments, a tripling of its external debt, a reduction in output, high unemployment and persistent inflation (Sandbrook 2005: 555). In the early 1980s the Mauritian government implemented a series of stabilization programmes with loans from the IMF and World Bank. It did so in an unorthodox manner counter to the typical Washington Consensus policies (see section on Sequenced and Pragmatic Liberalization). By 1983 it was well on its way to recovery.

Strong institutions contributed to maintaining Mauritius' macroeconomic stability, competitiveness and economic resilience. The government supported a pragmatic development strategy of heterodox liberalization and diversification and ensured that the quota rents from the EU and US markets were reinvested into strategic and productive sectors. They also promoted a strongly regulated and well-capitalized banking and financial system, thus shielding it from toxic assets prior to the 2008 global financial crisis.

In Viet Nam, economic reforms (Doi Moi) tamed the high inflation rates of the mid-1980s. Nominal interest rates also fell, generating a more stable and predictable investment climate. Complementary short-term macro policies (which were not necessarily part of *Doi Moi*) also played a role in taming hyper-inflation (Wood 1989). Economic reforms reflected

An agreement on the need for policy reforms aimed at reducing macroeconomic instability and accelerating economic growth, and that all economic levers (price, wages, fiscal and monetary policies) were to be used to achieve these objectives. (Van Arkadie and Mallon 2003)

Policies included in *Doi Moi* aimed at abolishing the system of bureaucratic centralised management (based on state subsidies) and replacing it with a multi-sector market-oriented economy, whereby the private sector had a role of competing with the state in non-strategic sectors (ibid.). The central components of this policy package were four-fold. They included: i) equitable land reform; ii) liberalization of the agricultural sector; iii) a pragmatic and sequenced liberalization that allowed Viet Nam to consistently attract and benefit from foreign investment; and iv) continued investment in human development. A strong sense of social solidarity and equity underpinned policy and implementation.

## *Sequenced and Pragmatic Trade Liberalization*

By most accounts, Mauritius and Viet Nam's progress in economic conditions was largely driven by their export oriented approach. Neither country went for a "big-bang" approach to liberalization. Rather they undertook a dual-track approach, complementing orthodox and heterodox trade policies, and strong government intervention. On the other hand, some sub-Saharan African and Latin American countries underwent comprehensive trade liberalization and made little progress in integrating into the global economy. Furthermore, as a result of opening their borders, they experienced little growth, and saw poverty and inequities increase. Underpinning Mauritius' and Viet Nam's liberalization approach were strong institutions and strong government intervention that enabled the implementation of a pragmatic and sequenced liberalization strategy and resisted international pressures to speed to re-sequence their liberalization.

Through a specific package and sequencing of both highly protectionist as well as export-oriented policies, Mauritius was able to segment the export and import competing sectors. These were led by strong government intervention[8] and their design and implementation were tailored to develop some competitive advantages.

Mauritius' liberalisation process occurred in phases and adapted to the country's evolving advantages on the international market. In a first phase (1970s) Mauritius profited from sugar rents and reinvested these into the manufacturing sector and established an export processing zone (EPZ). It also successfully attracted capital and foreign investment into its manufacturing sector. A second phase entailed the expansion of the EPZ, a significant increase in FDI and of tourism (1980–1990s). Rents accruing from preferential access to markets for sugar and clothing together amounted to 7 percent of GDP in the

1980s and 4.5 percent of GDP in the 1990s (Government of Mauritius 2008: 21). The capital and current accounts were liberalised, contributing to an investment and employment boom. The high inflow of FDI brought with it managerial know-how and market access. A third phase consisted of the broadening and deepening economic development through further diversification, liberalisation and investment (1990s–2010); broadening in terms of diversification of the economy and deepening in terms of ensuring more high value addition.

Rodrik (1999) and Subramanian (2009) contend that Mauritius' successful trade policies did not correspond to the orthodoxy of the Bretton Woods Institutions; "Mauritius is not the poster boy for the Washington Consensus. Mauritius had a highly restrictive trade regime" (Subramanian 2009: 10). After all, imports were limited through high trade barriers, while extensive and selective intervention occurred on the export side. Subramanian (2009) parallels this approach with the dirigiste approach of Korea, Taiwan, and Japan (ibid.: 11). As discussed above, the success of Mauritius' heterodox opening was bolstered by high levels of human capital and the preferential access to the markets of the United States and especially Europe.

Pragmatic and sequenced trade liberalisation enabled Viet Nam to attract and benefit from trade and foreign investment. A dual approach was followed, opening some sectors of the economy to international markets while protecting others. This course aimed to remove obstacles to foreign investors while protecting national industries, promoting industrial upgrading and retaining management and manufacturing skills within national borders and strengthening improvements in standard of living among the populace. Initial inflows of FDI were directed to labour-intensive sectors, for example, which provided an avenue out of poverty for unskilled workers (Rama 2008). The egalitarian redistribution of agricultural land in the early years of reform, combined with the liberalisation of trade in commodities, did much to boost rural living standards and reduce poverty (ibid.).

The Vietnamese state was involved in trade, maintaining import monopolies and retaining quantitative restrictions and high tariffs (30 percent to 50 percent) on agricultural and industrial imports (UNDP 2003). Despite high trade barriers, the country was able to integrate rapidly into the global economy. Indeed, contrary to economic orthodoxy, these strategies have been exceptionally successful—expanding trade at double-digit rates on an annual basis and attracting substantial FDI.

Malawi can learn from both of these examples. Mauritius, for example, reinvested rents from sugar into developing a competitive advantage in its manufacturing sector. Malawi cannot rely, *ad infinitum,* on its agricultural sector for poverty reduction and economic development. Rather it is important that rents from this growth (and other growth, such as from the mining sector) are reinvested into building the nation's human capital base, diversifying its economy or moving up the value chain. Mauritius' and Viet Nam's strategic and pragmatic liberalization progress was embedded in a nationally owned development plan—something that is beginning to emerge in Malawi. Policy coherence was maintained in these countries through strong institutional arrangements and consistent leadership.

# CHALLENGES UNDERMINING EQUITABLE GROWTH

A number of challenges have the potential to undermine equitable growth. The case studies point towards inequality and the environment having an important negative effect. Equity has been deprioritised in Viet Nam and Mauritius. Mauritius' and Malawi's vulnerability to climate change is severe and Viet Nam's environmental degradation is worrisome.

A reprioritization of equity and nation building strategy, as evidenced by rising inequality, particularly between wealth groups, urban-rural areas and ethnic groups, has been recent in Mauritius. In Viet Nam on the other hand, disparities have risen since the 1990s. Rising inequality not only undermines the poverty reduction potential of economic growth, it also undermines economic growth through several channels—such as social, financial and political instability.

Rising inequality appears to be a reflection of a vicious cycle where policy decisions reflect unequal constituencies, and in turn widen inequalities which propagate social tensions. Such widening disparities also shorten the time horizons of policy-makers, encouraging short-sighted policies and policies with a bias in the distribution of social spending (Woo 2005).

In Mauritius, the government's efforts at nation building and fostering cooperation between ethnicities appear to be weakening. Sriskandarajah (2005) writes "while official attention has been paid to the issue of social exclusion, it is unclear whether the government is willing or able to address the problem effectively...There is also concern that, even if there is commitment to redistribution, any efforts may be hampered if growth rates slow down and general economic conditions worsen. (ibid.:74).

Martin Rama, a former Lead Economist for the World Bank in Viet Nam, writes "Growing inequality and displays of conspicuous wealth could generate social resentment, especially if the prevalence of corruption casts doubts on the legitimacy of the new fortunes. Weak mechanisms to gather and process the demands of specific population groups, no matter how narrow, could encourage them to voice their frustration through unauthorized channels, resulting in political turmoil" (Rama 2008: 12). Concern over rising inequality is emerging as a priority in government discourse. If Viet Nam is to sustain progress in economic conditions it must ensure an equalizing growth pattern.

Risks from climate change include unreliable agricultural production, falling domestic food production with rising imports, climate-related disasters (Mauritius is vulnerable to cyclones), increased risk for foreign direct investment flight and more. According to UNEP (1997) "Mauritius is highly vulnerable to climate change impacts... [as] it suffers considerable damage at regular intervals from cyclones, and serious coastal erosion is already evident." Malawi is predominantly an agricultural economy which relies on favourable rainfall. Its main source of foreign exchange (aside from overseas development assistance) comes from tobacco exports. Eighty-eight percent of Malawians and 52 percent of poor people live in rural areas and a large majority of Malawians depend on agriculture for their livelihoods (World Bank 2009: 4). In 2006, agriculture contributed a third to GDP. A drought or untimely rains could put Malawi in a precarious economic situation.

Viet Nam has relied on economic expansion through extensive use of natural resources, including its land, water and forest resources. Although not unique to Viet Nam, the trend towards environmental degradation is worrisome. This type of economic development is environmentally unsustainable, will undermine economic growth and will aggravate disparities. Moreover, environmental damage often results in costs borne disproportionately by the poor and the poorest.

## Case Studies

### Mauritius

At independence, Mauritius did not appear predestined for the progress that followed. Challenges included extreme cultural diversity as well as racial inequality; power concentrated in a small elite; high unemployment; and high population growth. The country suffered from an economic crisis throughout the 1970s, was remote from world markets and was commodity dependent. It also exhibited low initial levels of human development.

Despite multiple factors stacked against it, Mauritius has achieved stellar progress in economic conditions and has been unique in its ability to take advantage of privileged access to international markets to develop in a sustained and equitable manner. This has been enabled and complemented by effective poverty reduction and equitable improvements in human development. These achievements have been made by means of a concerted strategy of nation building; strong and inclusive institutions; high levels of equitable public investment in human development; and a pragmatic development strategy.

## *What Has Been Achieved?*

Mauritius has achieved what few other sub-Saharan African countries have been able to achieve since independence—sustained progress in economic conditions. It has also accomplished what a minority of fast growing economies have achieved—reductions in inequality. Between 1977 and 2008, Mauritius averaged a 4.6 percent gross domestic product (GDP) growth rate, compared with a 2.9 percent average in sub-Saharan Africa (WDI database). Over the same time period, its GDP per capita averaged $2,921 (constant US$2,000), well above a sub-continent average of $540. It is also the only country in the region where household expenditure increased significantly between 1990 and 2008. Meanwhile, inequality (as measured by the Gini coefficient) fell from 45.7 to 38.9 between 1980 and 2006 (1980 data: *UNU-WIDER* WIID; 2006 data: Household Budget Survey 2006/07).

Mauritius has successfully translated economic growth into concrete poverty reduction and improvements in human development. Its poverty rates remain low by international standards, with less than 1 percent of the population estimated as living on less than $1 a day. Malaria has been successfully eradicated from the island, and life expectancy at birth increased from 61 overall in 1965 to 69.3 for men and 76.1 for women in 2008. The country has maintained a primary net enrolment ratio several points above 90 percent (93.3 percent in 1991 and 94.0 percent in 2009) (MDG database). Along with the Seychelles, it has the lowest under-five child mortality rate on the sub-continent (17 deaths out of 1,000 live births) and the highest rate of children immunized against measles (98 percent in 2008) (ibid.).

Its performance on the Human Development Index (HDI), which draws on data on income, education and health to form a composite index, has been exceptional—not only by sub-Saharan African standards but also by international standards when compared with South Asia, the Arab States and East Asia and the Pacific, performing on a par with Latin America and the Caribbean (figure 13.1).

**Figure 13.1. HDI trends, 1980–2005**

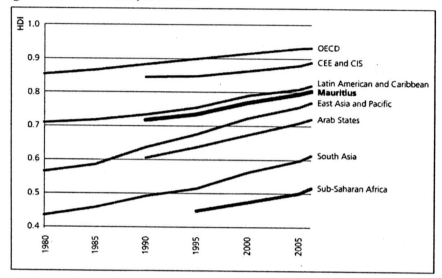

*Source*: UNDP (2008) *Human Development Report 2009*. New York: UNDP.

## What Has Driven Change?

### A Concerted Strategy of Nation Building

A concerted strategy of nation building in the early decades following independence set the foundations for sustained progress in Mauritius. This involved partnership among the major ethnic groups, a negotiated economic redistribution, a better balance of economic and political power and the building of strong institutions. These resulted in institutional arrangements that bolstered the emergence of a democratic developmental state to ensure economic redistribution and inclusive political representation. This included:

- A constitutional mandate for the inclusion of minorities;
- A system in the electoral process that ensures representation of ethnic groups, called the "best losers" system, whereby an independent electoral commission appoints up to eight losing candidates to each new National Assembly to represent "underrepresented" ethnic groups (Sandbrook 2005);
- An inclusive party system that does not have an ethnic basis; and
- A consultative approach to policy formation.

These institutional arrangements have contributed to the establishment of strong and inclusive institutions, to high levels of equitable investment in social welfare and human development and to a pragmatic development strategy of heterodox liberalisation and policy continuity. These have all enabled effective structural changes to the Mauritian economy.

## Strong and Inclusive Institutions Supporting a Social Consensus

A key outcome of the state's strategy was strong and inclusive institutions able to redistribute political and economic power away from the wealthy minority (Franco–Mauritian), in particular to disadvantaged ethnic groups (Sriskandarajah 2005). As one key informant from the Ministry of Finance noted:

> This important determinant of growth, that is, the ability of our domestic institutions to manage the distributional conflicts, triggered by local and external shocks, stands out markedly in the case of Mauritius. The quality of our domestic institutions seems to override the other primordial factors affecting growth.

Strong institutions have helped maintain the country's competitiveness, economic resilience and stability. They have supported development strategy and ensured that quota rents from international markets are reinvested in strategic and productive sectors. They have also promoted a strongly regulated and well-capitalized banking and financial system, thus shielding it from toxic assets prior to the 2008 global economic crisis.

## High Levels of Equitable Public Investment in Human Development

Independent Mauritius inherited a system of free education and health services. The government, as part of its strategy of nation building, avoided social and political tensions and supported solidarity and equity by investing in health and education, as well as in a non-contributory basic retirement pension and an extensive set of social security schemes. Since then, it has expanded these services, with the aim of expanding opportunities for its population and ensuring inclusive growth.

An unemployed, educated and easily adaptable labour force was an essential input into the success of the export-oriented strategy of the 1980s (Subramanian 2009). Mauritians responded to opportunities with vigour: after the establishment of the export processing zone (EPZ), around 90 percent of entrepreneurs there and in the manufacturing sector were Mauritian nationals (interview, Ministry of Finance). Businessmen and -women had the human capital, the know-how and the education to exploit market opportunities. The social infrastructure underpinning the flexible labor force was crucial to this.

## A Pragmatic Development Strategy of Heterodox Liberalization and Diversification

Progress in economic conditions was driven largely by the export-oriented approach, underpinned by a heterodox set of liberalization policies. The Mauritian government has played a strong and interventionist role: it has acted as facilitator (of the enabling environment for the private sector); as operator (to encourage competition); and as regulator (to protect the economy as well as vulnerable groups and sectors from shocks). A key strength is that leaders did not persist if a strategy did not work (ibid.). Of particular importance is that there has been consistency and stability in the approach to economic management, regardless of which political party is in government.

Liberalisation occurred in phases, adapted to the country's evolving advantages on the international market. In a first phase (1970s), Mauritius profited from sugar rents and established an EPZ. It also successfully attracted capital and foreign investment in manufacturing. A second phase (1980s–1990s) entailed the expansion of the EPZ and a significant increase in foreign direct investment (FDI) and tourism. Rents accruing from preferential access to markets for sugar and clothing together amounted to 7 percent of GDP in the 1980s and 4.5 percent in the 1990s (Subramanian 2009). Capital and current accounts were liberalised, contributing to an investment and employment boom, and the high inflow of FDI brought with it managerial know-how and market access. A third phase (1990s–2010) consisted in broadening and deepening economic development through further diversification, liberalization and investment.

## Viet Nam

Viet Nam's recent economic development has been exceptional. In the 1970s, the country was emerging from decades of war, which had decimated the country and its infrastructure and left many people dead and millions injured or displaced. Now Viet Nam is set to join the ranks of middle-income countries.

After the reunification of the North and the South in 1975, Viet Nam faced an economic crisis and declining standards of living, including serious food shortages. This undermined the legitimacy of the government. In response, the government began a process of economic reforms, shifting towards a market-based economy by using a trial-and-error approach. A decisive political shift came in the late 1980s, and this enabled significant economic, social and political reforms which contributed to noteworthy improvements in economic conditions and in human development. Equity and social cohesion were important components in these reform policies, as the government saw these as crucial to maintaining its legitimacy. The rate of economic growth and poverty reduction since the 1990s has been unsurpassed by most developing countries, although challenges remain with regard to rising inequality, environmental degradation and corruption.

## What Has Been Achieved?

Vietnam enjoyed an average annual growth rate of above 7 percent a year between 1990 and 2008. Exports rose by an exceptionally high average of 21 percent per year between 1991 and 2007. From having been a marginal player on the international market, Viet Nam has become a leading exporter of a number of commodities, such as rice and coffee, and this pattern has been repeated in the manufacturing sector. Overall, Viet Nam has been able to attract and take advantage of opportunities higher up the value chain (figure 13.2).

**Figure 13.2. Trends in Economic Conditions in Vietnam Between 1990 and 2008.**

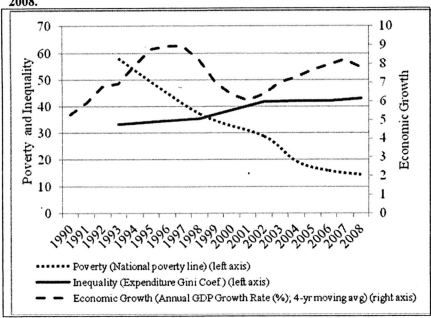

*Source*: Viet Nam Living Standards Survey (VLSS) 1993, 1998 (poverty); UNU-WIDER World Income Inequality Database (WIID) (inequality 1993, 1998); Viet Nam Household Living Standards Survey Results (VHLSS) (poverty and inequality 2002, 2004, 2006, 2008).

The expanding economy has enabled households to more than double per capita expenditure, from \$211 to \$439 between 1994 and 2008 (constant 2000 prices), and for the government to expand social investment.

Economic growth has been coupled with a dramatic reduction in poverty, from 58 percent to 14.5 percent between 1993 and 2008 (based on the national poverty line), and a fall in the proportion of people living under a dollar a day from 63 percent to 21.5 percent between 1993 and 2006. Viet Nam is one of the best performers in the world in reducing absolute poverty. Not all groups have benefited equally from progress, however. Inequality is on the rise: disparities in poverty levels, between ethnic minorities and the rest of the population, for example, widened from a factor of 1.6 in 1993 to 5.1 in 2006.

An expanding budget has enabled investment in services to improve health, education and water and sanitation. That is, improvements in economic conditions have accompanied improvements in human development. The under-five mortality rate declined from 56 to 14 deaths per 1,000 live births between 1990 and 2008. Underweight prevalence among children has fallen substantially, from 45 percent to 20 percent between 1994 and 2006, with Viet Nam ranked as the second best performer in terms of annual absolute reductions in underweight prevalence across 106 developing countries. Despite all income groups experiencing improvements, however, the poorest 40 percent have seen less progress when compared with the rest of the population, resulting in rising inequities in human development.

## What Has Driven Change?

### Pragmatic Leadership and Efective Policy Innovation

A key foundation for Viet Nam's economic progress has been pragmatic and non-dogmatic leadership by the Communist Party. "Democratic forces" enabled the adoption and implementation in 1986 of an effective set of policies, the *Doi Moi* ("renovation") policies, which represented a turning point in Viet Nam's economic development.

*Doi Moi* prioritised agricultural development, production of consumer goods and trade and foreign investment. The egalitarian redistribution of agricultural land, combined with the careful liberalisation of trade in commodities and targeted investment in agriculture, contributed to significant increases in agricultural yields. Viet Nam went from being a net importer of rice to being the second largest exporter in the world today. These reforms also led to high poverty reduction in rural areas through the diversification of rural incomes (e.g. Rama 2008).

Pragmatic and sequenced trade liberalization enabled Viet Nam to attract and benefit from foreign investment. A two-track approach was followed— opening some sectors of the economy to international markets while protecting others. This course aimed to remove obstacles to foreign investors while protecting national industries, promoting industrial upgrading and retaining management and manufacturing skills within national borders. Contrary to economic orthodoxy, these strategies have been exceptionally successful.

## Strong Institutions and a Wide-Reaching Foundation of Infrastructure

High-quality institutions are essential to a country's positive development outcomes. Viet Nam has seen improvements in its institutions, both informal and formal. In the agriculture sector, for instance, stronger institutions have enabled the expansion of the coffee subsector from being almost nonexistent to being among the largest global market players.

A tradition of resource mobilization at the local level in Viet Nam means that infrastructure has historically been built by local institutions, enhancing accountability and ownership, which contributes to their sustainability. Decentralized governance has also meant a wide spread of services, basic infrastructure and institutions around the country, in place to implement economic reform and to adopt new technologies and ideas.

## Equitable Initial Investment in Human Development

The pre-1990 government left a legacy of equitable investments in human development, most notably in formal education and health care. Equitable investment in education set the foundation for future improvements and contributed to a cultural interest in education among the population.

Continued investment in human development has led to significant improvements in human capital, enabling people to take advantage of market opportunities. In this way, Viet Nam's relatively high levels of literacy and quality of labour have contributed to its economic success. Completing the cycle, economic growth has allowed the government to go on to invest more resources in human development. Poor groups, such as ethnic minorities and those living in remote mountainous areas, have experienced fewer investments and benefited far less from these investments—explaining some of the widening disparities in the country.

## A Strong Sense of Social Solidarity and Equity

Both the population and the government of Viet Nam have a strong sense of equity, which is reflected in policies aimed at improving living standards and preventing the rise of inequality. These have included the egalitarian land distribution at the initial stages of *Doi Moi*, targeted investments in remote and poorer regions and the waiving of education and health fees for the poor. However, ambitions for more growth have gained precedence in recent years, and inequality is now increasing.

## Malawi

Malawi is a small and landlocked sub-Saharan African country, with a population of 13 million. Agriculture contributes about 35 percent to the country's gross domestic product (GDP) (AfDB and OECD 2010) and employs 85 percent of its workforce (GoM 2010). The national poverty rate in 2009 was 39 percent, with significant regional and rural–urban variation: about 14 percent of the population in urban areas lives in poverty compared with 43 percent in rural areas. Rural populations depend on agriculture for their livelihoods, mostly maize.

Between 2004 and 2009, Malawi averaged 7 percent annual GDP growth, well above the sub-Saharan African average. This has been paralleled by significant poverty reduction and a notable drop in inequality. Progress on human development is also evident: the country ranks among the top twenty performers on several Millennium Development Goal (MDG) indicators in terms of both absolute and relative progress (ODI 2010). However, the reality is more complicated than the macro indicators reflect.

Malawi is essentially a story of progress in economic conditions since 2004, thanks to macroeconomic stability, an effective input subsidy programme, improved policy coherence and strengthened national ownership of development policy. The UN Secretary-General has lauded Malawi for its progress: "In a few short years, Malawi has come from famine to feast; from food deficit to surplus; from food-importing country to food exporting country" (UN Multimedia).

## What has been achieved?

Since independence, Malawi's experience with development can be broadly categorised into three periods: the first, which lasted from 1964 to 1994, was one of uneven and inequitable growth; the second period of 1994–2004 is broadly considered a lost decade, as macroeconomic figures corroborate. Malawi's claim to progress in economic conditions thus relates to the most recent period; 2004 to date.

## Economic Growth

Malawi has sustained seven years of uninterrupted economic growth. Between 2004 and 2009, the annual GDP growth rate was 7 percent, reaching almost 8 percent in 2009. This was well above the sub-Saharan African average of 5 percent over the same period and the Latin American average of 4 percent. Since 2004, Malawi's GDP has increased over 40 percent, from $1.8 billion to $2.5 billion (2000 US$ prices) (WDI). Controlling for population growth, GDP per capita has risen by almost a third, from $130 to $166 (constant 2000 US$) (WDI). The country has also managed to sustain growth amid global recession, thanks to "generally sound macroeconomic policies" (Randall et al. 2010).

## More Equal Income Distribution

More equitable distribution of income began in the early 1990s (Tsoka et al. 2002): the Gini coefficient fell from 62 to 39 between 1993 and 2004 (UN WIDER database.[9] The income share held by the poorest 20 percent of the population increased from 4.8 percent in 1998 to 7 percent in 2004, and that of the richest 20 percent decreased from 56 percent to 46.5 percent. This indicates a redistribution of wealth to middle and poorer quintiles (World Bank data).[10]

## Falling Income Poverty

Poverty rates fell from 52 percent to 39 percent between 2004 and 2009, according to national statistics. Poverty reduction has been fastest in urban areas and in the north of the country, as table 13.1 shows.

**Table 13.1. Trends in Growth and Poverty across Malawi, 2004–2009**

|  | 2004 | 2005 | 2006 | 2007 | 2008 | 2009 |
|---|---|---|---|---|---|---|
| Economic growth |  |  |  |  |  |  |
| Annual GDP growth (%) | 5.7 | 2.6 | 8.2 | 8.6 | 9.7 | 7.7 |
| Poverty (%) |  |  |  |  |  |  |
| Malawi | 52 | 50 | 45 | 40 | 40 | 39 |
| Urban | 25 | 24 | 25 | 11 | 13 | 14 |
| Rural | .. | 53 | 47 | 44 | 44 | 43 |
| Northern region | 56 | 51 | 46 | 46 | 35 | 31 |
| Central region | 47 | 46 | 40 | 36 | 40 | 41 |
| Southern region | 64 | 60 | 55 | 51 | 51 | 51 |

Source: NSO 2009 data (2004 based on IHS2 data, 2005–2009 based on WMS surveys). Economic growth: WDI (accessed October 2010).

## Human Development

Between 1990 and 2008, the under-five mortality rate declined steadily and equitably, from 225 to 100 deaths per 1,000 live births (MDG database). Household survey data presented in Figure 13.3, show that this reduction was equitably spread across income groups (DHS and MICS data). Stunting and wasting of under-fives also fell from 6.4 percent and 6.8 percent in 2005, respectively, to 4.9 percent and 5.8 percent in 2007 (Waage et al. 2010).

**Figure 13.3A. Under-Five Mortality Rate in Malawi by Quintile, 2006**

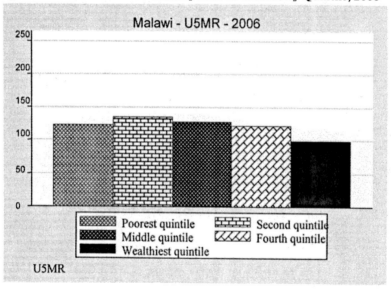

**Figure 113.3B: Under-five Mortality Rate in Malawi by Quintile, 1992**

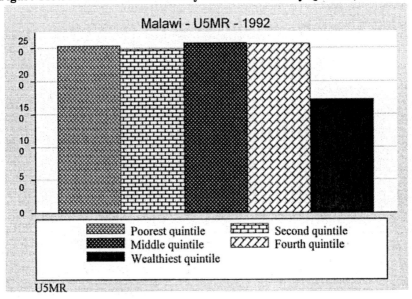

*Source*: Authors calculations based on DHS and MICS data.

The MDG Database shows a notable reduction in the proportion of 15–49 year olds living with HIV, from over 13 percent in 2001 to under 11 percent in 2007. Tuberculosis is also coming under control. Meanwhile, the number of children under five sleeping under an insecticide-treated bed net rose from 3 percent in 2000 to 25 percent in 2006.

Challenges remain in maternal health and in achieving universal primary education and improving education quality.

## What Has Driven Change?

### Effective and Appropriate Agricultural Input Subsidies

Malawi relies on agriculture for its economic development. In addition, a great majority of the 84 percent of Malawians living in rural areas depend on the sector for their well-being. Of these, 97 percent grow maize, and more than half of households grow no other crop (World Bank 2009). The country's large-scale input subsidy programme provides subsidized fertilizer and seed to farmers and this has increased maize production, farm incomes and household food security.

The success of the Fertilizer Input Subsidies Programme (FISP) lies in its design and in the broader context. Key factors include:

- High coverage and effective targeting through the use of vouchers;
- A large input package of fertiliser and seed;
- Strong government backing, as there is political mileage in increased food security; and
- Favourable rainfall since the programme's inception.

The FISP is not without its critics, who underline issues relating to potential regressivity, agro-ecological and fiscal unsustainability, market distortion, leakages and high opportunity costs. Moreover, some suggest that, while progress in economic conditions in Malawi is undeniable, the scale reported in national figures is inflated. These assessments reflect the controversial nature of the FISP among development practitioners, but also the inherent methodological challenges in measuring agricultural production in a largely smallholder farming sector.

## Macroeconomic Stability: Fiscal Balance and Reduced Interest and Inflation Rates

Between 2004 and 2010, national policy that focused on macroeconomic stability, low interest rates and the control of inflation both contributed to and benefited from increased agricultural production.

For example, Malawi's inflation rate, which reached levels above 70 percent in the 1990s and early 2000s, fell to 7 percent per year at the end of July 2010 (National Statistics Office, n.d.). As a large part of the Consumer Price Index basket is food (58 percent), and as maize is a large component of this, increased supply in recent years has helped control inflation and contributed to overall stability. It has also been one of the outcomes of increased macroeconomic stability in the country.

## National Ownership and Coherence of Development Policy

For a long time after independence, Malawi's national development planning was disjointed and donor-driven. Recently, though, policy coherence and national ownership have improved, and the Malawian government has become more assertive in terms of its own development programming.

This has contributed to progress in economic conditions in a number of ways. For example, it has reduced the inefficiencies that resulted from disjointed policymaking, it has ensured that development policy is coherent and it has enabled the government to prioritize allocations according to nationally agreed objectives.

# CONCLUSION: Lessons Learnt and Policy Recommendations

The overarching lesson of this chapter is that there is no single intervention that will lead to equitable growth. Rather, we see that in Mauritius, Viet Nam and Malawi a package of policies has been effective in harnessing and developing the country's competitive advantage. This has often been pioneered by a developmental state that has contributed effectively to enabling equitable growth. In each of the case study countries, achieving equitable growth took time and a concerted effort to stay the course. Sustained efforts by the state and policy continuity are important and not just in generating equitable growth but in maintaining positive momentum. We see that a relaxation of attention to equity can result in a shift in the pattern of growth and, although Mauritius and Viet Nam have both experienced equitable growth in the past, they are now seeing a much less equitable pattern emerging—the latter to a much greater degree—with possible implications for social cohesion and the sustainability of the growth process. Specific lessons include:

- *Reconciling high economic growth with reductions in inequality is possible,* as in the case of Mauritius, with resolute and pragmatic government leadership, complemented by strong institutions that enabled the redistribution of political and economic power across ethnic groups and equitable investments in human well-being.
- Equitable growth in Mauritius and Viet Nam was driven by a *pragmatic development strategy of heterodox liberalisation processes*. In Malawi it has been driven by growth in the agriculture sector, spurred by a fertilizer subsidy programme. Growth in the agriculture sector has absorbed labour, provided smallholder farmers with improved incomes and greater food security and generated export earnings.
- *Equitable investment in human capital*: A strong human capital foundation, achieved through consistent and equitable investment in human development, enabled Mauritius to exploit advantages and maintain competitiveness in a fast evolving international market. Equitable initial investment in human development, to maintain Viet Nam's comparative advantage of high-quality low-cost labour, has primed the populace to take advantage of economic opportunities and has contributed to the continued inflow of foreign investment.
- *Targeting and sequencing development policy* in the sectors and areas where poor people work and live contributes to poverty reduction, improved food security and equity. Malawi's government chose an input subsidy programme, Mauritius focused on its sugar sector, Viet Nam implemented equitable land reform, along with targeted investment in the agriculture

sector. Different countries may find that other policies are more sustainable and efficient in terms of achieving growth in agriculture, depending on their political, economic, environmental and social context.

- *Macroeconomic stability* can both contribute to and benefit from increased agricultural production. In Malawi, agricultural growth has contributed significantly to economic growth and to the control of inflation. Inversely, a stabilised macro-economy has stimulated private sector investment and donor assistance, enabling investment in agriculture.

## NOTES

1.    The authors acknowledge the Bill and Melinda Gates foundation for funding the production of the case studies as part of the *Progress in Development: A Library of Stories* project. These case studies are part of a larger project that includes twenty-four stories of progress on development across eight different        sectors.        For        more        information        see http://www.developmentprogress.org. Authors gratefully acknowledge Nisha Agarwal, Alison Evans, Liesbet Steer and Alan Winters for their review of respective case studies. All errors remain, and views in this chapter are, those of the authors alone. For comments on the chapter, please contact Milo Vandemoortele        (m.vandemoortele@odi.org.uk)        and        Kate        Bird (k.bird.ra@odi.org.uk).

2.    Qualitative information on potential case studies was sourced from key informant interviews and a review of literature.

3.    Criteria for choosing between the available quantitative indicators included: (1) data availability both across countries and over time (1990–2010); (2) prioritising observable indicators (e.g. U5MR, measles vaccination); (3) avoiding modelled indicators (e.g. $1 a day, percent people with inadequate calorie intake, completion rates), and  (4) avoiding perception indices (except for governance, where these were used to inform qualitative research).

4.    The MDGS provides a framework under which the international community can contribute to development planning. A government respondent stated that "*no donor in Malawi can talk about a programme or project without someone linking it with the MDGS.*" Key informants, across the board, indicate increasing national ownership of development policy processes, contending that both on paper and in practice the MDGS is more strongly nationally owned than previous development strategies. One donor respondent reports: "it is coming out clearly, the government is taking ownership of the development agenda."

5.    Prior to the 2004/05 programme, agricultural policy in Malawi was predominantly shaped by donor demands. Kumwenda and Phiri (2009) document that in the 1990s Bretton Woods Institutions advocated for the removal of agricultural subsidies and a complete liberalization of the agriculture

market. Authors also document donor influence in 2000 when the Targeted Inputs Programme (TIP) programme (which was demonstrating clear impact) was scaled down for sustainability, rather than technical, reasons. In 2005, DFID pulled out of its funding which led to the termination of the programme (Kumwenda and Phiri, 2009). Malawi's government then took the reins of the programme over in 2004/05.

6.    This consisted of free education, health services, a non-contributory basic retirement pension and an extensive set of social security schemes.

7.    Recent indications point to declining macro-economic management since the president replaced the minister of finance.

8.    The Mauritian government, for example, played a strong and interventionist role in the country's development policy and implementation. It has acted as a facilitator, operator and regulator in the liberalization process: a facilitator to provide the enabling environment for the private sector to thrive and develop; an operator in the economy, mostly to encourage competition; A regulator to protect the Mauritian economy from economic shocks, as well as seeking to protect vulnerable groups and sectors from changes in Mauritian economy.

9.    UN WIDER database. www.wider.unu.edu/research/Database/en_GB/wiid/. More recent data on Gini coefficients are not available.

10.    World Bank data, http://data.worldbank.org/. More recent data could not be found. Respondents indicate that the targeting of FISP and spill-over effects have contributed to reductions in inequality.

## BIBLIOGRAPHY

African Development Bank (AfDB) and Organization for Economic Co-operation and Development (OECD). *Malawi. African Economic Outlook.* Paris: OECD, 2010.

Government of Malawi (GoM). *A Medium-Term Plan for the Farm Input subsidy Programme 2011-2016.* Lilongwe: GoM, 2010.

Government of Mauritius (GoM). *Mauritius 40 Years After: New Goals, New Challenges.* Port Louis: Government of Mauritius, 2008.

Imam, Patrick, and Camelia Minoiu. *Mauritius: A Competitiveness Assessment.* Washington DC.: IMF, 2008.

Jaumotte, Florence, Subir Lall, and Chris Papageorgiou. *Rising Income Inequality: Technology, or. Trade and Financial Globalization?* Washington DC: IMF, 2008.

Kokko, Ari. *Vietnam: 20 Years for Doi Moi.* Ha Noi: The Gioi Publishers, 2008.

Kumwenda, Ian, and Horace Phiri. *Government Interventions in Fertilizer Market in Malawi: From 1994-2009: A Literature Review.* Lilongwe: Agriculture and Natural Resource Management (ANARMAC), 2009.

National Statistics Office - Malawi. "Malawi in Figures 2007." n.d. http://www.nso.malawi.net/data_on_line/general/malawi_in_figures/Malawi_in_Figure s.pdf (accessed November 2010).

National Statistics Office (NSO). "Welfare Monitoring Survey (2009)." http://www.nso.malawi.net/data_on_line/agriculture/wms_2009/wms2009.h tml (accessed November 2010).

Nguyen, Anh Ngoc, and Thang Nguyen. *Foreign Direct Investment in Vietnam: An Overview and Analysis the Determinants of Spatial Distribution Across Provinces.* Hanoi: Development and Policies Research Center, 2007.

Overseas Development Institute (ODI). *Millennium Development Goals (MDG) Report Card: Measuring Progress Across Countries Report.* London: ODI, 2010.

Rama, Martin. *Making Difficult Choices: Vietnam in Transition.* Washington DC.: The World Bank, 2008.

Randall, Ruby, Adelaide Matlanyane, Charles Amo Yartey, and Joe Thornton. "IMFSurvey Magazine: Countries and Regions Malawi's New IMF Loan Boosts Malawi's New IMF Loan Boosts Prospects for Sustained Growth." *IMF Survey Magazine: Countries & Regions,* 2010.

Rodrik, Dani. *The New Global Economy and Developing Countries: Making Openness Work.* London: Overseas Development Council, 1999.

Sandbrook, Richard. "Origins of the Democratic Developmental State: Interrogating Mauritius." *Canadian Journal of African Studies* 39, no. 3 (2005): 549-581.

Sriskandarajah, Dhananjayan. "Development, Inequality and Ethnic Accommodation: Clues from Malaysia, Mauritius and Trinidad and Tobago." *Oxford Development Studies* 33, no. 1 (2005): 63-79.

Stiglitz, Joseph, et al. "Report of the Commission of Experts of the President of the United Nations General Assembly on Reforms of the International Monetary and Financial System." *UN Conference on the World Financial and Economic Crisis and Its Impact.* New York: United Nations, 2009. 113.

Subramanian, Arvind. *The Mauritian Success Story and Its Lessons.* Helsinki: UNU-WIDER, 2009.

Tsoka, Maxton Grant, Nebert Nyirenda, Earnest Hayes, and Osten Chulu. *Millennium Development Goals: Malawi 2002 Report.* Lilongwe: Government of Malawi, 2002.

United Nations Development Programme (UNDP). *Making Global Trade Work for People.* London: Earthscan, 2003.

United Nations Environmental Programme (UNEP). "Mauritius." *The Newsletter of the UNEP Collaborating Centre on Energy and Environment -C2E2 News* 9 (May 1997).

Van Arkadie, Brian, and Ray Mallon. *Viet Nam: A Transition Tiger?* Canberra: ANU E Press, 2003.

Waage, Jeff, et al. "The Millennium Development Goals: A Cross-sectoral Analysis and Principles for Goal Setting after 2015." *The Lancet* 376, no. 9745 (2010): 991 - 1023.

Wilkinson, Richard, and Kate Pickett. *The Spirit Level: Why More Equal Societies Almost Always Do Better.* London: Allen Lane, 2009.

Woo, Jaejoon. "Social Polarization, Fiscal Instability and Growth." *European Economic Review* 49 (2005): 1451-77.

Wood, Adrian. "Deceleration of Inflation with Acceleration of Price Reform: Vietnam's Remarkable Recent Experience." *Cambridge Journal of Economics* 13, no. 4 (1989): 563-71.

World Bank (WB). *Malawi - Country Economic Memorandum: Seizing Opportunities for Growth through Regional Integration and Trade.* Washington DC: World Bank, 2009.

———. "World Development Indicators." http://data.worldbank.org/data-catalog/world-development-indicators (accessed March 2010).

# Index